T0402019

321

Topics in Current Chemistry

Topics in Current Chemistry
Recently Published and Forthcoming Volumes

EPR Spectroscopy: Applications in Chemistry and Biology
Volume Editors: Malte Drescher, Gunnar Jeschke
Vol. 321, 2012

Radicals in Synthesis III
Volume Editors: Markus R. Heinrich, Andreas Gansäuer
Vol. 320, 2012

Chemistry of Nanocontainers
Volume Editors: Markus Albrecht, F. Ekkehardt Hahn
Vol. 319, 2012

Liquid Crystals: Materials Design and Self-Assembly
Volume Editor: Carsten Tschierske
Vol. 318, 2012

Fragment-Based Drug Discovery and X-Ray Crystallography
Volume Editors: Thomas G. Davies, Marko Hyvönen
Vol. 317, 2012

Novel Sampling Approaches in Higher Dimensional NMR
Volume Editors: Martin Billeter, Vladislav Orekhov
Vol. 316, 2012

Advanced X-Ray Crystallography
Volume Editor: Kari Rissanen
Vol. 315, 2012

Pyrethroids: From Chrysanthemum to Modern Industrial Insecticide
Volume Editors: Noritada Matsuo, Tatsuya Mori
Vol. 314, 2012

Unimolecular and Supramolecular Electronics II
Volume Editor: Robert M. Metzger
Vol. 313, 2012

Unimolecular and Supramolecular Electronics I
Volume Editor: Robert M. Metzger
Vol. 312, 2012

Bismuth-Mediated Organic Reactions
Volume Editor: Thierry Ollevier
Vol. 311, 2012

Peptide-Based Materials
Volume Editor: Timothy Deming
Vol. 310, 2012

Alkaloid Synthesis
Volume Editor: Hans-Joachim Knölker
Vol. 309, 2012

Fluorous Chemistry
Volume Editor: István T. Horváth
Vol. 308, 2012

Multiscale Molecular Methods in Applied Chemistry
Volume Editors: Barbara Kirchner, Jadran Vrabec
Vol. 307, 2012

Solid State NMR
Volume Editor: Jerry C. C. Chan
Vol. 306, 2012

Prion Proteins
Volume Editor: Jörg Tatzelt
Vol. 305, 2011

Microfluidics: Technologies and Applications
Volume Editor: Bingcheng Lin
Vol. 304, 2011

Photocatalysis
Volume Editor: Carlo Alberto Bignozzi
Vol. 303, 2011

EPR Spectroscopy

Applications in Chemistry and Biology

Volume Editors: Malte Drescher · Gunnar Jeschke

With Contributions by

E. Bordignon · M. Drescher · B. Endeward · D. Hinderberger · I. Krstić · D. Margraf · A. Marko · D.M. Murphy · T.F. Prisner · E. Schleicher · S. Van Doorslaer · J. van Slageren · S. Weber

Editors
Dr. Malte Drescher
Department of Chemistry
University of Konstanz
Konstanz
Germany

Dr. Gunnar Jeschke
Laboratory of Physical Chemistry
ETH Zurich
Zurich
Switzerland

ISSN 0340-1022 e-ISSN 1436-5049
ISBN 978-3-642-28346-8 e-ISBN 978-3-642-28347-5
DOI 10.1007/978-3-642-28347-5
Springer Heidelberg Dordrecht London New York

Library of Congress Control Number: 2012932055

Printed on acid-free paper

Springer is part of Springer Science+Business Media (www.springer.com)

Volume Editors

Dr. Malte Drescher

Department of Chemistry
University of Konstanz
Konstanz
Germany

Dr. Gunnar Jeschke

Laboratory of Physical Chemistry
ETH Zurich
Zurich
Switzerland

Editorial Board

Topics in Current Chemistry Also Available Electronically

Topics in Current Chemistry is included in Springer's eBook package *Chemistry and Materials Science*. If a library does not opt for the whole package the book series may be bought on a subscription basis. Also, all back volumes are available electronically.

For all customers with a print standing order we offer free access to the electronic volumes of the series published in the current year.

If you do not have access, you can still view the table of contents of each volume and the abstract of each article by going to the SpringerLink homepage, clicking on "Chemistry and Materials Science," under Subject Collection, then "Book Series," under Content Type and finally by selecting *Topics in Current Chemistry*.

You will find information about the

- Editorial Board
- Aims and Scope
- Instructions for Authors
- Sample Contribution

at springer.com using the search function by typing in *Topics in Current Chemistry*.

Color figures are published in full color in the electronic version on SpringerLink.

Aims and Scope

The series *Topics in Current Chemistry* presents critical reviews of the present and future trends in modern chemical research. The scope includes all areas of chemical science, including the interfaces with related disciplines such as biology, medicine, and materials science.

The objective of each thematic volume is to give the non-specialist reader, whether at the university or in industry, a comprehensive overview of an area where new insights of interest to a larger scientific audience are emerging.

Thus each review within the volume critically surveys one aspect of that topic and places it within the context of the volume as a whole. The most significant developments of the last 5–10 years are presented, using selected examples to illustrate the principles discussed. A description of the laboratory procedures involved is often useful to the reader. The coverage is not exhaustive in data, but rather conceptual, concentrating on the methodological thinking that will allow the non-specialist reader to understand the information presented.

Discussion of possible future research directions in the area is welcome.

Review articles for the individual volumes are invited by the volume editors.

In references *Topics in Current Chemistry* is abbreviated *Top Curr Chem* and is cited as a journal.

Impact Factor 2010: 2.067; Section "Chemistry, Multidisciplinary": Rank 44 of 144

Preface

Electron paramagnetic resonance (EPR) spectroscopy [1-3] is the most selective, best resolved, and a highly sensitive spectroscopy for the characterization of species that contain unpaired electrons. After the first experiments by Zavoisky in 1944 [4] mainly continuous-wave (CW) techniques in the X-band frequency range (9-10 GHz) were developed and applied to organic free radicals, transition metal complexes, and rare earth ions. Many of these applications were related to reaction mechanisms and catalysis, as species with unpaired electrons are inherently unstable and thus reactive. This period culminated in the 1970s, when CW EPR had become a routine technique in these fields. The best resolution for the hyperfine couplings between the unaired electron and nuclei in the vicinity was obtained with CW electron nuclear double resonance (ENDOR) techniques [5].

Starting in the 1960s, stable free radicals of the nitroxide type were developed as spin probes that could be admixed to amorphous or weakly ordered materials and as spin labels that could be covalently attached to macromolecules at sites of interest [6,7]. In parallel, theory was developed for analyzing linewidths and lineshapes in CW EPR spectra in terms of molecular dynamics [8-10]. At about the same time a few select groups in the Soviet Union and the USA worked on pulse experiments, using spin echo phenomena to measure electron spin relaxation [11], to detect hyperfine couplings by electron spin echo envelope modulation (ESEEM) techniques [12], to acquire ENDOR spectra in a broader temperature range than with CW methods [13], and to measure distances between electron spins [14]. These developments were pursued by physicists, were heavily focused on methodology, and were hardly recognized by mainstream chemists even by the end of the 1980s when the groundwork was all done. As a result, EPR spectroscopy acquired the reputation of an old-fashioned, somewhat obscure technique applicable to only a small range of compounds. Many chemistry departments considered it as dispensable.

Several developments in the 1990s prepared the stage for the renaissance of EPR spectroscopy that we now experience. Concepts of pulse NMR were introduced into pulse EPR [15,16], which lead to a zoo of new experiments for the separation of different interactions of the electron spin with its environment [3]. After an induction period a new generation of EPR spectroscopists started using these techniques in the established application fields of transition metal catalysis and metalloenzymes. Within the same decade the application field in structural biology was extended tremendously by the introduction of site-directed spin labeling [17], which made diamagnetic proteins accessible to EPR spectroscopy, among them many that were difficult to study by x-ray crystallography or NMR spectroscopy. The third major development of the 1990s was the systematic combination of EPR measurements at multiple frequencies (multi-frequency EPR) to study more complex problems, and, in particular, the extension to higher fields and frequencies, made possible by new microwave technology and by the superconducting magnet technology developed for NMR spectroscopy [18,19].

This volume of *Topics in Current Chemistry* is devoted to the consequences that these three parallel developments have had on the application field of EPR spectroscopy. It is no exaggeration to state that the major part of the systems studied nowadays by EPR spectroscopy was inaccessible two decades ago and that for the remaining systems information can be obtained, which was inaccessible at that time. The scope of EPR spectroscopy arising from this combination has been hardly realized even by the most advanced practitioners.

This volume starts with three chapters that illustrate the wealth of information which can now be obtained in some of the traditional application fields of EPR spectroscopy. Chapter 1 by S. Van Doorslaer and D. Murphy is an in-depth review on work in catalysis focusing on the mechanistic information that can be obtained from EPR spectra. Work on radical enzymes is exemplified in Chapter 2 by S. Weber and E. Schleicher on the example of flavoproteins which play a role in both chemically and light-activated electron transfer processes. Chapter 3 on synthetic polymers by D. Hinderberger argues that careful analysis of mundane nitroxide spin label or spin probe CW EPR spectra can reveal a lot of information which is hard to obtain by any other characterization technique.

The following three chapters explore the opportunities provided by site-directed spin labeling of diamagnetic biomacromolecules. Intrinsically disordered proteins are one class of such biomacromolecules that is hard to characterize by established techniques. Chapter 4 by M. Drescher discusses how EPR spectroscopy can contribute to better understanding of these proteins. The main application of site-directed spin labeling techniques is on membrane proteins, which are more difficult to study by crystallography and high-resolution NMR spectroscopy than soluble proteins. EPR on membrane proteins is treated in Chapter 5 by E. Bordignon, with an emphasis on the nuts and bolts of the approach. During the past few years application of spin label EPR to nucleic acids has emerged, and Chapter 6 by I. Krstić, B. Endeward, D. Margraf, A. Marko, and T. Prisner provides a comprehensive overview of both spin labeling and EPR techniques applied in this field and on the information that can be obtained.

Finally, an emerging application field is discussed. The application to molecular magnets is a result of parallel development of new approaches in inorganic chemistry and new high-field and high-frequency EPR technologies. In Chapter 7 J. van Slageren discusses the newly emerging technologies of frequency-domain magnetic resonance and Terahertz spectroscopy and the importance of relaxation studies in the field of molecular nanomagnetism.

Due to space limitations only a selected range of systems can be covered. Recent good reviews exist about EPR spectroscopic studies on photosynthesis [20-22] and metalloproteins [23-26]. Strongly physics-related application fields, such as quantum computing [27], electrically detected [28] and optically [29] detected EPR spectroscopy of dopants and defects in solids are left out. Furthermore, this volume does not cover technical issues that are mainly of interest to method developers rather than chemists. Pulse EPR spectroscopy [3], high-field EPR spectroscopy [19], and quantum chemical computation of EPR parameters [30] were all subject of monographs. Note also that EPR distance measurements between spin labels in biological systems are covered by two forthcoming volumes of the series *Structure and Bonding* that are edited by C. Timmel and J. Harmer.

M. Drescher
G. Jeschke

References

1. J. A. Weil, J. R. Bolton, J. E. Wertz, *Electron Paramagnetic Resonance*, Wiley, New York (1994)
2. N. M. Atherton, Principles of Electron Spin Resonance, Ellis Horwood, New York (1993)
3. A. Schweiger, G. Jeschke, Principles of Pulse Electron Paramagnetic Resonance, Oxford University Press, Oxford, 2001.
4. E. Zavoisky, *J. Phys. USSR* **1945**, *9*, 211-216.
5. H. Kurreck, B. Kirste, W. Lubitz, Electron Nuclear Double Resonance Spectroscopy of Radicals in Solution, VCH Publishers, New York (1988)
6. A. K. Hoffmann, A. T. Henderson, *J. Am. Chem. Soc.* **1961**, 83, 4671-4672.
7. O. H. Griffith, A. S. Waggoner, *Chem. Rev.* **1969**, *2*, 17-24.
8. D. Kivelson, *J. Chem. Phys.* **1960**, *33*, 1094-1106.
9. J. H. Freed, G. K. Fraenkel, *J. Chem. Phys.* **1963**, 39, 326-348.
10. J. H. Freed, G. V. Bruno, C. F. Polnaszek, *J. Phys. Chem.* **1973**, *75*, 3385–3399.
11. A. M. Raitsimring, K. M. Salikhov, B. A. Umanskii, Y. D. Tsvetkov, *Fiz. Tverd. Tela* **1974**, *16*, 756-766.
12. W. B. Mims, *Phys. Rev. B* **1972**, *5*, 2409-2419.
13. W. B. Mims., *Proc. Roy. Soc. London A*, **1965**, *283*, 452-457.
14. A. D. Milov. K. M. Salikhov, M. D. Shirov, *Fiz. Tverd. Tela* **1981**, *23*, 975-982.
15. P. Höfer, A. Grupp, H. Nebenführ, M.Mehring, *Chem. Phys. Lett.* **1986**, *132*, 279-282.
16. J. M. Fauth, A. Schweiger, L. Braunschweiler, J. Forrer, R. R. Ernst, *J. Magn. Reson.* **1986**, 66, 74-85.
17. W. L. Hubbell, D. S. Cafiso, C. Altenbach, *Nat. Struct. Biol.* **2000**, *7*, 735-739.
18. A. K. Hassan, L. A. Pardi, J. Krzystek , A. Sienkiewicz, P. Goy, M. Rohrer, L. C. Brunel, *J. Magn. Reson.* **2000**, *142*, 300-312.
19. K. Möbius, A. Savitsky, *High-Field EPR Spectroscopy on Proteins and their Model Systems: Characterization of Transient Paramagnetic States*, RSC Publishing, Cambridge, 2008.
20. A. Savitsky, K. Möbius, *Photosynth. Res.* **2009**, *102*, 311-333.
21. A. van der Est, *Photosynth. Res.* **2009**, *102*, 335-347.

22. G. Kothe, M. C. Thurnauer, *Photosynth. Res.* **2009**, *102*, 349-365.
23. M. E. Pandelia, H. Ogata, W. Lubitz, *ChemPhysChem* **2010**, *11*, 1127-1140.
24. B. M. Hoffman, D. R. Dean, L. C. Seefeldt, *Acc. Chem. Res.* **2009**, *42*, 609-619.
25. R. Davydov, B. M. Hoffman, *Arch. Biochem. Biophys.* **2011**, *507*, 36-43.
26. J. Harmer, G. Mitrikas, A. Schweiger Biol. Magn. Reson. **2009**, *28(1)*, 13-61.
27. M. Mehring, J. Mende, W. Scherer, *Phys. Rev. Lett.* **2003**, *90*, 153001.
28. A. R. Stegner, C. Boehme, H. Huebl, M. Stutzmann, K. Lips, M. S. Brandt, *Nature Physics* **2006**, *2*, 835-838.
29. L. Childress, M. V. G. Dutt, J. M. Taylor, A. S. Zibrov, F. Jelezko, J. Wrachtrup, P. R. Hemmer, M. D. Lukin, *Science* **2006**, *314*, 281-285.
30. M. Kaupp, M. Bühl, V. G. Malkin, *Calculation of NMR and EPR Parameters. Theory and Applications*, Wiley-VCH, Weinheim, 2004.

Contents

Top Curr Chem (2012) 321: 1–40
DOI: 10.1007/128_2011_237

Published online: 17 September 2011

EPR Spectroscopy in Catalysis

Sabine Van Doorslaer and Damien M. Murphy

Abstract The modern chemical industry relies heavily on homogeneous and heterogeneous catalysts. Understanding the operational mode, or reactivity, of these catalysts is crucial for improved developments and enhanced performance. As a result, various spectroscopic techniques are inevitably used to characterize and interrogate the mechanistic details of the catalytic cycle. Where paramagnetic centres are involved, ranging from transition metal ions to defects and radicals, EPR spectroscopy is without doubt the technique of choice. In this review we will demonstrate the wealth and breadth of information that can be gleaned from this technique, in the characterization of homogenous and heterogeneous systems of catalytic importance, whilst illustrating the advantages that modern high-field and pulsed EPR methodologies can offer.

Keywords EPR · Heterogeneous catalysis · Homogeneous catalysis

Contents

S. Van Doorslaer (✉)
SIBAC Laboratory – Department of Physics, University of Antwerp, Universiteitsplein 1, 2610 Wilrijk, Belgium
e-mail: sabine.vandoorslaer@ua.ac.be

D.M. Murphy
School of Chemistry, Cardiff University, Main Building, Park Place, Cardiff CF10 3AT UK

1 Introduction

1.1 The Importance of Catalysis

Catalysis is an extremely important branch of science, which is vital in our modern society. It is estimated that about 90% of all processed chemical compounds have, at some stage of their production, involved the use of a catalyst. In general, catalytic reactions are more energy efficient and, at least in the case of highly selective reactions, lead to reduced waste and undesirable compounds, which is an important consideration with dwindling global reserves of raw materials [1]. Many of the catalysts used in the modern chemical industry are well established, particularly those utilized in bulk chemical production and energy processing. Because catalysts are so firmly embedded and established within the chemical industry, one could imagine that the research and development into new catalysts would present limited opportunities for industrial and academic research. However, nothing could be further from the truth. In the past 10 years alone the Nobel Prize in chemistry has recognized the outstanding achievement of nine scientists whose work has a strong bias in homogeneous catalysis: 2010 (R.F. Heck, E. Negishi and A. Suzuki) for palladium-catalysed cross couplings in organic synthesis[1]; 2005 (Y. Chauvin, R.H. Grubbs and R.R. Schrock) for the development of the metathesis method in organic synthesis[2]; 2001 (W.S. Knowles, R. Noyori and K.B. Sharpless) for chirally catalysed hydrogenation and oxidation reactions[3]. Their combined work has revolutionized the field of fine chemical synthesis and chiral feedstock production using well defined and discrete homogeneous organometallic catalysts. Classic

[1] http://nobelprize.org/nobel_prizes/chemistry/laureates/2010/
[2] http://nobelprize.org/nobel_prizes/chemistry/laureates/2005/
[3] http://nobelprize.org/nobel_prizes/chemistry/laureates/2001/

examples of this work include olefin methathesis using ruthenium carbene catalysts (Gubbs catalyst) [2], the Heck reaction using palladium-based catalysts [3] and the enantioselective preparation of 2,3-epoxyalcohols from primary and secondary allylic alcohols using a Ti tartrate catalyst [4–6].

1.2 Mechanistic Understanding of Catalysis: The Role of Electron Paramagnetic Resonance

Despite the phenomenal success of these homogeneous catalysts, further developments of new asymmetric catalysts, bio-catalysts and heterogeneous catalysts will benefit from a greater understanding of the mechanistic pathways involved in the catalytic reactions [7]. A good illustration of this process is the hydrolytic kinetic resolution of racemic epoxides using a Co-based Salen catalyst [8]. Detailed kinetic measurements of the catalytic reaction revealed, rather unexpectedly, that the mechanism was second order with respect to the catalyst [9]. This mechanistic insight served as the inspiration for the further development of cyclic oligomeric salen complexes, which displayed dramatically enhanced reactivities and higher enantioselectivities relative to the monomeric counterparts [8, 10]. Undoubtedly a greater understanding of the mechanism can lead to enhanced performance, even with well established catalytic systems.

In most cases, paramagnetic metal centres or reaction intermediates are involved in many catalytic cycles. For example, organometallic catalysts based on Ti, Mn, Cr, Fe, Co, Cu or Ru frequently undergo redox changes during the reaction whilst heterogeneous catalysts may involve a defect site or transient species [1], so that in principle Electron Paramagnetic Resonance (EPR) spectroscopy, a technique targeted at unravelling structure and dynamics information of paramagnetic compounds, becomes an important tool in their characterization. Indeed, EPR has been used for a number of years in the analysis of heterogeneous catalysts [particularly at conventional microwave frequencies (9.5/35 GHz) and in the continuous-wave (CW) mode], and this area of research has been reviewed several times over the years [11–15] and more recently from the perspective of in situ methodologies [16–18]. Surprisingly, the EPR technique has not been exhaustively or widely used to examine homogeneous catalysts [19, 20]. Considering that most homogeneous catalysts are based on paramagnetic transition metal complexes and organometallic systems, this absence is unusual. In the last 10 years, with the advent of higher microwave frequency capabilities [21–24], high-frequency/field (HF) EPR has begun to be applied in the realm of high-spin states in catalysis, notably in recent work by Telser and co-workers [25, 26] and Gatteschi and co-worker [27]. The large zero-field splitting (zfs) in some of these transition metal ions precluded their analysis in the past by lower microwave frequency spectrometers so that they were largely ignored or inaccessible until the advent of HF EPR.

But the modern EPR arsenal of techniques entails far more than just CW EPR alone; the plethora of old and new hyperfine methodologies, such as Electron

Nuclear Double Resonance (ENDOR), Electron Spin Echo Envelope Modulation (ESEEM) and HYperfine Sublevel CORrElation (HYSCORE), and the methods targeted at extracting inter-spin distance information, like Pulsed ELectron-electron DOuble Resonance (PELDOR) or Double Electron Electron Resonance (DEER), have all reinvigorated the field [28]. Most of these methodologies were rapidly adapted by many research groups to the study of the fields of macromolecular assemblies, enzymes and proteins (effectively nature's more advanced and elegant equivalent of the man-made asymmetric homogeneous catalysts) [29–36]. The success of these combined techniques in characterizing the biomolecular world should inspire the same approach to characterize the numerous classes of homogeneous catalysts. Homogeneous catalysis is not just about electronic structure and high-spin to low-spin transitions as probed by field-swept EPR techniques, such as CW EPR; it's also about changes to the ligand environment during the reaction, as potentially probed by these above-mentioned hyperfine techniques.

1.3 Scope of the Review

The aim of this review is therefore to provide a selective, rather than exhaustive, review of the literature over the past 10 years, primarily in the field of EPR applied to studies in catalysis. From an EPR perspective, the emphasis will be placed on the role of advanced EPR techniques to study the structure and reactivity of homogeneous and heterogeneous catalysts. We will begin with a 'case study' based on our recent work, demonstrating the role for weak outer-sphere forces in controlling asymmetric interactions. Next we will review recent developments in the preparation of model complexes with reactive active oxygen species, used as model systems for biocatalysts. Recent evidence has also shown how non-innocent organic ligands play an important role in modulating the redox properties of organometallic complexes, and how EPR is used to study these ligand-based radicals. We will then present some recent representative examples of more traditional areas of homogeneous catalysis where EPR has played an important characterization role, such as polymerization, selective oxidations, C–H activation and Diels–Alder reactions. Finally, we will turn our attention to a number of heterogeneous systems, specifically focusing on porous catalytic materials.

2 Origins of Selectivity in Asymmetric Homogeneous Catalysis

2.1 Non-Covalent Interactions in Asymmetric Complexes

There are two fundamental processes of key importance in asymmetric homogeneous catalysis: first the stabilization of the transition state and second the efficiency of 'chiral information transfer' between substrate and ligand [7].

The identification and investigation of both processes by spectroscopic techniques is not straightforward. In the latter case, the investigation requires the detection of weak inner- and outer-sphere substrate-ligand interactions, which are difficult to interrogate by most spectroscopic techniques as the perturbation to the metal centre can be quite small. These interactions can, however, be investigated by EPR techniques. By probing these key structure-reactivity relationships, one can build an accurate model for enantiomer discrimination and ultimately provide a fundamental basis for improvement in the operation of enantioselective catalysts.

Chromium and manganese complexes of ligand (**1**) (Fig. 1) have been shown to be highly effective catalysts for the epoxidation of alkenes [37, 38]. The cobalt derivative of (**1**) is also highly effective for the hydrolytic kinetic resolution of terminal epoxides [39]. Because chiral epoxides are formed or hydrolysed, respectively, in these two types of reactions at the metal complexes, we sought to investigate the nature of how chiral recognition occurs in the first place between the epoxide substrate and the asymmetric complex. Owing to the general reactivity of the Cr, Mn and Co ions towards epoxides, we utilised a less reactive

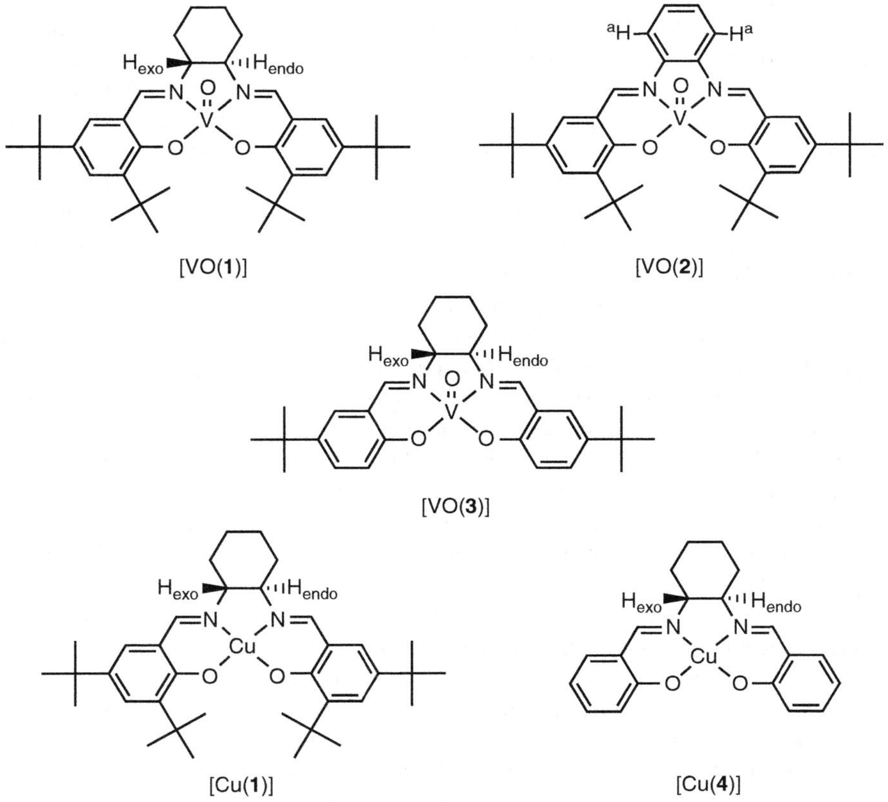

[VO(**1**)]

[VO(**2**)]

[VO(**3**)]

[Cu(**1**)]

[Cu(**4**)]

Fig. 1 Structures of complexes (**1**)–(**4**)

Lewis acid centre ([VO(**1**)]) simply to focus on the role of outer sphere interactions in the chiral transfer step (in the absence of unwanted ring opening reactions). Using CW-ENDOR spectroscopy, we observed the enantiomeric discrimination of chiral epoxides (specifically propylene oxide, C_3H_6O) by a chiral vanadyl salen-type complex [VO(**1**)] [40]. CW-EPR and ^{1}H-ENDOR spectra of *R*,*R*′-[VO(**1**)] and *S*,*S*′-[VO(**1**)] were systematically recorded in *R*-/*S*-propylene oxide. Whilst the EPR spectra of all enantiomeric combinations were virtually identical, the ^{1}H-ENDOR spectra were characteristically different; the heterochiral pairwise combinations of *R*,*R*-[VO(**1**)]+*R*-C_3H_6O and *R*,*R*-[VO(**1**)] +*S*-C_3H_6O yielded slightly different ^{1}H-ENDOR spectra, which was attributed to the presence of diastereomeric pairs [40]. This result showed for the first time how the subtle structural differences between the diastereomeric adducts in frozen solution could be detected by ENDOR [40]. Importantly, when racemic-[VO(**1**)] was dissolved in racemic-propylene oxide, the resulting ^{1}H-ENDOR spectrum was found to be identical to the spectrum of the homochiral enantiomeric pair *R*,*R*′-[VO(2)]+*R*-C_3H_6O. This result represented clear proof for the preferential binding of *R*-C_3H_6O by *R*,*R*′-[VO(**1**)] (and likewise of *S*-C_3H_6O by *S*,*S*-[VO(**1**)]).

Although CW-ENDOR revealed the presence of the diastereomeric adducts, it did not provide any evidence of how the adducts are actually formed and stabilised. Therefore we prepared two derivatives of ligand (**1**), [VO(**2**)] and [VO(**3**)] (Fig. 1), and studied their interactions with simple epoxides [41–44]. CW ENDOR was used to identify the role of H-bonds responsible for the stabilisation of the [VO(**1**)]+*cis*-2,3-epoxybutane adduct [41]. By comparison, no evidence for binding of the *trans*-2,3-epoxybutane isomer was found. In combination with DFT, a series of weak H-bonds, formed between the vanadyl complex and the epoxide substrate were identified. Notably, an H-bond was observed between the epoxide oxygen atom, O_{ep}, and one of the methine protons (H_{exo}) of the cyclohexyl group in [VO(**1**)]. Two additional H-bonds were also found to exist between the vicinal epoxide protons and each of the two phenoxide O atoms of the salen ligand. Crucially these combined H-bonds were proposed to facilitate the overall orientation of the more symmetrical *cis*-epoxide between the metal centre and the chiral salen backbone [41]. The role of these H-bonds in orientating the substrate was furthermore confirmed using the phenylene derivative [VO(**2**)] [42]. In the absence of the key H_{exo} proton in [VO(**2**)] (Fig. 1), the H-bonding between the epoxide and the complex was weakened [42], as evidenced not only by ENDOR but also by CW EPR.

Other weak outer-sphere forces, such as electrostatic interactions [43] and steric contributions [44], between the substrate and the VO-complexes [VO(**1**)] and [VO(**3**)] were also shown to contribute to the mode of chiral binding in the asymmetric adducts. In the specific case of [VO(**3**)], removal of the bulky inner *tert*-butyl groups from the 3,3′ positions was not found to moderate the electronic properties of the VO centre (revealed via EPR) or its interactions with the surrounding ligand ^{14}N nuclei (revealed via HYSCORE), but was found to reverse the stereoselectivity of epoxide binding [44]. Whilst homochiral enantiomeric adducts were preferentially formed in [VO(**1**)]+propylene oxide, EPR unambiguously proved that the opposite heterochiral enantiomeric adducts were formed in [VO(**3**)]+propylene oxide [44].

2.2 Chiral Amine Recognition

Observation of the stereoselective manner of chiral substrates binding to these asymmetric metal-salen complexes was not confined to [VO(**1**,**3**)] or chiral epoxides. Recently we showed how asymmetric copper salen complexes, [Cu(**1**)] and [Cu(**4**)] (Fig. 1), could also discriminate between chiral amines (*R*-/*S*-methylbenzylamine, MBA) as evidenced by multi-frequency CW and pulsed EPR, ENDOR, HYSCORE and DFT [45]. The discrimination of the MBA enantiomers was directly observed by W-band EPR. By simulating the W-band EPR spectra of the individual diastereomeric adduct pairs (i.e. *R*,*R'*-[Cu(**4**)]+*R*-MBA and *R*,*R'*-[Cu(**4**)]+*S*-MBA), accurate spin-Hamiltonian parameters could be extracted for each adduct. The EPR spectrum of the racemic combinations (i.e. *rac*-[Cu(**4**)]+*rac*-MBA) was then simulated using a linear combination of the *g*/*A* parameters for the homochiral (*R*,*R'*-[Cu(**4**)]+*R*-MBA) and heterochiral (*R*,*R'*-[Cu (**4**)]+*S*-MBA) adducts. An analogous series of measurements was performed for the [Cu(**1**)] complex. This revealed an 86:14 preference for the heterochiral adducts (*RR*-*S* and *SS*-*R*) compared to the homochiral adducts (*RR*-*R* and *SS*-*S*) in [Cu(**1**)], diminishing to 57:43 in favour of the heterochiral adducts in [Cu(**4**)] [45].

DFT also sheds light on the origins of this selectivity. The computational results revealed that the bulky phenyl ring of MBA destabilises the formation of any adduct whereby the MBA-phenyl ring is placed over the *tert*-butyl groups at positions 3,3′ and 5,5′ of the complex (Fig. 1). Instead, the steric hindrance between the complex and MBA is minimised when the MBA-phenyl ring is positioned over the phenyl rings of the [Cu(**1**)] complex. Two stabilisation sites could be identified in the homochiral adduct *R*,*R'*-[Cu(**1**)]+*R*-MBA with one structure slightly preferred by 2 kJ mol^{-1}, due to the small unfavourable steric interactions between the MBA-phenyl ring and the ligand cyclohexyl ring that occurred in the other structure. However, the most stable site was found for the heterochiral adduct *R*,*R'*-[Cu(**1**)]+*S*-MBA, in agreement with the experiments. This site was found to be slightly preferred by 5 kJ mol^{-1} compared to the homochiral adduct sites. In this heterochiral case, the α-proton of *S*-MBA was found to point away from the ligand methine proton; the reverse situation occurred in the homochiral adducts.

2.3 Chiral Recognition and the Role of the Outer-Sphere

Homogeneous asymmetric catalysts often deliver enantiomeric excesses (e.e.s) of greater than 99%. The structural features of the ligand are clearly important to achieve these high e.e.s, since the ligand not only stabilises the transition metal ion and associated transition state, but also modulates the trajectory of the incoming chiral substrate. Bulky framework substituents (such as *tert*-butyl groups) are known to prevent stabilisation of transition states, since these have similar energies for the two diastereomers. This is particularly true in chiral metal-salen complexes, whereby the efficiency of the catalyst depends on the nature of the bulky substituents at the 3,3′ and 5,5′ positions and regulates the orientation of the incoming substrates, creating a high diastereofacial preference.

It is clear from the work summarised above [40–45] that diastereomeric discrimination of chiral substrates occurs in asymmetric complexes. On the one hand in [Cu (**1**)], W-band EPR revealed a strong preference for the *heterochiral* adducts of [Cu (**1**)] compared to the homochiral adducts [45]. On the other hand, X-band ENDOR revealed an exclusive preference for the *homochiral* adducts of [VO(**1**)] with chiral propylene oxide [40, 43]. The origin of these selectivities was shown to arise from a combination of weak outer sphere interactions including H-bonding [40, 41], electrostatic influences of the substrate [43] and the subtle steric perturbations of key functional groups on the asymmetric ligand [45]. Whilst the bulky *tert*-butyl groups may affect the stability of the transition states during the reaction, these groups undoubtedly also affect the stereo-discrimination of chiral substrates. It is important to note that, whilst the presence of such diastereomeric adducts are often presumed as mechanistic intermediates, they are rarely observed directly in cases of weak complex–substrate interactions. The results reported here therefore demonstrate the useful role of EPR techniques in probing such diastereomeric adducts, which may be of direct relevance to studies in homogeneous asymmetric catalysis.

3 Active Catalytic Oxygen Species: Model Bio-Mimetic Systems

Nature has evolved iron enzymes, like non-heme iron oxygenases, capable of carrying out hydrocarbon oxidations with high degrees of selectivity under mild conditions [46–50]. Significant efforts have therefore been made recently to reproduce these reactions for fine chemical production by synthesis of low molecular weight (homogeneous) analogues. A particularly active field of research is the preparation of artificial metalloenzymes for enantioselective catalysis [36, 51]. In the specific case of Fe-based systems, one key part of this effort is to determine the reactivity of non-heme iron-(hydro)peroxo species in oxidation reactions [52] and thus to understand and mimic the key structural and functional properties of the natural enzymes. Paramount to these investigations is the availability of synthetic analogues that can react with molecular oxygen (O_2) and its reduced forms (O_2^- and H_2O_2). A number of groups have synthesised metal complexes that mimic the iron site in superoxide reductase (SOR) and thus developed bio-inspired iron-based oxidation catalysts [53, 54]. The ferrous site in SOR is based on one cysteine and four histidine ligands bound to the iron centre in 5-coordinate, square pyramidal geometry. The mechanism of O_2^- reduction is not well known, but participation of a (hydro)peroxo-iron^{III} intermediate [Fe^{III}-OO(H)] is considered likely [55–57]. Identification of such an intermediate is not, however, straightforward.

Jiang et al. [58] therefore prepared the [Fe^{II}([15]aneN_4)(SC_6H_4-*p*-Cl)]BF_4 complex (**5**), which reacts with molecular oxygen at low temperatures to yield the Fe-OOH centre (**6**) (Fig. 2). The X-band CW-EPR spectrum of the centre was attributed to a low-spin species with principal g parameters of [2.347(2), 2.239 (2), 1.940(2)], consistent with the π-bonding between Fe^{III} and the hydroperoxo π* orbitals, which destabilizes the d_{xy} orbital relative to the other t_2 orbitals and results in the unpaired electron being predominantly in d_{xy} [58]. A possible assignment assuming a peroxo bridge di-Fe^{III} complex was ruled out based on quantitative measurements. Analysis of the EPR spectrum also gave the ligand-field splitting parameters $|\Delta/\xi| = 7.85$ and $|V/\xi| = 2.6$. These values were found to be comparable to those of other Fe^{III}-OOH (or R) complexes [59]. This ferric-hydroperoxo complex (**6**) was unable to oxidise PPh_3 to $OPPh_3$. This lack of reactivity in electrophilic oxidations has been seen before for non-heme Fe^{III}-OOH complexes [58]. However, complex (**6**) was found to react towards weak acids.

Ferric-hydroperoxo and -peroxo centres based on tris-(pyridylmethyl)ethane-1,2-diamine ligands have also been studied by EPR [60]. The two complexes displayed in Fig. 2 were characterized in solution as the purple low-spin Fe^{III}-OOH complex (**7**), which converts upon addition of base to a blue Fe^{III}-η^2-peroxo species (**8**). The low-temperature CW-EPR spectrum of (**7**) displayed a characteristic low-spin ferric spectrum, with $\boldsymbol{g} = [2.12, 2.19, 1.95]$. However, the low-temperature CW-EPR spectrum of (**8**) was quite different, with strong signals at $\boldsymbol{g} = [7.4, 5.7, 4.5]$ typical of high-spin ferric species ($S = 5/2$). The high-spin EPR signal of (**8**) had almost axial zero-field splitting. The signal at $g \approx 7.4$ and $g \approx 4.5$ were assigned to the effective values g'_y and g'_x of the $m_S = \pm 1/2$ Kramers' doublet, respectively. The signal at $g \approx 5.7$ corresponded to the effective g'_z of the middle doublet $m_S = \pm 3/2$. Furthermore, this peroxo complex (**8**) exhibited well resolved magnetic hyperfine patterns in the Mössbauer spectra that matched the EPR results.

The chemistry of heme and non-heme iron in high-valent states is also of great interest. A problem encountered in this chemistry is that the porphyrin ligand itself can also be oxidized to form a π radical. Reactive Fe^{IV}-oxo units can then be coordinated to an oxidized porphyrin radical. In such complexes the oxidation state of the iron would actually be higher if it were not for the oxidation state of the porphyrin. In non-heme iron systems, the ligands bound to iron are generally considered to be redox innocent, and intermediates containing Fe^{IV} and Fe^{V} have been postulated. An experimental and theoretical approach was taken by Berry et al. [61] in the investigation of electron-transfer processes for three ferric complexes of the pentadentate ligand 4,8,11-trimethyl-1,4,8,11-tetraazacyclotetradecane-1 acetate (Me_3cyclam acetate) with axial chloride, fluoride and azide ligands. This provided a unique opportunity to observe iron-centred redox processes in Fe^{II}, Fe^{III} and Fe^{IV} complexes with essentially the same ligand sphere.

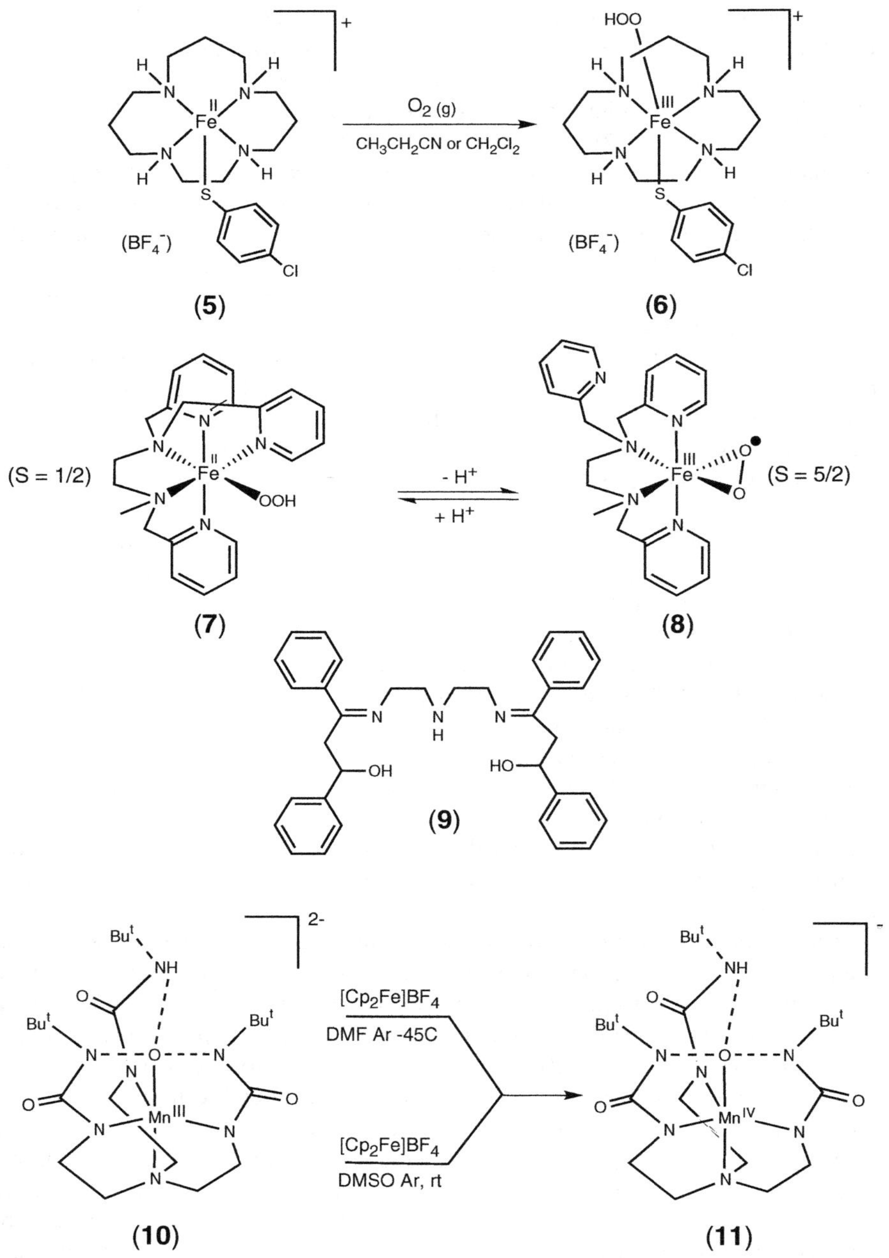

H H N N Fe II N N H H S (BF_4^-) Cl +

O_2 (g)
CH_3CH_2CN or CH_2Cl_2

HOO H H N N Fe III N N H H S (BF_4^-) Cl +

(**5**) (**6**)

(S = 1/2) N N Fe II N N N OOH

- H^+
+ H^+

N N N Fe III O O N N (S = 5/2)

(**7**) (**8**)

N N H N OH HO

(**9**)

Bu^t NH O Bu^t Bu^t N O N N O O N Mn^{III} N N 2-

$[Cp_2Fe]BF_4$
DMF Ar -45C

$[Cp_2Fe]BF_4$
DMSO Ar, rt

Bu^t NH O Bu^t Bu^t N O N N O O N Mn^{IV} N N -

(**10**) (**11**)

Fig. 2 Structures of complexes (**5**)–(**11**)

Whereas the above cited works of Jiang et al. [58] and Simaan et al. [60] were heavily focused on the spectroscopic (EPR) characterization of the model Fe^{III}-hydroperoxo and Fe^{III}-peroxo complexes, Bilis et al. [62] investigated the catalytic oxidation of hydrocarbons (cyclohexane) by homogeneous and heterogeneous non-heme Fe^{III} centres using H_2O_2. The Fe^{III} complex was based on 3-{2-[2-(3-hydroxy-1,3-diphenyl-allylideneamino)-ethylamino]-ethylimino}-1,3-diphenyl-propen-1-ol [Fig. 2; complex (**9**)]. CW EPR initially revealed the presence of a high-spin Fe^{III} ($S = 5/2$) centre in a rhombic field characterized by $E/D \sim 0.33$. This signal was, however, found to be heavily solvent based, since in the presence of CH_3CN a new Fe^{III} signal with g parameters [2.052, 2.005, 1.80] was observed to form at the expense of the high-spin EPR signal. This low-spin Fe^{III} centre was proposed to be Fe^{III}-OOH. In situ EPR revealed that this low-spin EPR signal is progressively lost during the catalytic reaction, whereas the high-spin EPR signal remains unaffected, indicating the role of Fe^{III}-OOH in the catalysis.

The investigation of reactive metal centres bearing oxygen intermediates has not been confined to iron. Manganese [63, 64] and copper [65] have also attracted significant attention. Parsell et al. [64] investigated the properties of a Mn^{IV} complex bearing a terminal oxo ligand, which converted some phosphines to phosphine oxides. This complex was formed starting from an $[Mn^{III}H_3buea(O)]^{2-}$ ($[H_3buea]^{3-}$, tris[(N'-*tert*-butylureaylato)-N-ethylene]aminato) (**10**) (Fig. 2), a monomeric Mn^{III}-O complex in which the oxo ligand derived from dioxygen cleavage or deprotonation of water. This $[H_3buea]^{3-}$ ligand is important since it regulates the secondary coordination sphere by providing a sterically constrained H-bond network around the Mn^{III}-O unit. The Mn^{IV}-oxo species (**11**) (Fig. 2) was then formed using a mild oxidant at low temperatures. The low-temperature (4 K) X-band EPR spectrum of (**11**) revealed principal g values of [5.15, 2.44, 1.63], corresponding to a system having an $S = 3/2$ state with an $E/D = 0.26$ [64]. The temperature dependence of the EPR spectra and simulation of the signals indicated a value of $D = 3.0$ cm^{-1} and a ^{55}Mn hyperfine coupling of 190 MHz, comparable to other Mn^{IV} complexes. Complex (**11**) did not react with PPh_3 or PCy_3 in DMSO, but did react with $PMePh_2$ via O-atom transfer to produce $O{=}PMePh_2$ in 50–70% yields. Oxygen-atom transfer is normally a two-electron process and would yield phosphine oxide and the corresponding Mn^{II} complex. Evidence for the formation of this reduced complex came from the X-band EPR spectra.

4 Ligand- Versus Metal-Centred Redox Reactions

For the past 10 years, chemists have been interested in how to define the charge of a metal ion in complexes bearing non-innocent ligands, most notably through the works of Bill and Wieghardt [66]. As Jörgensen [67] suggested many years ago, an oxidation number that is derived from a known d^n electron configuration should be specified as the *physical* (or spectroscopic) oxidation number. However, this is not always straightforward. When organic radicals with open-shell electron configurations are coordinated to a transition metal ion, the oxidation state of the metal is less well defined because the oxidation may be ligand- or metal-centred:

$$M^{n+} - O - R \xrightarrow{-e} M^{n+} - O^{\bullet} - R \quad \text{or} \quad M^{(n+1)+} - O - R \tag{1}$$

For example, in the Fe (d^5) coordinated phenoxyl-radical complex (Fe^{III}–$O^{\bullet}$–Ph), the formal oxidation state of the metal is classed as +IV, since a closed shell phenolato anion would have to be removed. However, in many cases spectroscopic measurements, amongst others EPR, have proven the presence of a high-spin d^5 electron configuration at the iron and a phenoxyl ligand in such complexes. In this case, the iron ion has a *physical oxidation* number of +III even though the *formal oxidation state* would be classed as +IV. As a result of these potential confusions, several research groups have prepared numerous examples of metal-coordinated ligand-radical complexes, particularly coordinated phenoxyl radicals, in order to examine the nature of the metal oxidation states and the extent of spin delocalisation in such complexes.

There is also another important reason for studying these coordinated (phenoxyl) radicals. The inter-conversion and synergism between the redox active metal centres and proximal organic cofactors is very important in many biological reaction centres, particularly those including ET reactions [68]. A good example is the two-electron oxidation of primary alcohols with O_2 to produce aldehydes and H_2O_2 as catalysed by galactose oxidase (GAO). The active site in GAO is a Cu centre ligated by a cysteine-modified tyrosine group. Owing to the growing number of metal-phenoxyl radical systems, chemists have developed strategies to prepare and characterise model compounds containing coordination Cu-phenoxyl radicals [68]. Studies of these model complexes have provided important insights into the structure and function of the GAO enzyme, and, equally important, have acted as a cornerstone in the recent development of bio-inorganic catalysts [69]. The inactive site of GAO is EPR active, and produces an EPR spectrum with well defined parameters [70]. Unfortunately the active and reduced forms of GAO are EPR silent. Therefore EPR is of limited use in the characterization of true GAO mimics, since the magnetically coupled spins of the active state produce a diamagnetic ground state. Whilst EPR has naturally been used to investigate many of these model Cu^{II}-coordinated phenoxyl radical complexes, significant attention has also been given to the EPR characterization of phenoxyl radicals of Cr^{III}, Mn^{III}, Fe^{III}, Co^{III} and Ni^{II} [71–77].

Coordinated phenoxyl radicals in Schiff bases and phenolate ligands have understandably attracted the most attention in the past 10 years because of the reversible redox states that can be achieved [78–82]. The most striking example of this redox chemistry was demonstrated by the recent findings with Ni^{II}-salen complexes [77, 78]. Shimazaki et al. [78] studied the electrochemical oxidation of [Ni(**1**)] (Fig. 3). The Ni^{II}-radical valence isomer could be obtained in CH_2Cl_2 and converted into the Ni^{III}-phenolate valence isomer simply by changing the

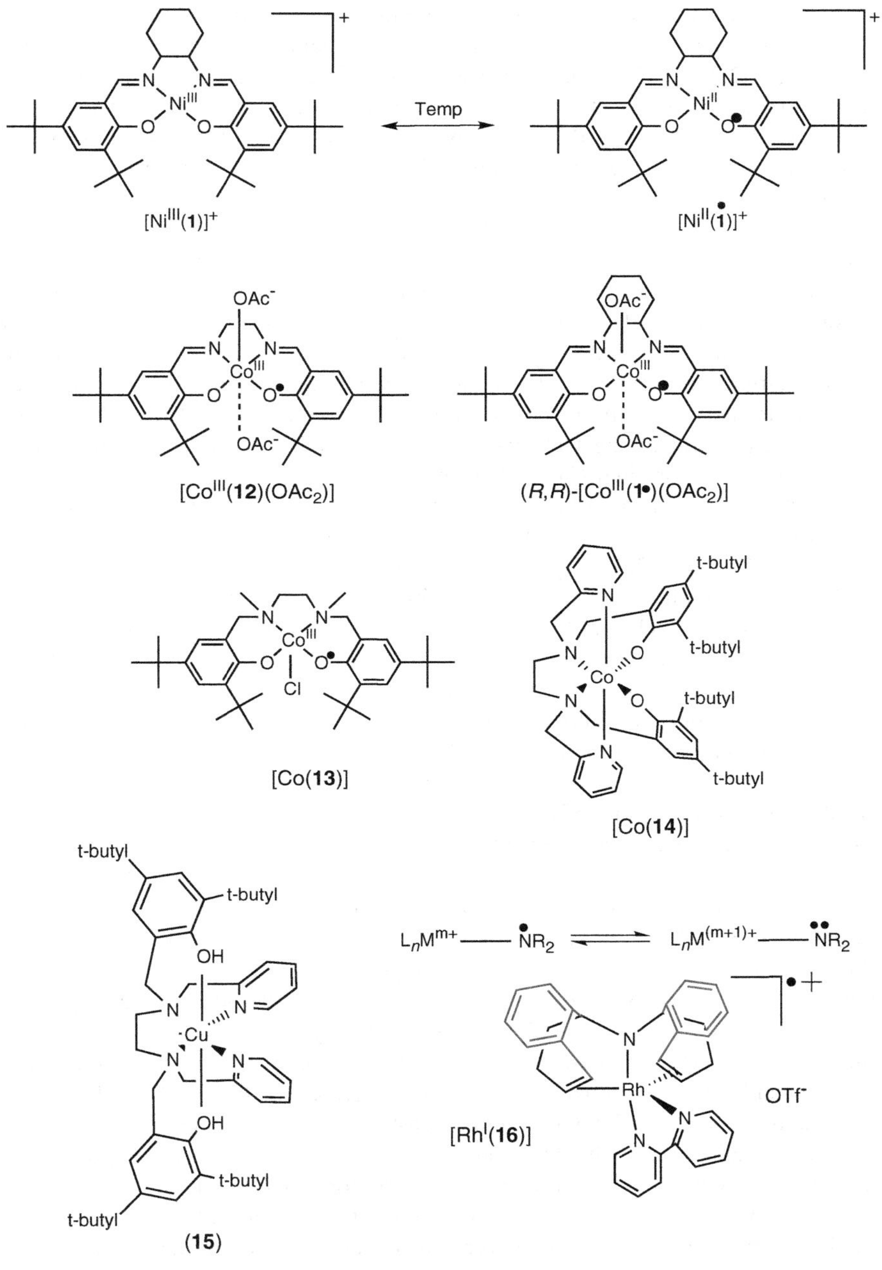

N N Ni^{III} O O +

Temp

N N Ni^{II} O O +

$[Ni^{III}(\mathbf{1})]^+$

$[Ni^{II}(\dot{\mathbf{1}})]^+$

OAc⁻ N N Co^{III} O O OAc⁻

OAc⁻ N N Co^{III} O O OAc⁻

$[Co^{III}(\mathbf{12})(OAc_2)]$

(R,R)-$[Co^{III}(\mathbf{1}^\bullet)(OAc_2)]$

N N Co^{III} O O Cl

$[Co(\mathbf{13})]$

N t-butyl N O t-butyl Co N O t-butyl N t-butyl

$[Co(\mathbf{14})]$

t-butyl t-butyl OH N N Cu N N OH t-butyl t-butyl

(**15**)

L_nM^{m+}—$\dot{N}R_2$ ⇌ $L_nM^{(m+1)+}$—$\ddot{N}R_2$

N Rh N N •+ OTf⁻

$[Rh^I(\mathbf{16})]$

Fig. 3 Structures of complexes (**1**), (**12**)–(**16**)

temperature and the solvent. At low temperature (123 K), a Ni^{III}-phenolate complex was easily identified by EPR, based on the characteristic rhombic $\boldsymbol{g}$ tensor ($\boldsymbol{g}$ = [2.30, 2.23, 2.02]) typical of a low-spin |z^2, $^2A_1$ > ground state. However, at elevated temperatures (158–173 K), this rhombic signal evolves into an isotropic EPR signal with $g_{iso} = 2.04$, which was assigned to the Ni^{II}-phenoxyl radical, indicating the tautomerism that can exist between the two redox states ($[Ni^{III}(\mathbf{1})]^+$ or $[Ni^{II}(\mathbf{1}^{\bullet})]^+$) (Fig. 3).

Most recently, Rotthaus et al. [80, 81] extended this study to a range of closely related Schiff base Ni^{II} complexes, proving the formation of the Ni^{II}-phenoxyl radical with partial delocalisation of the SOMO onto the metal orbitals. Pratt and Stack [83, 84] have also generated and characterised the coordinated phenoxyl radical in $[Cu^{II}(\mathbf{1})]$ (Fig. 1) (labelled $[Cu^{II}(\mathbf{1}^{\bullet})]^+$) as a bio-mimetic model for galactose-oxidase complexes. The EPR data reported in their study was consistent with the formation of an anti-ferromagnetically coupled Cu^{II}-phenoxyl complex, whereby oxidation of the Cu^{II} complex to Cu^{II}-phenoxyl simply resulted in an attenuation of the original Cu^{II} EPR signal by ~15%.

In a more unusual case of a coordinated phenoxyl radical bearing a Schiff-base-type ligand, we recently reported the identification of $[Co^{II}(\mathbf{1}^{\bullet})(OAc)_n](OAc)_m$ ($n = m = 1$ or $n = 2$, $m = 0$), simply by treatment of $[Co(\mathbf{1})]$ with acetic acid under aerobic conditions [85]. These conditions are analogous to those employed in the activation of $[Co(\mathbf{1})]$ for the widely used hydrolytic kinetic resolution (HKR) of epoxides [8, 9]. Initially we investigated the electronic properties of the parent pre-catalyst $[Co(\mathbf{1})]$ in the absence of acetic acid and subsequently followed the changes to this catalyst after the addition of acetic acid under anaerobic conditions [86]. The pre-catalyst complex produced a CW-EPR spectrum typical for a species possessing an |yz, $^2A_2$ > ground state. Upon acetic acid addition under anaerobic conditions, new high-spin [87] and low-spin [86] centres were generated. The latter low-spin centre was characterized by the parameters $\boldsymbol{g}$ = [2.41, 2.27, 2.024]; $\boldsymbol{A}$ = [±100, ±70, ±310] MHz indicative of a |z^2, $^2A_1$ > ground state, induced by acetate ligation to the Co^{II} complex. When molecular oxygen was introduced into this system (or alternatively when the acetic-acid addition was conducted directly under aerobic conditions) a new signal assigned to the phenoxyl radical was observed. This signal was characterized by the parameters of $\boldsymbol{g}$ = [2.0060, 2.0031, 1.9943]; $\boldsymbol{A}$ = [±17, ±55, ±14] MHz, readily identifiable at X- and W-band microwave frequencies (Fig. 4) [85]. A combination of HYSCORE, Resonance Raman and DFT results proved conclusively the presence of a coordinated phenoxyl radical, as opposed to a bound (ligating) substrate based radical [85].

The formation of the phenoxyl radical was proposed to occur in the presence of acetic acid by coupling the two-electron, two-proton reduction of molecular oxygen to H_2O_2 [85]. In some way this process is reminiscent of the half reaction observed in GAO. The unusual aspect, however, was its identification in the activated HKR reaction system, although there was no evidence for its involvement in the hydrolysis of epoxides [85]. As Wieghardt noted in earlier works, bulky substituents are required for stabilization of metal-coordinated phenolate radicals [66]. We confirmed this by activation of a Co-salen derivative, $[Co(\mathbf{12})]$ (Fig. 3). In the absence

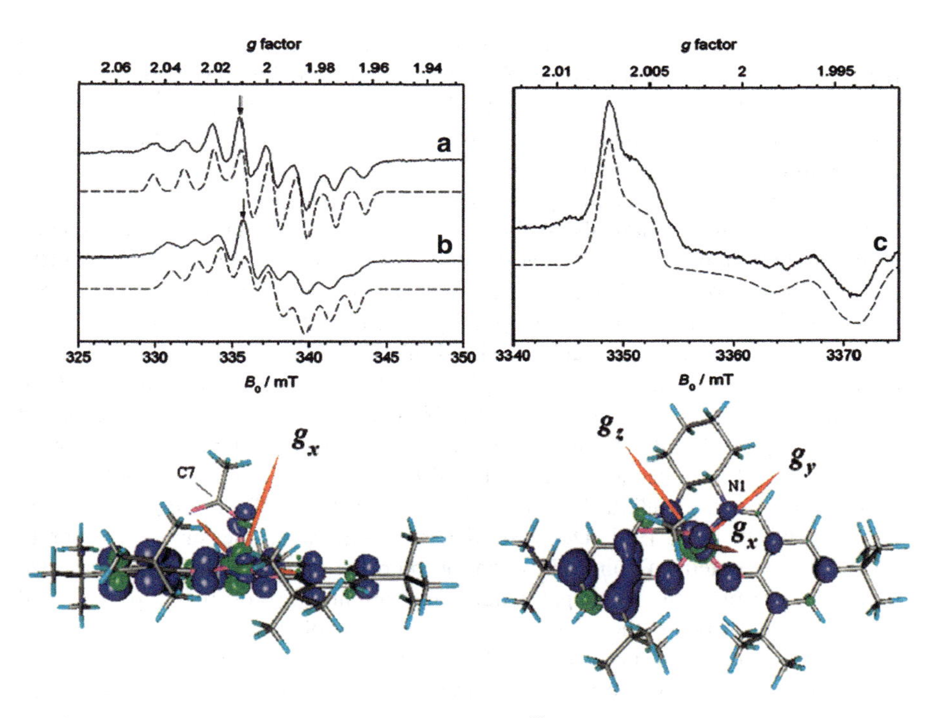

Fig. 4 Schematic illustration of the coordinated Co^{II}-phenoxyl radical, bearing coordinated acetate groups, derived from [Co(**1**)] after addition of acetic acid under aerobic conditions. *Top*: (**a**, **c**) the X- and W-band CW EPR spectra of $[Co^{II}(\mathbf{1}^{\bullet})(OAc)_n](OAc)_m$ ($n = m = 1$ or $n = 2$, $m = 0$) and (**b**) the X-band CW-EPR spectrum of $[Co^{II}(\mathbf{1}^{\bullet})(Py)_2]$. *Bottom*: the DFT-computed spin densities of $[Co(\mathbf{1}^{\bullet})(OAc)]^+$ shown from the side and top elevation. *Blue* is positive spin density, *green* represents negative spin density. Adapted and reprinted with permission from [85]. Copyright 2011 American Chemical Society

of the cyclohexyl backbone, but importantly in the presence of the *tert*-butyl groups, we confirmed (using EPR, HYSCORE and DFT) that coordinated phenoxyl radicals could indeed be generated and stabilized in [Co(**1**)] [88].

Thomas et al. [74] have also studied the one- and two-electron-oxidized (electrochemically generated) radicals of [Co(**13**)], [Co(**14**)] and [Cu(**15**)] (Fig. 3). The X-band CW-EPR spectra of [Co(**13**)] and [Co(**14**)] were shown to be typical of an $S = 1/2$ ground state. The di-radical of [Co(**14**)] was also prepared electrochemically, and in this case the X-band CW-EPR spectrum of the di-radical (recorded at 100 K) was replaced by a signal typical of an $S = 1$ spin state observable at 9 K. The Curie plot revealed that the triplet corresponded to the ground state and thus coupling between the two radical fragments was ferromagnetic [74]. This coupling generally arises from a lack of overlap between the SOMOs and thus a co-planarity between the two rings. This interpretation was also in good agreement with the structure based on the bis-phenolate precursor [Co(**15**)], thereby revealing that the global arrangement of the complex did not change significantly upon two-electron oxidation. The behaviour was also in good agreement with the $S = 1$ ground state recently reported for the di-radical Zn^{II} analogue of (**14**) [89]. As anticipated, the oxidized [Cu(**15**)] complex was EPR silent (a residual, 30%, Cu^{II} EPR signal remained in the electrochemically oxidized solution) due to the magnetic interaction between the phenoxyl radicals ($S = 1/2$) and the Cu^{II} spin.

The study of coordinated ligand radicals in metal complexes has not been confined to phenoxyl radicals. Although much rarer, the analogous coordinated aminyl radicals have also being investigated. Büttner et al. [90] revealed how a Rh-based coordination complex could also support an aminyl radical, $[Rh^{I}(trop_2N^{\bullet})(bipy)]^{+}OTf^{-}$ (trop = 5-*H*-dibenzo[*a,d*]cycloheptene-5-yl) labelled $[Rn(\mathbf{16})]^{\bullet+}$ (Fig. 3). The complex essentially consisted of a $trop_2N^{\bullet}$ radical coordinated to the cationic Rh centre. Just like phenoxyl radicals, the unpaired electron in these coordinated radicals may either be at the N centre or at the metal, but advanced EPR techniques and DFT demonstrated the existence of the ligand-based paramagnetic centre. The experimental S-band CW-EPR spectrum was simulated using a rhombic $\boldsymbol{g}$ matrix [2.0822, 2.0467, 2.0247] and a large ^{14}N principal hyperfine coupling of 98 MHz [90]. Davies-ENDOR spectra were measured in order to determine the extent of spin delocalization onto the ligand. HYSCORE was also used to determine the principal values and orientations of the hyperfine and quadrupole tensors for the strongly and weakly coupled nitrogens. A large A_{iso} for the apical nitrogen atom of 45.1 MHz was identified. This nitrogen has pronounced anisotropy, indicating the unpaired electron resides in an orbital with high p-character. These experimental results were fully supported by the DFT calculations and proved the paramagnetic centre was best described as the aminyl Rh^{I} complex, rather than the Rh^{II}-amide complex (Fig. 3).

The chemical reactivity of $[Rh^{I}(\mathbf{16})]^{\bullet+}$ was also examined in reactions with H-atom donors, where the complex behaves as a nucleophilic radical [90]. The reaction rates were dependent on the X–H bond dissociation energies, reacting rapidly with stannane (Bu_3Sn–H) and thiophenol, more slowly with *tert*-butyl thiol and thioglycolic acid methyl ester and not at all with phenol and triphenylsilane [90].

5 Mechanistic Insights into Ethylene Polymerization

Linear alpha olefins (LAOs) are useful intermediates for a range of important commodity chemicals (including surfactants, lubricants, plasticizers, etc.). They are produced via ethylene oligomerisation, using transition metal catalysts. A major problem associated with these catalysts is the formation of a broad chain length distribution of α-olefins. One approach to solving this problem, operating via a uniquely different mechanism, is ethylene trimerization and tetramerization to 1-hexene and 1-octene, respectively [91]. Recent developments have focussed on designing highly selective catalysts and, to date, Cr catalysts account for >90% of the literature of ethylene oligomerization, with Ti, Co, Ta and Fe catalysts also available. The mechanism is thought to follow a metallacyclic route, involving oxidative addition of two ethylene molecules to the metal followed by insertion of another to yield a metallacycloheptane species [91, 92]. Concerning the Cr oxidation state, there is evidence for $Cr^{I/III}$ and $Cr^{II/IV}$ couples [93, 94]. Attempts have been made to determine the oxidation states by experimental and computational studies [95]. The debate still continues, and, while the precise nature of the redox states is not known, very little consideration has been given to the spin states involved.

As stated above, chromium has been used mostly extensively as a catalyst for ethylene polymerisation, and several groups have studied the catalytic system by EPR [96–99]. For example, Brückner et al. [97] has used in situ EPR to monitor the changes to the oxidation state of a $Cr(acac)_3$/PNP mixed catalyst [PNP = Ph_2PN(*i*-Pr)PPh_2] and also in a [(PNP)$CrCl_2$(μ-Cl)]$_2$ complex for ethylene oligomerization. The authors demonstrated that, for the $Cr(acac)_3$/PNP system, the initial Cr^{III} centre was reduced to a low-spin Cr^{I} centre upon activation with methyl aluminoxane (MMAO). It was proposed that the major active species in the reaction were EPR-silent centres, possibly anti-ferromagnetic Cr^{I} dimers. The authors also reported the decomposition of the [(PNP)$CrCl_2$(μ-Cl)]$_2$ complex into a bis-tolyl centre, reported to be $[Cr(\eta^6\text{-}CH_3C_6H_5)_2]^+$, which may deactivate the catalytic system [97].

Skobelev et al. [98] also examined the nature of the redox changes in $Cr(acac)_3$/pyrrole/$AlEt_3$/$AlEt_2Cl$ and $Cr(EH)_3$/pyrrole/$AlEt_3$/$AlEt_2Cl$ catalytic systems by EPR (acac = acetylacetonate). These systems were chosen specifically to model the Philips ethylene trimerization catalyst. The initial pre-catalysts, $Cr(acac)_3$, had reported parameters of $g_x = g_y = g_z = 1.97$, $D = 0.413$ and $E = 0.011$ cm^{-1}, whilst $Cr(EH)_3$ had parameters of $g_x = g_y = g_z = 1.982$, $D = 0.052$ cm^{-1} and $E = 0.008$ cm^{-1} [98]. Following activation with $AlEt_3$ two types of EPR-active chromium species were identified in both catalytic systems, including Cr^{III} and Cr^{I} complexes with proposed structures of $Cr^{III}(Pyr)_xCl_yEt_zL$ and $Cr^{I}(Pyr)L$ respectively, where L was (an) unidentified ligand(s). The authors also proposed that the major part of the chromium probably existed in the form of EPR-silent Cr^{II} species. The ethylene trimerization activity of the catalyst systems studied correlated with the concentration of Cr^{I} species in the reaction solution. The data obtained therefore supported the $Cr^{I,III}$ mechanism of ethylene trimerization [98].

Whilst many groups investigated the role of the $Cr^{I,III}$ couple in ethylene trimerization, starting from the air-stable Cr^{III} precursor, McDyre et al. [99] have taken an alternative approach to probe the mechanism by starting from the Cr^{I} precursor. Using CW-EPR and ENDOR, a series of structurally related Cr^{I}-carbonyl complexes $[Cr(CO)_4L]^+$ [L = ($Ph_2PN(R)PPh_2$)/($Ph_2P(R)PPh_2$)] complexes were investigated (complex (**17**) in Fig. 5). The EPR spectra were dominated by the ***g*** anisotropy, with notably large hyperfine couplings from the two equivalent ^{31}P nuclei. The spin Hamiltonian parameters [$g_\perp$ ($g_x = g_y$) > g_e > $g_\parallel$ (g_z)] were shown to be consistent with a low-spin d^5 system possessing C_{2v} symmetry, with a SOMO where the metal contribution was primarily d_{xy} for all complexes [99]. Attempts to correlate trends in EPR-derived parameters with the catalysis data revealed no obvious connections. However, upon activation of the $[Cr(CO)_4L]^+$ complex using $AlEt_3$, the spectra changed dramatically. No evidence was found for the

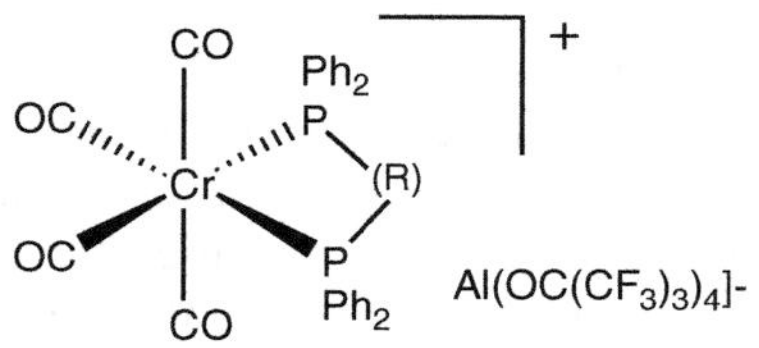

Fig. 5 Structures of complexes (17)–(20)

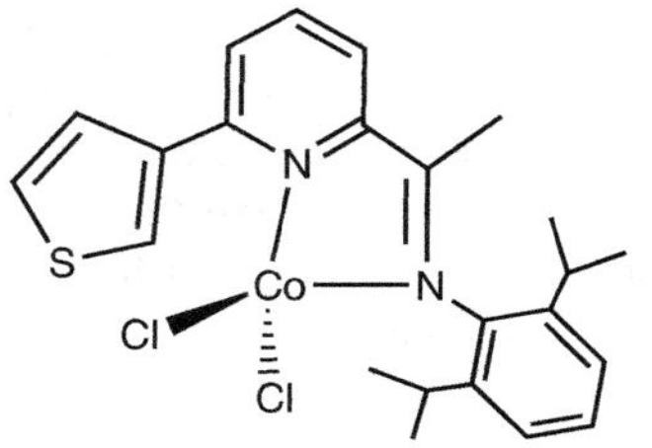

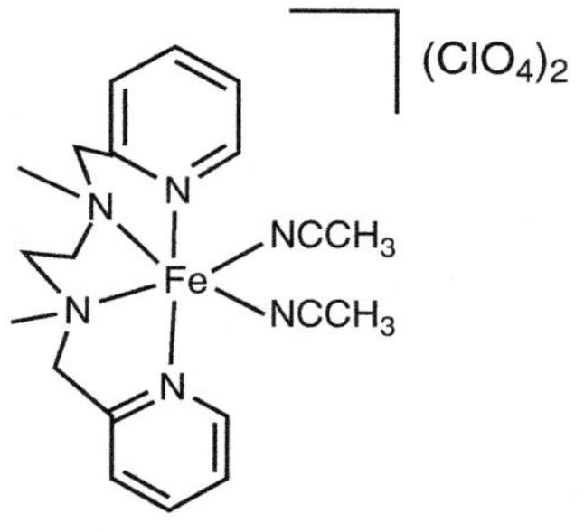

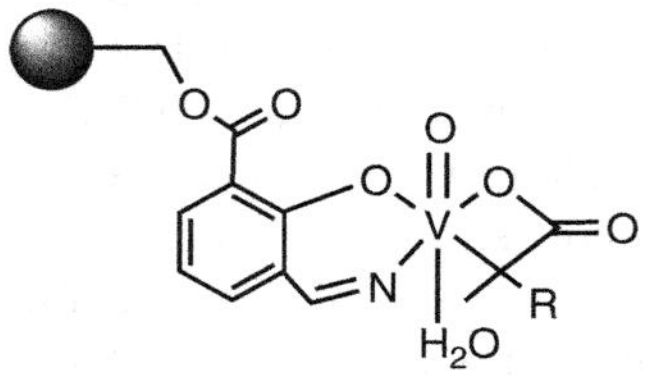

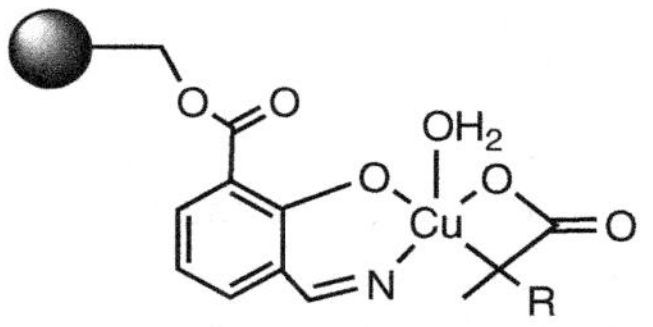

formation of any Cr^{III} centres. Instead a new series of Cr^{I} complexes was formed, some of which retained the coordinated PNP or P(R)P backbone (i.e. $[Cr(CO)_xL]^+$) whilst in other cases new Cr^{I} complexes having undergone a ligand slippage process were identified [100].

A significant number of oligomerisation studies have also been devoted to cobalt, as investigated by CW EPR [27, 101, 102]. Bianchini et al. [27] showed how the position of the sulphur atom in the thienyl groups of 6-(thienyl)-2-(imino) pyridine ligands ((**18**) in Fig. 5) strongly affects the catalytic activity of the corresponding tetrahedral high-spin dihalide Co^{II} complex following activation with methylaluminoxane (MAO). In catalytic experiments, Co^{II} complexes bearing a sulphur atom in the 3-position of the thienyl ring were found to catalyse the selective conversion of ethylene to 1-butene. From an EPR perspective, in situ experiments revealed the occurrence of a spin-state changeover from high-spin tetrahedral Co^{II} to low-spin Co^{II} following activation with MAO. The tetrahedral (imino)pyridine cobalt complexes were EPR-silent at room temperature. However, at low temperatures (10 K), a broad signal was detected for these high-spin states. After MAO addition the signal changed dramatically, reverting from high-spin to low-spin ($S = 1/2$). The most likely coordination geometry of the Co^{II} centres in the activated species was proposed to be square planar with two nitrogen atoms from the (imino)pyridine ligand, a carbon atom from a methyl group released by MAO, and a fourth ligand that might be provided by the organyl group in the 6-position of the pyridine ring. Indeed, in the absence of either ring, as in $CoCl_2N_2Br$, no EPR signal appeared upon treatment of the complex in toluene with an excess of MAO in the temperature range from 293 to 20 K. It is therefore very likely that the organyl group in the 6-position interacted with the cobalt centre.

The nature of the active sites responsible for ethylene polymerisation was also examined in a series of related Fe and Co bis(imino)pyridine complexes [102, 103]. In these studies EPR was used as a complementary characterisation technique, in conjunction with Mössbauer and NMR, revealing a change in oxidation state upon activation with MAO and triethylenealuminium.

6 Catalytic Activations and Transformations

The catalytic activation of small molecules is an extremely important area of research. For example, the ability to activate CO and CO_2 or the activation of C–H and C–C bonds as a potential source of chemical feedstocks, is a strong motivation in today's chemical industry. Recently cationic Ir- and Rh-carbonyl complexes were shown to be effective for CO activation [104], whilst Pd, Mo and Ru polyoxometalates were also shown to be very good candidates for CO [105] and CO_2 [106] activation. In the particular case of alkene epoxidation, the highly desirable epoxide derivatives are used as valuable chemicals in organic synthesis and in the manufacture of commodity chemicals. Transition metal-catalysed epoxidation is one of the most efficient approaches for this transformation. Metals such as manganese and chromium complexes, with ligand frameworks such as salens [37, 38, 107, 108], porphyrins [109–111] and aromatic N-donors [112–114], are among the most widely used and investigated systems. Not surprisingly, EPR has played an important role in understanding the mechanistic details of all these small molecule activation processes, as illustrated in the following.

6.1 Selective Oxidation

Whilst [MnCl(**1**)] complexes are effective in the epoxidation of *Z*-alkenes [38, 107], the [CrCl(**1**)] complex is also very suitable for epoxidation of alkenes. In order to identify the paramagnetic intermediates involved in the [CrCl(**1**)] epoxidation reaction, Bryliakov et al. [115, 116] studied two structurally related complexes, [CrCl(**1**)] and *racemic*-*N*,*N*′-bis(3,4,5,6-*tetra*-deutero-salicylidene)-1,2-cyclohexanediamino chromium(III) chloride. The effective g values of $g \approx 4$ and 2 identified in the spectra of both complexes were assigned to an $S = 3/2$ spin system. Zero-field-splitting parameters revealed moderately large D values (~0.7–0.8 cm^{-1}) and a small rhombicity parameter, E (~0.108–0.042 cm^{-1}). The addition of pyridine (Py) as an activator produced a decrease of D (~0.6–0.67 cm^{-1}) and an increase in E (~0.119–0.150 cm^{-1}), indicating a noticeable structural change upon complexation with Py.

The identification of the reaction intermediates formed between [CrCl(**1**)] and iodosylbenzene (PhIO) were also investigated by EPR [115, 116]. The first intermediate was characterised by the reported spin-Hamiltonian parameters of $g = 1.970–1.974$, $A_{Cr} = 54$ MHz, $A_N = 4.5–5.6$ MHz, whilst the second species produced the parameters of $g = 1.976–1.980$, $A_{Cr} = 54$ MHz, $A_N = 5.6–6.4$ MHz. Based on the CW-EPR and ^{1}H-NMR investigation, the first intermediate was identified as a reactive mononuclear oxochromium(V) intermediate, labelled [CrVO(**1**)L] where L = Cl^- or solvent molecule. The second intermediate was identified as an inactive mixed-valence binuclear [L(**2**)Cr^{III}OCr^V(**2**)L] complex. Bryliakov et al. [115, 116] thereby proposed that the [CrCl(**1**)]-catalysed epoxidation of alkenes proceeds in accordance with a modified "oxygen rebound cycle".

In the case of [MnCl(**1**)], Campbell et al. [117] sought to provide direct spectroscopic evidence for the formation of the Mn^V=O species, widely believed to be responsible for the high enantioselectivity of the epoxidation reaction, by applying dual-mode EPR during the epoxidation of *cis*-β-methylstyrene. The parallel mode EPR spectrum of [Mn^{III}Cl(**1**)] consisted of many (>16)^{55}Mn hyperfine lines possessing distinct temperature dependencies, arising from two or more coupled Mn^{III} centres [117]. In the presence of NMO or 4-phenylpyridine-*N*-oxide (4-PPNO), the EPR spectrum showed six well-resolved hyperfine lines. The authors assigned the EPR lines to transitions between the $M_S = \pm 2$ levels, which are the lowest energy levels for the [Mn^{III}Cl(**1**)] when the axial zfs parameter is negative. The results indicated that NMO and 4-PPNO alter the ligand field around the initially five-coordinate [Mn^{III}Cl(**2**)] complex by binding to Mn^{III}, forming an axially elongated six-coordinate complex, before the formation of any reaction species. Following addition of the oxidant NaOCl to the [Mn^{III}Cl(**1**)], no parallel-mode EPR signal was observed; this implied oxidation of all Mn^{III} species. Subsequently signals assigned to an $Mn^{III,IV}$ dinuclear complex were identified during the course of the reaction. The authors thus demonstrated that the use of dual-mode EPR techniques provide a sensitive probe to changes in the ligand environment of the Mn^{III} centre and enables one to observe reactants, Mn intermediates and Mn by-products simultaneously.

As non-toxic chiral Fe^{III} complexes have recently been used as catalysts [118–120], increased knowledge of their structure-reactivity relationships becomes pertinent. X-band CW-EPR spectra of [Fe^{III}Cl(**1**)], reported by Bryliakov et al. [121], were found to be typical of high-spin $S = 5/2$ Fe^{III} complexes with $E/D \approx 0.15$. Using this complex, the conversion and selectivity of the asymmetric sulphide oxidation reaction was investigated in a variety of solvents. In previous studies [122], the active site was proposed to be the [Fe^{IV}=O(1)]$^{+\bullet}$ species. However an alternative active species was proposed [121]. Oxo-ferryl π-cation radicals are expected to have typical $S = 3/2$ spectra with resonances at $g_{eff} \approx 4$

and $g_{eff} \approx 2$. Treatment of the [Fe^{III}Cl(**1**)] complexes with PhIO and *m*-CPBA did not lead to formation of $S = 3/2$ type spectra; instead a sharp peak at $g = 4.2$ belonging to an unidentified $S = 5/2$ species was found. The species associated with this signal did not contribute to the catalytic cycle, and the intensity of its EPR signal accounted for only 10% of the total Fe concentration. From this data, the authors proposed a new catalytic system for the asymmetric oxidation of sulphides where the active species was shown to be an [Fe^{III}Cl(**2**)]+PhIO complex [121].

Iron complexes with aminopyridine ligands are also known to catalyze selective olefin oxidation efficiently using H_2O_2 or CH_3CO_3H as terminal oxidants. Duban et al. [123] used EPR spectroscopy to identify the intermediates formed during the reaction cycle of [Fe^{II}(**19**)$(CH_3CN)_2$]$(ClO_4)_2$ (Fig. 5). EPR spectra recorded after the onset of the reaction of [Fe^{II}(**19**)$(CH_3CN)_2$]$(ClO_4)_2$ with CH_3CO_3H showed a signal at $g = 4.23$, an axially anisotropic signal with $\boldsymbol{g} = [2.42, 2.42, 2.67]$ and a broad signal at $g \approx 2$. On warming the sample to room temperature, the signal at $g = 4.23$ decayed and was replaced by a weaker, sharper signal with the same g factor that remained stable over several hours. The species causing the axial signal was tentatively assigned to a mixed-valence $Fe^{III}Fe^{IV}$ complex, [(**19**)Fe^{III}-O-Fe^{IV}=O(**19**)(S)]$^{3+}$.

In contrast, the EPR spectrum recorded after onset of the reaction between [Fe^{II}(**19**)$(CH_3CN)_2$]$(ClO_4)_2$ and H_2O_2 showed several signals. Low-spin ferric hydroperoxo intermediates [Fe^{III}(**19**)(OOH)(CH_3CN)]$^{2+}$ ($\boldsymbol{g} = [2.218, 2.178, 1.967]$) and [$Fe^{III}$(**19**)(OOH)$(H_2O)$]$^{2+}$ ($\boldsymbol{g} = [2.195, 2.128, 1.970]$) were observed, in addition to the dinuclear mixed valence $Fe^{III}Fe^{IV}$ complex described above [124]. Different reaction intermediates were therefore observed under the different catalytic conditions and coincided with the differing reactivities and selectivities of the epoxidation of olefins. Whilst it was not clear if the dinuclear $Fe^{III}Fe^{IV}$ complex could possibly act as an active species in the corresponding catalytic systems, it was certainly clear that in the [Fe^{II}(**19**)$(CH_3CN)_2$]$(ClO_4)_2$/H_2O_2/CH_3COOH systems the mono-nuclear Fe^{IV} species [Fe^{IV}=O(**19**)(S)]$^{2+}$ did play an important role. More recently this group has followed up this work by investigating the oxidation reactions of a series of iron complexes with aminopyridine ligands [123, 125].

Although much less common for selective oxidation, catalytic vanadium- and copper-based complexes have also been investigated by EPR [126, 127], including vanadium complexes based on salen derivatives [128]. These catalysts are particularly active using H_2O_2 as a readily available oxidant. Maurya et al. [126] examined the oxidation of *p*-chlorotoluene and cyclohexene catalysed by the polymer anchored oxovandium^{IV} and copper^{II} complexes of amino derived tridentate ligands (**20a**, **20b**) (Fig. 5). CW-EPR was primarily used to characterise the VO and Cu EPR signals before and after the reaction. Whilst the VO catalysts displayed minimal changes, significant changes were detected in the Cu^{II} system after the catalytic reaction. The structural nature of the recovered Cu^{II} catalysts was not however assigned [126]. Di-nuclear and tri-nuclear copper clusters, derived from the enantiomeric octadentate ligand *S*-**21** (a 1,1′-binaphthyl-2,2′diamine ligand), were also successfully used in the oxidation of L-/D-Dopa derivatives to quinones [127]. High enantioselectivities were observed in the oxidation of L-/D-Dopa methyl

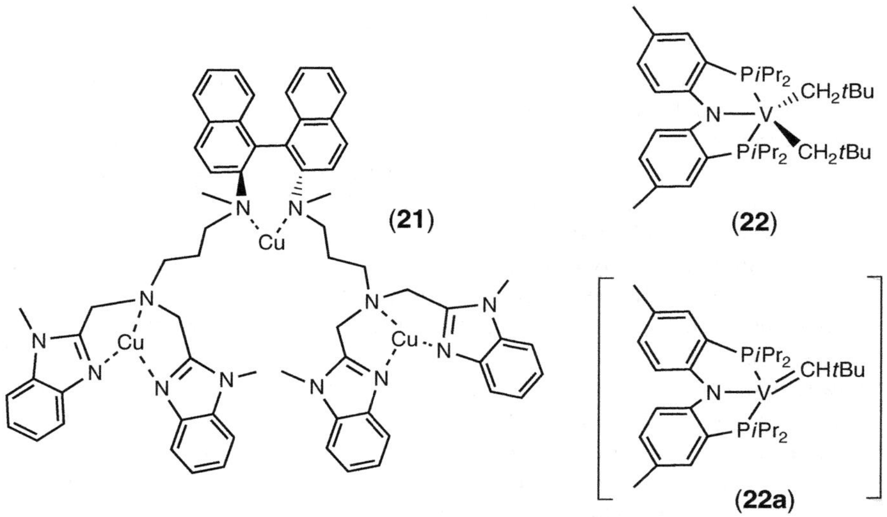

Fig. 6 Structure of complexes (**21**)–(**22**)

ester catalysed by the dinuclear Cu complex, which exhibited strong preference for the D enantiomer. The enantioselectivity was largely lost for the trinuclear Cu complex. A detailed X-band CW-EPR study was undertaken on these copper complexes. The EPR analysis of the trinuclear complex, $[Cu_3(S)\text{-}\mathbf{21}]^{6+}$ (Fig. 6), had some remarkable similarities to the copper cluster found in multicopper oxidases such as laccase. The dinuclear Cu^{II} complex exhibited the most interesting behaviour [127] because it allowed stronger chiral recognition by the binaphthyl residue. As the authors demonstrated convincingly, the origin of the enantioselectivity in their complex was indeed ligand induced.

6.2 C–H Bond Activation

As mentioned earlier, activation of C–H bonds is an important area of research owing to the dwindling reserves of global hydrocarbon feedstocks. Andino et al. [129] therefore investigated the activation of benzene to benzyne, a very useful intermediate for a variety of selective transformations in organic chemistry. The homogeneous catalyst used was the vanadium-based [V(**22**)] complex, a vanadium-alkyl complex of the form $(PNP)V(CH_2tBu)_2$ (where PNP=$N(2\text{-}P(CHMe_2)_2\text{-}4\text{-}methylphenyl)_2$) (Fig. 6). The inherent reactivity of these complexes arises from the formation of a transient alkylidene (PNP)V=CH*t*Bu [V(**22a**)] (Fig. 6), active towards two-electron oxidants. High-frequency and -field EPR (HF EPR), from 50 to 300 GHz, was then used to study the oxidation state of the [V(**22**)] complex. The 224-GHz spectrum produced a readily recognizable spin triplet ($S = 1$) state of near axial symmetry. Simulation revealed the parameters $S = 1, D = +3.93\ \text{cm}^{-1}$, $E = +0.145\ \text{cm}^{-1}$, $g_x = g_y = 1.955$, $g_z = 1.99$. The relatively large magnitude of D for [V(**22**)] was consistent with a system best described as V^{III} rather than as an organic (ligand centred) di-radical, expected to produce zero-field splittings much below 1 cm^{-1}. The observed D value was also larger than that for five-coordinate V^{III} centres, but lower than that expected for classical octahedral $S = 1$ complexes [130, 131]. The spectroscopic data in this study showed nicely how high frequencies combined with high resonant magnetic fields allows observation of EPR resonances from systems traditionally regarded as "EPR-silent". This is especially useful in homogeneous catalysis where multiple oxidation states may be involved in the reaction.

6.3 Diels–Alder Transformations

A number of EPR techniques have also been used to investigate the mechanistic details of the catalytic Diels–Alder reaction. Using CW EPR, HYSCORE and pulsed ENDOR, Bolm et al. [132] examined the changes in the ligand sphere surrounding their homogeneous Cu^{II} catalyst, a chiral bis-sulfoxime Cu^{II} complex bearing labile triflate groups. Introduction of the dienophile [*N*-(1-oxoprop-2-en-1-yl) oxazolidin-2-one] resulted in the formation of a new complex with well defined hyperfine spectra. The geometry of the complex at the different stages of the catalytic reaction was determined by EPR. In solvent-free conditions, the initial bis-sulfoxime Cu^{II} complex possessed square planar geometry but upon addition of the dienophile, the EPR parameters were found to be typical of a distorted, non-symmetric square pyramidal geometry. Mims ENDOR also revealed that at least one triflate anion directly participates in the first coordination sphere of the Cu^{II} by occupying an axial site [132].

Although the choice of the anion (such as triflate, TfO^-) in the Cu^{II} complex did not affect the overall conversion in Diels–Alder catalysis, the stereoselectivity of the reaction was considerably influenced by the choice of anion used. The authors used EPR to investigate a series of Cu^{II} catalysts bearing TfO^-, SbF_6^-, Cl^- and Br^- anions in the presence and absence of the dienophile [133]. The profile of the EPR spectra were significantly different for the complexes bearing the halide anions compared to those bearing the bulkier TfO^- and SbF_6^- anions, and the spectra were also notably dependent on the presence of the dienophile.

The authors concluded that using the anions TfO^- or SbF_6^-, a complex was formed involving an asymmetric coordination sphere around the Cu^{II} bis-sulfoxime complex (bound via two non-equivalent nitrogens). The dienophile was suggested to replace the two equatorially bound counterions (bound via two non-equivalent oxygens) and the weakly bound counterions in an axial position [133]. In contrast to this, two distinct complexes were established using the anions Cl^- or Br^-. The first revealed Cu^{II}–Cu^{II} electron–electron interactions via the halogen atoms. Upon addition of the dienophile, this orientation was changed towards a distorted arrangement with strongly tetra-coordinated counterions and N atoms coming from the bis-sulfoxime ligand. The authors concluded that high levels of stereo-selectivity requires weakly coordinating counter-ions that are able to move to axial positions during the catalytic cycle, thus allowing the substrate to occupy equatorial positions [133].

7 Nanoporous Catalysts

7.1 Microporous Zeotype Materials

Zeolites, i.e. microporous aluminosilicate materials with pores smaller than 2 nm, play key roles in the fields of sorption and catalysis [134, 135]. The global annual market for zeolites is several million tons. In the past few decades a large variety of zeolites and related zeotype materials have been produced, whereby transition metal incorporation is extensively used to modulate the catalytic characteristics of these materials. Since the catalytic properties depend on the structure and accessibility of the transition metal sites, a lot of effort is put into probing these sites. Nevertheless, the exact nature of the transition metal incorporation is often strongly debated, since most spectroscopic evidence for isomorphous substitution is indirect.

If the transition metal ion is paramagnetic, EPR techniques offer an unambiguous way to unravel the nature of the metal incorporation. This is convincingly demonstrated in the work of D. Goldfarb and co-workers, who used high-field ENDOR in combination with DFT to probe the isomorphous substitution of Mn^{II} into aluminophosphate zeotypes [136, 137]. They studied a large variety of zeotype structures, serving as examples for different channel morphologies and framework densities of the zeolite family. In all cases, the observation of strong ^{31}P hyperfine interactions proved that Mn^{II} can replace the framework Al. Furthermore, high-field ENDOR allowed these researchers to map out the full process of Mn^{II} incorporation into different aluminophospate zeotypes during synthesis [138].

A careful X-band ^{27}Al HYSCORE and W-band ^{1}H ENDOR analysis showed that from the three Cu^{II} species found in Cu-containing Si:Al zeolite Y (Si:Al $=$ 12 and 5), only one Cu^{II} was bound to the framework oxygens [139]. The other species consisted of a copper ion with a complete coordination sphere of water and no direct bonding with the zeolite framework. In a similar way, combined CW-EPR and ^{27}Al HYSCORE provided evidence of the interaction of Cu^{II} with the framework in copper-doped nanoporous calcium aluminate (mayenite) [140]. In mayenite, the positively charged calcium aluminate framework is counter-balanced by extra-lattice O^{2-} ions. Such free oxide ions are responsible for the ion conductivity of the materials and are readily replaced by various guest anions, such as O_2^- and OH^-. A native O_2^- species could indeed be identified with EPR in the Cu-doped mayenite materials [140].

Hyperfine techniques also allow for the probing of the accessibility of the paramagnetic transition metal sites to gases and small molecules. When ammonia is adsorbed to vanadium (VO^{2+})-exchanged ZSM-5, ^{14}N HYSCORE features typical of equatorial ammonia ligation to the vanadyl site are observed [141].

Copper-exchanged zeolites have been known for a long time to be active in NO_x decomposition. The decomposition of NO is shown to occur via formation of a Cu^I-$(NO)_2$ dimer with a paramagnetic Cu^I-NO monomer as a precursor [142]. This has prompted A. Pöppl and co-workers to use multi-frequency pulsed EPR and ENDOR techniques to investigate Cu^I-NO adsorption complexes in a range of copper-exchanged zeolites prepared by both solid- and liquid-state ion exchange [143–145]. ^{27}Al HYSCORE and ENDOR analyses of these complexes in Cu-L and Cu-ZSM-5 zeolites allowed the estimation of hyperfine parameters of an aluminium nucleus found near the Cu^I-NO site. The data showed that the Al atom is located in the third coordination sphere of the adsorbed NO [144], and hence supporting the O_2–Al–O_2–Cu^I–NO structure proposed earlier on the basis of quantum-chemical computations [146].

In contrast, zeolites with poor de-NO_x properties may be very promising as materials to store and deliver NO under controlled conditions in clinical applications. Indeed, nitric oxide is a crucial biomolecule in the cardiovascular, nervous and immune systems. The non-toxic zinc-exchanged Linde type A (LTA) zeolite has a relatively high storage capacity for NO and is hence a promising material for clinical applications. EPR revealed that the NO monomer is interacting more strongly with the metal sites in Zn-LTA than in the corresponding Na-LTA [147].

7.2 *Mesoporous Siliceous Materials*

Despite the current importance of microporous zeotype materials, they have the drawback that the size of their pores limits their applicability to smaller molecules. To overcome this, many efforts have led to the production of a wide variety of mesoporous (2 nm $< \emptyset <$ 50 nm) siliceous and non-siliceous materials [148]. In this respect, the discovery of the M41S family by researchers from Mobil Oil Corporation has played a crucial role [149, 150]. The synthesis process of mesoporous materials is based on a self-assembly process of organic–inorganic composites where the organic self-organized structures serve as a template for the inorganic skeleton. One of the most intensively studied M41S materials is MCM-41 that possesses hexagonally packed uni-dimensional cylindrical pores with pore diameters between 2 and 10 nm. This silica material is synthesized using the ionic surfactant cetyltrimethylammonium bromide as template. Conversely, mesoporous materials with larger pores and higher (hydro)thermal stability can be obtained using non-ionic block co-polymers. In this class of materials, SBA-15 with large tailorable uniform pores (3–15 nm) is found to be particularly promising [151–153].

The formation of mesoporous materials can be followed via EPR using nitroxide spin probes (i.e. nitroxide radicals) [154]. By introducing such a spin probe into the system, or by labelling a molecule with this nitroxide, EPR can be used to monitor the direct environment of this radical. The spin probe thus acts as a 'spy' that keeps track of the changing environment during the formation reaction. Spin-labelled surfactants, silane-based spin-labels and spin-labelled pluronics are ideal to monitor the formation of templated mesoporous materials [154]. By varying the type of probes added in the reaction mixture, different regions in the forming mesostructure can be studied. CW-EPR experiments give a direct insight in the polarity and microviscosity of the local environment, while ESEEM experiments reveal the water content and the presence of additives or ions in the proximity of the label [155–157]. Variations in the size of the micelles can be probed during the initial stages of the reaction by using DEER spectroscopy, a pulsed-EPR technique targeted at determining inter-spin distances [158].

Transition metal-based redox centres render the mesoporous silica materials catalytically active. The transition metals can be introduced during synthesis or postsynthetic (e.g. by impregnation) and the local structure of the transition metal site will determine the catalytic properties. As also demonstrated for the zeolite cases, EPR offers a unique tool to monitor these local sites. The observation of strong ^{29}Si hyperfine couplings in the HYSCORE spectra of vanadium-doped MCM-41 unambiguously proved the framework incorporation of vanadium [159]. The material was obtained via a direct synthesis method at room temperature. When vanadia was post-synthetically deposited on the surface of MCM-41 by the molecular designed dispersion (MDD) method using vanadyl acetylacetonate complexes, the vanadyl ions were, in contrast, found to be fully hydrated with no binding to the silica walls. This agrees with our earlier findings that the vanadyl acetylacetonate complexes have a great tendency to increase their coordination sphere by coordinating waters when deposited on SBA-15 [160]. Interestingly, this tendency is highly reduced when a Ti environment (e.g. a TiO_x layer) is present [160].

7.3 Plugged Hexagonal Template Silica: Combined Zeolite-Mesoporous Systems

While mesoporous materials have the advantage of larger pores, the crystalline microporous materials are much more stable. When the silica/surfactant ratio is increased during the SBA-15 synthesis, microporous amorphous nanoparticles are formed inside the mesoporous channels [161]. This leads to mechanically more stable SBA-15 materials that are named plugged hexagonal templated silica (PHTS). When vanadium-activated zeolitic nanoparticles are deposited inside the mesoporous channels of SBA-15 via a post-synthetic dry impregnation with the zeolite nanoparticles, a catalytically active PHTS is formed [162]. The detailed CW and pulsed EPR analysis of these materials not only revealed valuable information

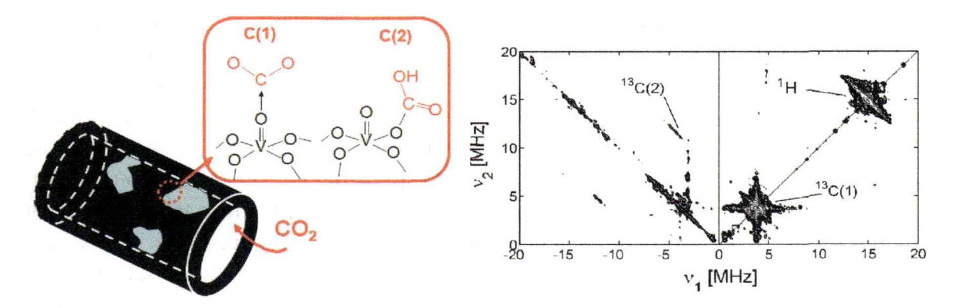

Fig. 7 The observation of strong ^{13}C [C(2)] and weak [C(1)] hyperfine couplings in the HYSCORE spectra after adsorption of $^{13}CO_2$ to a PHTS formed by deposition of vanadium silicate-1 nanoparticles in SBA-15 reveals equatorial and axial ligation of CO_2 to the vanadyl-sites. Adapted and reprinted with permission from [164]. Copyright 2011 American Chemical Society

about the stability, H_2O, CO_2 and NH_3 accessibility of the vanadium sites [163, 164] (Fig. 7) – it also gave new insight into the formation process of vanadium silicalites [163]. In order to obtain the silicalite nanoparticles, the zeolite growth needs to be stopped by acidification of the solution. Although this has been known for some time, it was never clear what triggers this process. Our results clearly showed that acidification affects the stability of the tetrapropylammonium-hydroxide template molecules such that one propyl ligand is lost by acid hydrolysis. It was also shown that at this stage of the zeolite synthesis, the vanadyl species are not fixed completely within the hydrophilic network. The true incorporation of the vanadyl species in the zeolite only occurs when the solvated vanadyl species get trapped between the aggregating nanoparticles during the hydrothermal growth of the zeolite.

7.4 Porous Titania for Photo-Catalytic Applications

Titanium dioxide-based photocatalysts promise to be excellent materials for alternative water and air treatments, because the photo-induced oxidation processes may achieve a complete mineralization of most pollutants [165, 166]. One of the limiting factors is, however, the wide band gap (3.0–3.2 eV) [167] of most conventional TiO_2 materials, so that the transfer of electrons from the valence band to the conduction band, the key step in the photocatalysis process, needs light in the UV region (wavelength less than 380 nm). Considerable effort is therefore put in the synthesis of TiO_2 materials with a lower bandgap via different doping of the materials [168] or by synthesis of stable TiO_2-based mesoporous materials or nanoparticles [169]. The latter materials have the added advantage that they have a large active surface. In order to come to a targeted synthesis of TiO_2 materials, a thorough understanding of the photocatalytic working plays a key role. Upon light-irradiation, a large number of (EPR detectable) paramagnetic centres are formed that stem from the light-induced electron-hole pair. Trapping of an electron by a Ti^{4+} ion leads to different Ti^{3+} sites [170–172]. Combination of surface Ti^{3+} with O_2 can result in formation of O_2^- [173] that in turn can break down organic molecules. Trapping of a valence-band hole by O^{2-} can lead to the paramagnetic O^- that can further react with O_2 to form O_3^- [173]. The conduction-band electron can also be trapped in an oxygen vacancy, leading to an F^+ colour centre [174]. In addition, the dopants may also be paramagnetic, either before and/ or after light illumination [175, 176]. It therefore will not come as a surprise that EPR in combination with light illumination has been used as a prime technique to understand the mechanistic steps involved in this type of photocatalysis [170–173, 175–182].

One of the problems of using EPR to study these materials lies in the fact that a large amount of paramagnetic centres are typically formed. At X-band, this tends to lead to strongly overlapping EPR signals, which may hamper interpretation in terms of the individual centres. This is clearly demonstrated by our recent study of TiO_2 nanotubes [180]. In this work, we combined a catalytic study with an EPR analysis of multi-walled hydrogen trititanate nanotubes. A calcination of the latter material leads to the formation mixed-phase anatase/trititanate nanotubes. Although clear spectral differences could be observed for the two types of nanotubes that could be linked to the catalytic activity under UV and visible-light illumination, a detailed analysis of the paramagnetic oxygen-based species formed could not be achieved due to strong spectral overlap in the $g = g_e$ region. This is a common problem also found with the analysis of TiO_2 nanoparticles [173, 178, 179].

The high resolution of HF EPR may allow a more detailed analysis of these oxygen-based species, although first attempts [177, 182] have shown that this technique will have to be combined with other variations, such as variation in the atmosphere (O_2, N_2, . . .), the calcination and measurement temperature and the irradiation conditions, to allow a full disentanglement of the spectra. Combination of EPR on samples treated under controlled circumstances with DFT computations promises to enhance further our understanding of the photocatalytic systems as shown by one of us for the interaction of molecular oxygen with oxygen vacancies on thermally reduced TiO_2 [181]. HF EPR also offers added possibilities for studying doped TiO_2 photocatalysts. Fittipaldi and co-workers used W-band CW EPR to disentangle the EPR contributions of different paramagnetic species in B-doped TiO_2, including a paramagnetic species mainly centered on the B atom [176].

Up till now, the use of hyperfine techniques has been very limited in the analysis of light-induced species in TiO_2-based materials. Nevertheless, a recent combined CW-EPR, HYSCORE and ENDOR study of UV-photo-induced Ti^{3+} in titanium oxide-based gels shows the potential of such an approach [183]. Not only could the kinetics of the formation of different Ti^{3+} centres under UV-irradiation be followed with CW-EPR – the hyperfine techniques revealed strong hyperfine couplings with nearby protons stemming from organic ligands. It can be expected that a combination of both HF EPR and an in-depth pulsed EPR/ENDOR analysis of TiO_2-based photocatalysts will give paramount information about the different paramagnetic species formed following the initial electron-hole creation. Studies of relevant model systems [184] and theoretical analyses will further extend these possibilities.

7.5 *Metal-Organic Framework Compounds*

Metal organic framework (MOF) compounds are novel three-dimensional, crystalline solids that are based on metal ions linked by bridging ligands containing organic carbon [185]. Depending on the metal–ligand combination, (meso)porous materials can be obtained with different physical and chemical properties. The probing of these properties is not trivial. Although the field is still relatively young, the group around A. Pöppl has shown that EPR offers an excellent tool to study these materials [186–189].

Using a combination of Q-band CW-EPR and Davies and Mims ENDOR, they were able to study the incorporation of chromium in the MOF compound MIL-53 [187]. The data convincingly showed the transformation from an open to a closed pore structure upon change of the temperature.

The EPR analysis of the MOF compound $Cu_3(BTC)_2(H_2O)_3 \bullet H_2O$ (BTC = benzene 1,3,5-tricarboxylate) revealed the presence of cupric ions in two different chemical environments [188]: Cu_2^{II} clusters in the paddle-wheel building blocks of the MOF giving rise to an anti-ferromagnetically coupled spin state and Cu^{II} monomeric species accommodated in the pores of the system. In a next step [189], the authors substituted Cu^{II} ions with Zn^{II}, thus forming paramagnetic binuclear Cu–Zn clusters that allowed EPR monitoring of the interaction of the Cu^{II} ions with adsorbates such as methanol.

8 Conclusions and Perspectives

With the continuing presence of paramagnetic transition metal ions or defect sites in homogeneous and heterogeneous catalysis, the important role played by advanced EPR techniques to uncover the mechanistic details of the catalytic cycle is clear and justified. The electronic structure and symmetry of the metal centre are crucial elements of the catalytic active site. These properties can be easily probed by EPR, whilst the additional resolving power of HF EPR, ENDOR, ESEEM and HYSCORE offers further insights into the extended (ligand) structure around the catalytic site. In many ways, when paramagnetic centres are involved, these combined techniques provide an unparalleled glimpse into the actions of the working catalyst. The coupling of the traditional CW-EPR techniques with the more advanced methodologies that have been developed over the past 20 years facilitates the improved characterisation and indeed the detection of paramagnetic and radical intermediates of relevance to catalysis. Although this family of EPR techniques offers tremendous opportunities to investigate mechanistic pathways and transition states in future catalysis research, they are often underused in the field.

Acknowledgments DMM would like to thanks EPSRC for funding (EP/H023879/1). SVD thanks the University of Antwerp for support (via NOI-BOF funding).

References

1. Gates BC (1992) Catalytic chemistry. Wiley, New York
2. Nguyen ST, Johnson LK, Grubbs RH et al (1992) Ring-opening metathesis polymerization (ROMP) of norbornene by a group VIII carbene complex in protic media. J Am Chem Soc 114:3974–3975
3. Heck RF, Nolley JP Jr (1972) Palladium-catalyzed vinylic hydrogen substitution reactions with aryl, benzyl, and styryl halides. J Org Chem 37:2320–2322
4. Katsuki T, Sharpless KB (1980) The 1st practical method for asymmetric epoxidation. J Am Chem Soc 102:5974–5976
5. Rossiter BE, Katsuki T, Sharpless KB (1981) Asymmetric epoxidation provides shortest routes to four chiral epoxy alcohols which are key intermediates in syntheses of methymycin, erythromycin, leukotriene C-1 and disparlure. J Am Chem Soc 103:464–465
6. Martin VS, oodard SS, Katsuki T et al (1981) Kinetic resolution of racemic allylic alcohols by enantioselective epoxidation – a route to substances of absolute enantiomeric purity. J Am Chem Soc 103:6237–6240
7. Walsh PJ, Kozlowski MC (2009) Fundamentals of asymmetric catalysis. University Science, Sausalito
8. Yoon TP, Jacobsen EN (2003) Privileged chiral catalysts. Science 299:1691–1693
9. Nielsen LPC, Stevenson CP, Blackmond DG et al (2004) Mechanistic investigation leads to a synthetic improvement in the hydrolytic kinetic resolution of terminal epoxides. J Am Chem Soc 126:1360–1362
10. Ready JM, Jacobsen EN (2002) A practical oligomeric [(salen)Co] catalyst for asymmetric epoxide ring-opening reactions. Angew Chem Int Ed 41:1374–1377
11. Lunsford JH (1972) Electron spin resonance in catalysis. Adv Catal 22:265–344
12. Howe R (1982) EPR spectroscopy in surface chemistry: recent developments. Adv Colloid Interface Sci 18:1–55
13. Che M, Taarit YB (1985) Applications of electron paramagnetic resonance to heterogeneous systems. Adv Colloid Interface Sci 23:235–255
14. Che M, Giamello E (1987) Electron paramagnetic resonance. Stud Surf Sci Catal 57: B265–B332
15. Murphy DM (2008) Electron paramagnetic resonance spectroscopy of polycrystalline oxide systems. In: Jackson SD, Hargreaves J (eds) Metal oxide catalysis. Wiley-VCH, New York
16. Hunger M, Weitkamp J (2001) In situ IR, NMR, EPR, and UV/Vis spectroscopy: tools for new insight into the mechanisms of heterogeneous catalysis. Angew Chem Int Ed 40:2954–2971
17. Bruckner A (2010) In situ electron paramagnetic resonance: a unique tool for analyzing structure-reactivity relationships in heterogeneous catalysis. Chem Soc Rev 39:4673–4684
18. Stosser R, Marx U, Herrmann W et al (2010) In situ EPR study of chemical reactions in Q-band at higher temperatures: a challenge for elucidating structure-reactivity relationships in catalysis. J Am Chem Soc 132:9873–9880

19. Van Doorslaer S, Caretti I, Fallis IA et al (2009) The power of electron paramagnetic resonance to study asymmetric homogeneous catalysts based on transition-metal complexes. Coord Chem Rev 253:2116–2130
20. Carter E, Murphy DM (2009) Structure – function relationships and mechanistic pathways in homogeneous enantioselective catalysis as probed by ENDOR spectroscopy. In: Douthwaite R, Duckett S (eds) Spectroscopic properties of inorganic and organometallic compounds, vol 40. RSC, Cambridge
21. Mobius K, Savitsky A (2008) High field EPR spectroscopy on proteins and their model systems. RSC, Cambridge
22. Goldfarb D (2006) High field ENDOR as a characterization tool for functional sites in microporous materials. Phys Chem Chem Phys 8:2325–2343
23. Bennati M, Prisner TF (2005) New developments in high field electron paramagnetic resonance with applications in structural biology. Rep Prog Phys 68:411–448
24. Andersson KK, Schmidt PP, Katterle B et al (2003) Examples of high-frequency EPR studies in bioinorganic chemistry. J Biol Inorg Chem 8:235–247
25. Fortman GC, Kegl T, Li QS et al (2007) Spectroscopic detection and theoretical confirmation of the role of Cr2(CO)5 (C5R5)2 and.Cr(CO)2(ketene)(C5R5) as intermediates in carbonylation of N=N=CHSiMe3 to O=C=CHSiMe3 by.Cr(CO)3 (C5R5) (R = H, CH3). J Am Chem Soc 129:14388–14400
26. Krzystek J, Ozarowsk A, Telser J (2006) Multi-frequency, high-field EPR as a powerful tool to accurately determine zero-field splitting in high-spin transition metal coordination complexes. Coord Chem Rev 250:2308–2324
27. Bianchini C, Gatteschi D, Giambastiani G et al (2007) Electronic influence of the thienyl sulfur atom on the oligomerization of ethylene by cobalt(II) 6-(thienyl)-2-(imino)pyridine catalysis. Organometallics 26:726–739
28. Schweiger A, Jeschke G (2001) Principles of pulse electron paramagnetic resonance. Oxford University Press, Oxford
29. Zein S, Kulik LV, Yano J et al (2008) Focusing the view on nature's water-splitting catalyst. Philos Trans R Soc B 363:1167–1177
30. Hoffman BM (2003) Electron-nuclear double resonance spectroscopy (and electron spin-echo envelope modulation spectroscopy) in bioinorganic chemistry. Proc Natl Acad Sci USA 100:3575–3578
31. Peloquin JM, Britt RD (2001) EPR/ENDOR characterization of the physical and electronic structure of the OEC Mn cluster. Biochim Biophys Acta Bioenerg 1503:96–111
32. Lubitz W, Reijerse E, Van Gastel M (2007) [NiFe] and [FeFe] hydrogenases studied by advanced magnetic resonance techniques. Chem Rev 107:4331–4365
33. Lubitz W, Lendzian F, Bittl R (2002) Radicals, radical pairs and triplet states in photosynthesis. Acc Chem Res 35:313–320
34. Schiemann O, Prisner TF (2007) Long-range distance determinations in biomacromolecules by EPR spectroscopy. Quart Rev Biophys 40:1–53
35. Prisner T, Rohrer M, MacMillan F (2001) Pulsed EPR spectroscopy: biological applications. Annu Rev Phys Chem 52:270–313
36. Rosati F, Roelfes G (2010) Artificial metalloenzymes. ChemCatChem 2:916–927
37. Palucki M, Finney NS, Pospisil PJ et al (1998) The mechanistic basis for electronic effects on enantioselectivity in the (salen)Mn(III)-catalyzed epoxidation reaction. J Am Chem Soc 120:948–954
38. Katsuki T (1995) Catalytic asymmetric oxidations using optically active (salen)-manganese (III) complexes as catalysts. Coord Chem Rev 140:189–214
39. Schaus SE, Brandes BD, Larrow JF et al (2002) Highly selective hydrolytic kinetic resolution of terminal epoxides catalyzed by chiral (salen)Co-III complexes. Practical synthesis of enantioenriched terminal epoxides and 1,2-diols. J Am Chem Soc 124:1307–1315
40. Fallis IA, Murphy DM, Willock DJ et al (2004) Direct observation of enantiomer discrimination of epoxides by chiral salen complexes using ENDOR. J Am Chem Soc 126:15660–15661

41. Murphy DM, Fallis IA, Willock DJ et al (2008) Discrimination of geometrical epoxide isomers by ENDOR & DFT – the role of H-bonds. Angew Chem Int Ed 47:1414–1416
42. Carter E, Murphy DM, Fallis IA et al (2010) Probing the role of weak outer sphere interactions (H-bond) in VO(3,5-$^{t}Bu_2$-salophen) – epoxide adducts by EPR, ENDOR and HYSCORE. Chem Phys Lett 486:74–79
43. Murphy DM, Fallis IA, Landon J et al (2009) Enantioselective binding of structural epoxide isomers by a chiral vanadyl salen complex: a pulsed EPR, CW-ENDOR and DFT investigation. Phys Chem Chem Phys 11:6757–6769
44. Carter E, Fallis IA, Kariuki BM et al (2011) Structure and EPR characterization of N,N'-bis (5-tert-butylsalicylidene)-1,2-cyclohexanediamino-vanadium (IV) oxide and its adducts with propylene oxide. Dalton Trans 40:7454–7462
45. Murphy DM, Caretti I, Carter E et al (2011) Visualising diastereomeric interactions of chiral amine-chiral copper Salen adducts by EPR/ENDOR/HYSCORE spectroscopy and DFT. Inorg Chem (submitted)
46. Merkx M, Kopp DA, Sazinsky MH et al (2001) Dioxygen activation and methane hydroxylation by soluble methane monooxygenase: a tale of two irons and three proteins. Angew Chem Int Ed 40:2782–2807
47. Wallar BJ, Lipscomb JD (1996) Dioxygen activation by enzymes containing binuclear non-heme iron clusters. Chem Rev 96:2625–2657
48. Pulver SC, Froland WA, Lipscomb JD et al (1997) Ligand field circular dichroism and magnetic circular dichroism studies of component B and substrate binding to the hydroxylase component of methane monooxygenase. J Am Chem Soc 119:387–395
49. Wolfe MD, Parales JV, Gibson DT et al (2001) Single turnover chemistry and regulation of O_2^- activation by the oxygenase component of naphthalene 1,2-dioxygenase. J Biol Chem 276:1945–1953
50. Karlsson A, Parales JV, Parales RE et al (2003) Crystal structure of naphthalene dioxygenase: side-on binding of dioxygen to iron. Science 299:1039–1042
51. Ward TR (2011) Artificial metalloenzymes based on the biotin-avidin technology: enantioselective catalysis and beyond. Acc Chem Res 44:47–57
52. Park MJ, Lee J, Suh Y et al (2006) Reactivities of mononuclear non-heme iron intermediates including evidence that iron(III) – hydroperoxo species is a sluggish oxidant. J Am Chem Soc 128:2630–2634
53. Que L, Ho RYN (1996) Dioxygen activation by enzymes with mononuclear non-heme iron active sites. Chem Rev 96:2607–2624
54. Costas M, Chen K, Que L (2000) Biomimetic nonheme iron catalysts for alkane hydroxylation. Coord Chem Rev 200–202:517–544
55. Katona G, Carpentier P, Nivière V et al (2007) Raman-assisted crystallography reveals end-on peroxide intermediates in a nonheme iron enzyme. Science 316:449–453
56. Dey A, Jenney FE, Adams MK et al (2007) Sulfur K-edge X-ray absorption spectroscopy and density functional theory calculations on superoxide reductase: role of the axial thiolate in reactivity. J Am Chem Soc 129:12418–12431
57. Clay MD, Yang TC, Jenney FE et al (2006) Geometries and electronic structures of cyanide adducts of the non-heme iron active site of superoxide reductases: vibrational and ENDOR studies. Biochemistry 45:427–438
58. Jiang YB, Telser J, Goldberg DP (2009) Evidence for the formation of a mononuclear ferric-hydroperoxo complex via the reaction of dioxygen with an (N_4S(thiolate))iron(II) complex. Chem Commun 6828–6830
59. Girerd JJ, Banse F, Simaan AJ (2000) Characterization and properties of non-heme iron peroxo complexes. In: Metal-oxo and metal peroxo species in catalytic oxidations. Structure and Bonding, vol 97. Springer, Hidelberg, pp 145–177
60. Simaan AJ, Banse F, Girerd JJ (2001) The electronic structure of non-heme iron(III)-hydroperoxo and iron(III)-peroxo model complexes studied by Mossbauer and electron paramagnetic resonance spectroscopies. Inorg Chem 40:6538–6540

61. Berry JF, Bill E, Bothe E et al (2006) Octahedral non-heme oxo and non-oxo Fe(IV) complexes: an experimental/theoretical comparison. J Am Chem Soc 128:13515–13528
62. Bilis G, Christoforidis KC, Deligiannakis Y et al (2010) Hydrocarbon oxidation by homogeneous and heterogeneous non-heme iron(III) catalysts with H_2O_2. Catal Today 157:101–106
63. Shi W, Liu Y, Liu B et al (2006) Synthesis and characterization of a six-coordinate monomeric Mn(III) complex with SOD-like activity. J Coord Chem 59:119–130
64. Parsell TH, Behan RK, Green MT et al (2006) Preparation and properties of a monomeric Mn^{IV}-oxo complex. J Am Chem Soc 128:8728–8729
65. Donoghue PJ, Gupta AK, Boyce DW et al (2010) An anionic, tetragobal copper(II) superoxide complex. J Am Chem Soc 132:15869–15871
66. Chaudhuri P, Wieghardt K (2001) Phenoxyl radical complexes. Prog Inorg Chem 50:151–216
67. Jörgensen CK (1969) Oxidation numbers and oxidation states. Springer, Heidelberg
68. Jazdzewski BA, Tolma WB (2000) Understanding the copper-phenoxyl radical array in galactose oxidase: contributions from synthetic modeling studies. Coord Chem Rev 200–202:633–685
69. Krüger H-J (1999) What can we learn from nature about the reactivity of coordinated phenoxyl radicals? A bioinorganic success story. Angew Chem Int Ed 38:627–631
70. Bereman RD, Kosman DJ (1977) Stereo-electronic properties of metalloenzymes 5. Identification and assignment of ligand hyperfine splittings in electron spin resonance spectrum of galactose oxidase. J Am Chem Soc 99:7322–7325
71. Müller J, Kikuchi A, Bill E et al (2000) Phenoxyl radical complexes of chromium(III), manganese(III), cobalt(III) and nickel(II). Inorg Chimica Acta 297:265–277
72. dos Anjos A, Bortoluzzi AJ, Osorio R et al (2005) New mononuclear Cu^{II} and Zn^{II} complexes capable of stabilizing phenoxyl radicals as models for the active form of galactose oxidase. Inorg Chem Comm 8:249–253
73. Mukherjee A, Lioret R, Mukherjee R (2008) Synthesis and properties of diphenoxo-bridged Co^{II}, Ni^{II}, Cu^{II} and Zn^{II} complexes of a new tripodal ligand: generation and properties of M^{II}-coordinated phenoxyl radical species. Inorg Chem 47:4471–4480
74. Thomas F, Arora H, Philouze C et al (2010) Co^{III} and Cu^{II} complexes of reduced Schiff bases: generation of phenoxyl radical species. Inorg Chimica Acta 363:3122–3130
75. Chaudhuri P, Verani CN, Bill E et al (2001) Electronic structure of bis(o-iminobenzosemiquinonato)metal complexes (Cu, Ni, Pd). The art of establishing physical oxidation states in transition metal complexes containing radical ligands. J Am Chem Soc 123:2213–2223
76. Weyhermüller T, Paine TK, Bothe E et al (2002) Complexes of an aminebis(phenolate) [O, N, O] donor ligand and EPR studies of isoelectronic, isostructural Cr(III) and Mn(IV) complexes. Inorg Chimica Acta 337:344–356
77. Telser J (2010) Overview of ligand *versus* metal centred redox reactions in tetraaza macrocyclic complexes of nickel with a focus on electron paramagnetic resonance. J Braz Chem Soc 21:1139–1157
78. Shimazaki Y, Tani F, Fukui K et al (2003) One-electron oxidized nickel(II)-(disalicylidene) diamine complex: temperature-dependent tautomerism between Ni(III)-phenolate and Ni(II)-phenoxyl radical states. J Am Chem Soc 125:10512–10513
79. Shimazaki Y, Yajima T, Tani F et al (2007) Synthesis and electronic structures of one-electron-oxidized group 10 metal(II) – (disalicylidene)diamine complexes (metal = Ni, Pd, Pt). J Am Chem Soc 129:2559–2568
80. Rotthaus O, Jarjayes O, Del Valle CP et al (2007) A versatile electronic hole in one-electron oxidized Ni-II bis-salicylidene phenylenediamine complexes. Chem Commun 4462–4464
81. Rotthaus O, Labet V, Philouze C et al (2008) Pseudo-octahedral schiff base nickel(II) complexes: does single oxidation always lead to the nickel(III) valence tautomer? Eur J Inorg Chem 4215–4224

82. Storr T, Wasinger EC, Pratt RC et al (2007) The geometric and electronic structure of a one-electron-oxidized nickel(II) bis(salicylidene)diamine complex. Angew Chem Int Ed 46:5198–5201
83. Storr T, Verma P, Pratt RC et al (2008) Defining the electronic and geometric structure of one-electron oxidized copper-bis-phenoxide complexes. J Am Chem Soc 46:15448–15459
84. Pratt RC, Stack TDP (2005) Mechanistic insights from reactions between copper(II)-phenoxyl complexes and substrates with activated C-H bonds. Inorg Chem 44:2367–2375
85. Vinck E, Murphy DM, Fallis IA et al (2010) Formation of a cobalt(III)-phenoxyl radical complex by acetic acid promoted aerobic oxidation of a Co(II)salen complex. Inorg Chem 49:2083–2092
86. Vinck E, Van Doorslaer S, Murphy DM et al (2008) The electronic structure of N,N′-bis(3,5-di-tert-butylsalicylidene)-1,2-cyclohexane-diamino cobalt(II). Chem Phys Lett 464:31–37
87. Vinck E, Van Doorslaer S, Murphy DM et al (2011) (in prep)
88. Vinck E, Murphy DM, Fallis IA et al (2010) A pulsed EPR and DFT investigation of the stabilization of coordinated phenoxyl radicals in a series of cobalt Schiff-base complexes. Appl Magn Reson 37:289–303
89. Orio M, Jarjayes O, Philouze C et al (2010) Spin interaction in octahedral zinc complexes of mono- and diradical Schiff and Mannich bases. Inorg Chem 49:646–658
90. Büttner T, Geier J, Frison G et al (2005) A stable aminyl radical metal complex. Science 307:235–238
91. Dixon JT, Green MJ, Hess FM et al (2004) Advances in selective ethylene trimerisation – a critical overview. J Organomet Chem 689:3641–3668
92. Emrich R, Heinemann O, Jolly PW et al (1997) The role of metallacycles in the chromium-catalyzed trimerization of ethylene. Organometallics 16:1511–1513
93. Manyik RM, Walker WE, Wilson TP (1997) Soluble chromium based catalysts for ethylene trimerization and polymerization. J Catal 47:197–209
94. Köhn RD, Haufe M, Mihan S et al (2000) Triazacyclohexane complexes of chromium as highly active homogeneous model systems for the Phillips catalyst. Chem Commun 1927–1928
95. Kohn R (2008) Reactivity of chromium complexes under spin control. Angew Chem Int Ed 47:245–247
96. Moulin JO, Evans J, McGuinness DS et al (2008) Probing the effects of ligand structure on activity and selectivity of Cr(III) complexes for ethylene oligomerisation and polymerization. Dalton Trans 1177–1185
97. Brückner A, Jabor JK, McConnell AEC et al (2008) Monitoring structure and valence state of chromium sites during catalyst formation and ethylene oligomerization by in situ EPR spectroscopy. Organometallics 27:3849–3856
98. Skobelev IY, Panchenko VN, Lyakin OY et al (2010) In situ EPR monitoring of chromium species formed during Cr-pyrrolyl ethylene trimerization catalyst formation. Organometallics 29:2943–2950
99. McDyre LE, Hamilton T, Murphy DM et al (2010) A CW EPR and ENDOR investigation on a series of Cr(I) carbonyl complexes with relevance to alkene oligomerization catalysis: [Cr $(CO)_4L]^+$ (L = $Ph_2PN(R)PPh_2$, $Ph_2P(R)PPh_2$, $Ph_2PN(R)NPPh_2$). Dalton Trans 39:7792–7799
100. McDyre LE, Carter E, Murphy DM et al (2011) (in prep)
101. Bianchini G, Giambastiani IG, Rios A et al (2007) Synthesis of a new polydentate ligand obtained by coupling 2,6-bis(amino)pyridine and (imino)pyridine moieties and its use in ethylene oligomerisation in conjunction with iron(II) and cobalt(II) bis-halides. Organometallics 26:5066–5078
102. Soshnikov IE, Semikolenova NV, Bushmelev AN et al (2009) Investigating the nature of the active species in bis(imino)pyridine cobalt ethylene polymerisation catalysts. Organometallics 28:6003–6013
103. Britovsek GJP, Clentsmith GKB, Gibson VC et al (2002) The nature of the active site in bis (imino)pyridine iron ethylene polymerisation catalysts. Catal Commun 3:207–211

104. Dzik WI, Smits JMM, Reek JNH et al (2009) Activation of carbon monoxide by (Me(3)tpa) Rh and (Me(3)tpa)Ir. Organometallics 28:1631–1643
105. Goldberg H, Kaminker I, Goldfarb D et al (2009) Oxidation of carbon monoxide cocatalyzed by palladium(0) and the $H_5PV_2Mo_{10}O_{40}$ polyoxometalate probed by electron paramagnetic resonance and aerobic catalysis. Inorg Chem 48:7947–7952
106. Khenkin AM, Efremenko I, Weiner L et al (2010) Photochemical reduction of carbon dioxide catalyzed by a ruthenium substituted polyoxometalate. Chem A Eur J 16:1356–1364
107. Che C-M, Huang J-S (2003) Metal complexes of chiral binaphthyl Schiff-base ligands and their application in stereoselective organic transformations. Coord Chem Rev 242:97–113
108. Shitama H, Katsuki T (2007) Synthesis of metal-(pentadentate-salen) complexes: asymmetric epoxidation with aqueous hydrogen peroxide and asymmetric cyclopropanation (salenH(2): N,N-bis(salicylidene)ethylene-1,2-diamine). Chem Eur J 13:4849–4858
109. Groves JT, Stern MK (1988) Synthesis, characterization, and reactivity of oxomanganese(IV) porphyrin complexes. J Am Chem Soc 110:8628–8638
110. Ostovic D, Bruice TC (1992) Mechanism of alkene epoxidation by iron, chromium and manganese higher valent oxo-metalloporphyrins. Acc Chem Res 25:314–320
111. Collman JP, Zhang X, Lee VJ et al (1993) Regioselective and enantioselective epoxidation catalyzed by metalloporphyrins. Science 261:1404–1411
112. Murphy A, Dubois G, Stack TDP (2003) Efficient epoxidation of electron-deficient olefins with a cationic manganese complex. J Am Chem Soc 125:5250–5251
113. Murphy A, Pace A, Stack TDP (2004) Ligand and pH influence on manganese-mediated peracetic acid epoxidation of terminal olefins. Org Lett 6:3119–3122
114. de Boer JW, Browne WR, Brinksma J et al (2007) Mechanism of cis-dihydroxylation and epoxidation of alkenes by highly H_2O_2 efficient dinuclear manganese catalysts. Inorg Chem 46:6353–6372
115. Bryliakov KP, Lobanova MV, Talsi EP (2002) EPR and ^1H NMR spectroscopic study of the Cr^{III}(salen)Cl catalysts. J Chem Soc Dalton Trans 2263–2265
116. Bryliakov KP, Talsi EP (2003) Cr^{III}(salen)Cl catalyzed asymmetric epoxidations: insight into the catalytic cycle. Inorg Chem 42:7258–7265
117. Campbell KA, Lashley MR, Wyatt JK (2001) Dual-mode EPR study of Mn(III) salen and the Mn(III) salen-catalyzed epoxidation of cis-beta-methylstyrene. J Am Chem Soc 123:5710–5719
118. Scarpellini M, Casellato A, Bortoluzzi AJ et al (2006) EPR and semi-empirical studies as tools to assign the geometric structures of Fe-III isomer models for transferrins. J Braz Chem Soc 17:1617–1626
119. Dyers L, Que SY, Van Derveer D et al (2006) Synthesis and structures of new salen complexes with bulky groups. Inorg Chimica Acta 359:197–203
120. Tani F, Matsu-ura M, Nakayama S et al (2001) Synthesis and characterization of alkanethiolate-coordinated iron porphyrins and their dioxygen adducts as models for the active center of cytochrome P450: direct evidence for hydrogen bonding to bound dioxygen. J Am Chem Soc 123:1133–1142
121. Bryliakov KP, Talsi EP (2004) Evidence for the formation of an iodosylbenzene(salen)iron active intermediate in a (salen)iron (III)-catalyzed asymmetric sulfide oxidation. Angew Chem Int Ed 43:5228–5230
122. Sivasubramanian VK, Ganesan M, Rajagopal S et al (2002) Iron(III)-salen complexes as enzyme models: mechanistic study of oxo(salen)iron complexes oxygenation of organic sulfides. J Org Chem 67:1506–1514
123. Ottenbacher RV, Bryliakov KP, Talsi EP (2010) Nonheme manganese-catalyzed asymmetric oxidation. A Lewis acid activation versus oxygen rebound mechanism: evidence for the third oxidant. Inorg Chem 49:8620–8628
124. Duban EA, Bryliakov KP, Talsi EP (2007) The active intermediates of non-heme-iron-based systems for catalytic alkene epoxidation with H_2O_2/CH_3COOH. Eur J Inorg Chem 852–857

125. Lyakin OY, Bryliakov KP, Britovsek GJP et al (2009) EPR spectroscopic trapping of the active species of nonheme iron-catalyzed oxidation. J Am Chem Soc 131:10798–10799
126. Maurya MR, Kumar M, Pessoa JC (2008) Oxidation of *p*-chlorotoluene and cyclohexene catalyzed by polymer-anchored oxovanadium(IV) and copper(II) complexes of amino acid derived tridentate ligands. Dalton Trans 4220–4232
127. Mimmi MC, Gullotti M, Santagostini L (2004) Models for biological trinuclear copper clusters. Characterization and enantioselective catalytic oxidation of catechols by the copper(I) complexes of a chiral ligand derived from (*S*)-(−)-1,1′binaphthyl-2,2′-diamine. Dalton Trans 2192–2201
128. Adão P, Maurya MR, Kumar U et al (2009) Vanadium-salen and -salan complexes: characterization and application in oxygen-transfer reactions. Pure Appl Chem 81:1279–1296
129. Andino JG, Kilgore UJ, Pink M et al (2010) Intermolecular C-H bond activation of benzene and pyridines by a vanadium(III) alkylidene including a stepwise conversion of benzene to a vanadium-benzyne complex. Chem Sci 1:351–356
130. Ye S, Neese F, Ozarowski A et al (2010) Family of V(III)-tristhiolato complexes relevant to functional models of vanadium nitrogenase: synthesis and electronic structure investigations by means of high-frequency and -field electron paramagnetic resonance coupled to quantum chemical computations. Inorg Chem 49:977–988
131. Krzystek J, Fiedler AT, Sokol JJ et al (2004) Pseudooctahedral complexes of vanadium(III): electronic structure investigation by magnetic and electronic spectroscopy. Inorg Chem 43:5645–5658
132. Bolm C, Martin M, Gescheidt G et al (2003) Spectroscopic investigation of bis(sulfoximine) copper(II) complexes and their relevance in asymmetric catalysis. J Am Chem Soc 125:6222–6227
133. Bolm C, Martin M, Gescheidt G et al (2007) Mechanistic insights into stereoselective catalysis – the effects of counterions in a Cu^{II}-bissulfoximine-catalyzed Diels-Alder reaction. Chem Eur J 13:1842–1850
134. Tatsumi T (2004) Zeolites: catalysis. In: Atwood JL, Steed JW (eds) Encyclopedia of supramolecular chemistry, vol 2. Marcel Dekker, New York, pp 1610–1616
135. Čejka J (2004) Zeolites: structures and inclusion properites. In: Atwood JL, Steed JW (eds) Encyclopedia of supramolecular chemistry, vol 2. Marcel Dekker, New York, pp 1623–1630
136. Arieli D, Delabie A, Vaughan DEW et al (2002) Isomorphous substitution of Mn(II) into aluminophosphate zeotypes: a combined high-field ENDOR and DFT study. J Phys Chem B 106:7509–7519
137. Arieli D, Prisner TF, Hertel M et al (2004) Resolving Mn framework sites in large cage aluminophosphate zeotypes by high field EPR and ENDOR spectroscopy. Phys Chem Chem Phys 6:172–181
138. Arieli D, Delabie A, Groothaert M et al (2002) The process of Mn(II) incorporation into aluminophosphate zeotypes through high-field ENDOR spectroscopy and DFT calculations. J Phys Chem B 106:9086–9097
139. Carl PJ, Vaughan DEW, Goldfarb D (2002) Interactions of Cu(II) ions with framework Al in high Si:Al zeolite Y as determined from X- and W-band pulsed EPR/ENDOR spectroscopies. J Phys Chem B 106:5428–5437
140. Maurelli S, Ruszak M, Witkowski S et al (2010) Spectroscopic CW-EPR and HYSCORE investigations of Cu2+ and O2− species in copper doped nanoporous calcium aluminate (12CaO.7Al2O3). Phys Chem Chem Phys 12:10933–10941
141. Woodworth J, Bowman MK, Larsen SC (2004) Two-dimensional pulsed EPR studies of vanadium-exchanged ZSM-5. J Phys Chem B 108:16128–16134
142. Giamello E, Murphy D, Magnacca G et al (1992) Interaction of NO with copper ions in ZSM5. An EPR and IR investigation. J Catal 136:510–520
143. Umamaheswari V, Hartmann M, Pöppl A (2005) EPR spectroscopy of Cu(I)-NO adsorption complexes formed over Cu-ZSM-5 and Cu-MCM-22 zeolites. J Phys Chem B 109:1537–1546

144. Umamaheswari V, Hartmann M, Pöppl A (2005) Pulsed ENDOR study of Cu(I)-NO adsorption complexes in Cu-L zeolite. J Phys Chem B 109:10842–10848
145. Umamaheswari V, Hartmann M, Pöppl A (2005) Critical assessment of electron spin resonance studies on Cu(I)-NO complexes in Cu-ZSM-5 zeolites prepared by solid- and liquid-state ion exchange. J Phys Chem B 109:19723–19731
146. Pietrzyk P, Piskorz W, Sojka Z et al (2003) Molecular structure, spin density and hyperfine coupling constants of the $\eta^1\{CuNO\}^{11}$ adduct in the ZSM-5 zeolite:DFT calculations and comparison with EPR data. J Phys Chem B 107:6105–6113
147. Pal C, Wheatley PS, El Mkami H et al (2010) EPR on medically relevant NO adsorbed to Zn-LTA. Appl Magn Reson 37:619–627
148. Soler-Illia J, Sanchez C, Lebeau B et al (2002) Chemical strategies to design textured materials: from microporous and mesoporous oxides to nanonetworks and hierarchical structures. Chem Rev 102:4093–4138
149. Kresge CT, Leonowicz ME, Roth WJ et al (1992) Ordered mesoporous molecular sieves synthesized by a liquid-crystal template mechanism. Nature 359:710–712
150. Beck JS, Vartuli JC, Roth WJ et al (1992) A new family of mesoporous molecular sieves prepared with liquid crystal templates. J Am Chem Soc 114:10834–10843
151. Zhao D, Feng J, Huo Q et al (1998) Triblock copolymer syntheses of mesoporous silica with periodic 50 to 300 Angstrom pores. Science 279:548–552
152. Kruk M, Jaroniec M, Ko CH et al (2000) Characterization of the porous structure of SBA-15. Chem Mater 12:1961–1968
153. Bennadja Y, Beaunier P, Margolese D et al (2001) Fine tuning of the interaction between pluronic surfactants and silica walls in SBA-15 nanostructured materials. Microporous Mesoporous Mater 44–45:147–152
154. Ruthstein S, Goldfarb D (2008) An EPR tool box for exploring the formation and properties of ordered template mesoporous materials. Electron Paramagn Reson 21:184–215
155. Zhang J, Carl PJ, Zimmermann H et al (2002) Investigation of the formation of MCM-41 by electron spin-echo envelope modulation spectroscopy. J Phys Chem B 106:5382–5389
156. Ruthstein S, Frydman V, Kababya S et al (2003) Study of the formation of the mesoporous material SBA-15 by EPR spectroscopy. J Phys Chem B 107:1739–1748
157. Ruthstein S, Frydman V, Goldfarb D (2004) Study of the initial formation stages of the mesoporous material SBA-15 using spin-labeled block co-polymer templates. J Phys Chem B 108:9016–9022
158. Ruthstein S, Goldfarb D (2008) Evolution of solution structures during the formation of the cubic mesoporous material, KIT-6, determined by double electron-electron resonance. J Phys Chem C 112:7102–7109
159. Zamani S, Meynen V, Hanu AM et al (2009) Direct spectroscopic detection of framework-incorporated vanadium in mesoporous silica materials. Phys Chem Chem Phys 11:5823–5832
160. Van Doorslaer S, Segura Y, Cool P (2004) Structural investigation of vanadyl-acetylacetonate-containing precursors of TiO_x-VO_x mixed oxides on SBA-15. J Phys Chem B 108:19404–19412
161. Van Der Voort P, Ravikovitch PI, De Jong KP et al (2002) Plugged hexagonal templated silica: a unique micro- and mesoporous composite material with internal silica nanocapsules. Chem Commun 1010–1011
162. Meynen V, Beyers E, Cool P et al (2004) Post-synthesis deposition of V-zeolitic nanoparticles in SBA-15. Chem Commun 898–899
163. Chiesa M, Meynen V, Van Doorslaer S et al (2006) Vanadium silicalite-1 nanoparticles deposition onto mesoporous walls of SBA-15. Mechanistic insights from a combined EPR and Raman study. J Am Chem Soc 128:8955–8963
164. Zamani S, Chiesa M, Meynen V et al (2010) Accessibility and dispersion of vanadyl sites of vanadium silicate-1. J Phys Chem C 114:12966–12975

165. Carp O, Huisman CL, Reller A (2004) Photoinduced reactivity of titanium dioxide. Prog Solid State Chem 32:33–177
166. Ayoub K, van Hullebusch ED, Cassir M et al (2010) Application of advanced oxidation processes for TNT removal: a review. J Hazard Mater 178:10–28
167. Pascual J, Camassel J, Mathieu H (1977) Resolved quadrupole transition in TiO_2. Phys Rev Lett 39:1490–1493
168. Asahi R, Morikawa T, Ohwak T et al (2001) Visible-light photocatalysis in nitrogen-doped titanium oxides. Science 293:269–271
169. Du GH, Chen Q, Peng LM (2001) Preparation and structure analysis of titanium oxide nanotubes. Appl Phys Lett 79:3702–3704
170. Naccache C, Meriaude P, Che M et al (1971) Identification of oxygen species adsorbed on reduced titanium dioxide. Trans Faraday Soc 67:506–512
171. Hurum DC, Agrios AG, Crist SE et al (2006) Probing reaction mechanisms in mixed phase TiO_2 by EPR. J Electron Spectrosc Relat Phenomena 150:155–163
172. Micic OI, Zhang YN, Cromack KR et al (1993) Trapped holes on TiO_2 colloids studied by electron paramagnetic resonance. J Phys Chem 97:7277–7283
173. Berger T, Sterrer M, Diwals O et al (2005) Charge trapping and photoadsorption of O_2 on dehydroxylated TiO_2 nanocrystals – an electron paramagnetic resonance study. ChemPhysChem 6:2104–2112
174. Serpone N (2006) Is the band gap of pristine TiO_2 narrowed by anion- and cation-doping of titanium dioxide in second generation photocatalysts? J Phys Chem B 110:24287–24293
175. Livraghi S, Paganini MC, Giamello E et al (2006) Origin of photoactivity of nitrogen-doped titanium dioxide under visible light. J Am Chem Soc 128:15666–15671
176. Fittipaldi M, Gombac V, Montini T et al (2008) A high-frequency (95 GHz) electron paramagnetic resonance study of B-doped TiO_2 photocatalysts. Inorganica Chim Acta 361:3980–3987
177. Riss A, Berger T, Stankic S et al (2008) Charge separation in layered titanate nanostructures: effect of ion exchange induced morphology transformation. Angew Chem Int Ed 47:1496–1499
178. Berger T, Sterrer M, Diwald O et al (2005) Light-induced charge separation in anatase TiO_2 particles. J Phys Chem B 109:6061–6068
179. Kumar CP, Gopal NO, Wang TC et al (2006) EPR investigation of TiO_2 nanoparticles with temperature-dependent properties. J Phys Chem B 110:5223–5229
180. Ribbens S, Caretti I, Beyers E et al (2011) Unraveling the photocatalytic activity of multiwalled hydrogen trititanate and mixed-phase anatase/trititanate nanotubes: a combined catalytic and EPR study. J Phys Chem C 115:2302–2313
181. Green J, Carter E, Murphy DM (2009) Interaction of molecular oxygen with oxygen vacancies on reduced TiO_2: site specific blocking by probe molecules. Chem Phys Lett 477:340–344
182. Riss A, Elser MJ, Bernardi J et al (2009) Stability and photoelectronic properties of layered titanate nanostructures. J Am Chem Soc 131:6198–6206
183. Pattier B, Henderson M, Pöppl A et al (2010) Multi-approach electron paramagnetic resonance investigations of UV-photoinduced Ti^{3+} in titanium oxide-based gels. J Phys Chem B 114:4424–4431
184. Maurelli S, Livraghi S, Chiesa M et al (2011) Hydration structure of the Ti(III) cation as revealed by pulse EPR and DFT studies: new insights into a textbook case. Inorg Chem 50:23852394
185. Hertzsch T, Hulliger J, Weber E et al (2004) Organic zeolites. In: Atwood JL, Steed JW (eds) Encyclopedia of supramolecular chemistry, vol 2. Marcel Dekker, New York, pp 996–1004
186. Meilikhov M, Yusenko K, Torrisi A et al (2010) Reduction of a metal-organic framework by an organometallic complex: magnetic properties and structure of the inclusion compound $[(\eta^5\text{-}C_5H_5)_2Co]0.5$@MIL-47(V). Angew Chem Int Ed 49:6212–6215

187. Mendt M, Jee B, Stock N et al (2010) Structural phase transitions and thermal hysteresis in the metal-organic framework compound MIL-53 as studied by electron spin resonance spectroscopy. J Phys Chem C 114:19443–19451
188. Pöppl A, Kunz S, Himsl D et al (2008) CW and pulsed ESR spectroscopy of cupric ions in the metal-organic framework compound $Cu_3(BTC)_2$. J Phys Chem C 112:2678–2684
189. Jee B, Eisinger K, Gul-E-Noor F et al (2010) Continuous wave and pulsed electron spin resonance spectroscopy of paramagnetic framework cupric ion in the Zn(II) doped porous coordination polymer $Cu_{3-x}Zn_x(btc)_2$. J Phys Chem C 114:16630–16639

Top Curr Chem (2012) 321: 41–66
DOI: 10.1007/128_2011_301

Published online: 19 November 2011

Radicals in Flavoproteins

Erik Schleicher and Stefan Weber

Abstract Current technical and methodical advances in electron paramagnetic resonance (EPR) spectroscopy have proven to be very beneficial for studies of stationary and short-lived paramagnetic states in proteins carrying organic cofactors. In particular, the large number of proteins with flavins as prosthetic groups can be examined splendidly by EPR in all its flavors. To understand how a flavin molecule can be fine-tuned for specific catalysis of different reactions, understanding of its electronic structure mediated by subtle protein-cofactor interactions is of utmost importance. The focus of this chapter is the description of recent research progress from our laboratory on EPR of photoactive flavoproteins. These catalyze a wide variety of important photobiological processes ranging from enzymatic DNA repair to plant phototropism and animal magnetoreception. Whereas increasing structural information on the principal architecture of photoactive flavoproteins is available to date, their primary photochemistry is still largely undetermined. Interestingly, although these proteins carry the same light-active flavin chromophore, their light-driven reactions differ significantly: Formations of photoexcited triplet states and short-lived radical pairs starting out from triplet or singlet-state precursors, as well as generation of stationary radicals have been reported recently. EPR spectroscopy is the method of choice to characterize such paramagnetic intermediates, and hence, to assist in unravelling the mechanisms of these inimitable proteins.

Keywords BLUF domains · Cryptochrome · ENDOR · EPR · ESR · Flavoprotein · Photolyase · Phototropin · TREPR

E. Schleicher and S. Weber (✉)
Institut für Physikalische Chemie, Albert-Ludwigs-Universität Freiburg, Albertstr. 21, 79104 Freiburg, Germany
e-mail: Stefan.Weber@physchem.uni-freiburg.de

Contents

1 Introduction

During the past two decades an increasing number of enzymatic reactions were found to proceed via mechanisms involving radicals as intermediate states (see, e.g., [1–5]). Hence, a paramagnetic transient state is formed for a diamagnetic resting state of an enzyme. Radical reactions in enzymology are most often associated with redox-active cofactors, which can be divided into three subgroups. (1) Enzymes carrying a transition metal, such as iron or copper, are often involved in one-electron transfer processes. However, it has to be noted that, depending on the type and redox state of the metal ion, the resting state of the metalloenzyme could be intrinsically paramagnetic. (2) A few enzymatic reactions that involve molecular oxygen as one substrate proceed without cofactor involvement if the other substrate possesses appropriate chemical properties [6]. (3) Most radical reactions in enzymology utilize redox-active organic cofactors, such as the hetero-aromatic heme group, flavins, or quinones. To comprehend the reactivity of these enzymes on a molecular level, it is necessary to localize and characterize the paramagnetic center and its close vicinity, and to collect chemical and kinetic information on reaction intermediates. This also includes electronic and geometric information of traits that control the reactivity of enzymes. Protein systems are usually quite complex; hence, the complementary use of various molecular and structure determining methods is mandatory to obtain such details. However, for the characterization of the radical state itself, EPR spectroscopy is usually the method of choice that provides the most reliable and detailed information. To illustrate how various EPR techniques can be used for this purpose, we focus in this review on proteins carrying the flavin cofactor, which is the most ubiquitous redox-active organic coenzyme.

Flavins (Fl) catalyze many different bioreactions of physiological importance [7–9]. Riboflavin, flavin mononucleotide (FMN), and flavin adenine dinucleotide (FAD) have the 7,8-dimethyl isoalloxazine ring in common but differ in the side chain attached to N10. With their five redox states, fully oxidized, one-electron reduced semiquinoid ($FlH^{\bullet}$ and $Fl^{\bullet-}$), and fully reduced hydroquinone (FlH_2 and FlH^-), flavins are involved in one-electron and two-electron transfer reactions [10].

Flavin semiquinones were one of the first enzymatic radicals identified in electron-transfer reactions of flavoproteins and reactions of dihydroflavins with molecular oxygen [11]. Flavin semiquinones function in terminal electron transport complexes by facilitating electron transfer between obligatory two-electron-reducing agents (e.g., NADH), and one-electron acceptors (e.g., iron sulfide centers). These processes are often further mediated by quinone cofactors such as the coenzyme Q [1]. Moreover, flavin semiquinones play important roles as intermediates in many light-activated processes ranging from blue-light photoreception [12–16] and magnetoreception [17, 18] to DNA photorepair [19–21].

This review is divided into the following sections. First, EPR measurements of **g**-tensors of flavoproteins and the modulation of the principal values of **g** by the protein surroundings of the cofactor are discussed. Then, two recent examples of application of pulsed ENDOR spectroscopy will be reviewed, and, finally, time-resolved EPR spectroscopy, that is most favorably used to study photo-excited triplet states and radical pairs, will be introduced.

2 EPR Investigations of Flavoproteins

In studies of paramagnetic flavin species application of EPR has traditionally been very valuable to distinguish the protonation state of flavin semiquinones by means of the signal width of its typical inhomogeneously broadened EPR resonance centered at $g_{\mathrm{iso}} = 2.0034$ [22, 23]. Anion flavin radicals (deprotonated at N5) show peak-to-peak line widths (of the EPR signals in the first derivative) of around 1.2–1.5 mT, whereas neutral flavin radicals (protonated at N5) exhibit significantly larger spectral widths (around 1.8–2.0 mT) due to the presence of the additional large (and anisotropic, see below) hyperfine coupling from the H5 proton of the 7,8-dimethyl isoalloxazine moiety [24–27]. However, because hydrogen bonding of variable strength of surrounding amino acids to either N5 in anion flavin radicals ($\mathrm{Fl}^{\bullet-}$) or from NH5 in neutral flavin radicals ($\mathrm{FlH}^{\bullet}$) contributes to the EPR signal width, a clear-cut assignment of a flavin semiquinone signal to either a neutral or an anion flavin radical is often not possible based on the peak-to-peak EPR line width alone. Therefore, recent studies have been targeted on precisely measuring the **g**-tensor of protein-bound flavin radicals to correlate this quantity to the chemical structure of flavin semiquinones [28–35]. Because in most cases the principal values of **g** only deviate marginally from the free-electron value, $g_{\mathrm{e}} \approx 2.00232$, rather large magnetic fields and correspondingly high microwave frequencies are required to resolve the very small **g** anisotropies of flavin radicals. With the recent availability of powerful EPR instrumentation operating at high magnetic fields and high microwave frequencies, it is nowadays possible to perform such precision measurements that are not feasible at standard X-band frequencies where strong hyperfine inhomogeneities typically obscure the **g**-anisotropy [36]. In Fig. 1, characteristic high-magnetic-field/high-microwave-frequency EPR spectra of three

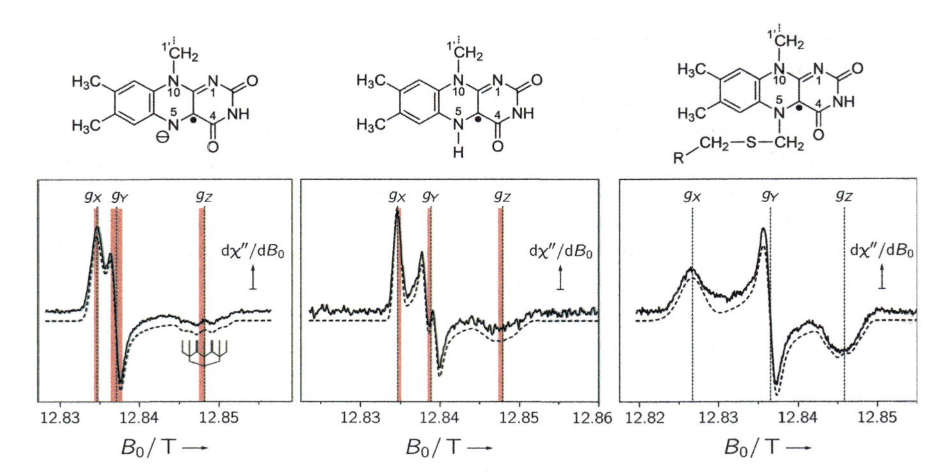

Fig. 1 High-magnetic-field/high-microwave-frequency continuous-wave EPR spectra (first derivatives) of various flavin radicals. *Left*: 360.04-GHz EPR spectrum of the stable anionic FAD radical of *Aspergillus niger* glucose oxidase (pH 10) recorded at 140 K [33]. *Middle*: 360.03-GHz EPR spectrum of the neutral FAD radical of *E. coli* CPD photolyase [28]. *Right*: 360.03 GHz EPR spectrum an FMN radical, which is bound at its N5 position to the LOV1 protein domain (C57M mutant) of *C. reinhardtii* phototropin [32]. Experimental and calculated EPR spectra are shown as *solid* and *dashed lines*, respectively. The *red shaded* areas show ranges for the respective principal values of **g** that have been measured by high-magnetic-field/high-microwave-frequency EPR experiments ($\geq$3.5 T/95 GHz)

different flavin radical species, namely neutral, anionic, and 5-methionine flavin radicals, are depicted and the ranges of typical **g**-principal values are shown.

For protein-bound flavin radicals it turned out that the **g**-tensor reflects the overall electronic structure on the redox-active isoalloxazine ring, and thus is potentially a valuable probe by which chemically different flavin radicals (e.g., non-covalently vs covalently bound at specific isoalloxazine ring positions, and neutral radical vs anion radical) may be distinguished [32]. This is because the **g**-principal values of non-covalently bound flavins seem rather unaffected towards changes in the isoalloxazine's local surroundings, unlike other organic cofactors (such as quinones, tyrosines, or tryptophans) or the nitroxides frequently used in spin-label EPR studies [37], which show broader distributions of g_X-, g_Y-, and g_Z-principal values, depending, e.g., on the polarity and/or the hydrogen-bonding situation of the cofactor binding pocket. Flavin neutral radicals in quite dissimilar protein surroundings, however, render **g**-principal values that are quite "robust," and hence the subtle differences of g_i, $i \in \{X, Y, Z\}$, in different flavoproteins, can only be observed by EPR performed in very strong magnetic fields.

So far, no single-crystal EPR studies of flavin radicals have been reported. However, the orientations of the **g**-principal axes relative to the molecular frame of the flavin's isoalloxazine moiety have been derived from orientation-selection effects of the quite anisotropic hyperfine coupling of H5 (or D5 in an isotope-exchange experiment) of the neutral flavin radical, both with EPR [28] and ENDOR

[30, 38]. As expected, the Z-principal axis of $\mathbf{g}$ is oriented perpendicular to the π-plane of the flavin ring. Analyses of EPR and ENDOR data revealed angles of $(-29 \pm 4)°$ and $(-14 \pm 2)°$ between the X axis of $\mathbf{g}$ and the N5–H5 (or N5–D5) bond in (6–4) photolyase [38] and cyclobutane pyrimidine dimer (CPD) photolyase [28, 30], respectively. The factors that cause the reorientation of the X and Y axes of $\mathbf{g}$ of a neutral flavin radical in the two highly homologous cofactor binding pockets of CPD photolyase and (6–4) photolyase remain elusive. Also, the orientations of the principal axes of $\mathbf{g}$ relative to the molecular frame of the isoalloxazine ring of a flavin anion radical still need to be determined experimentally.

In high-field EPR studies of nitroxide spin labels, quinones, and tyrosines, the g_X principal component of the $\mathbf{g}$-tensor is usually most sensitive towards changes in the hydrogen-bonding situation or the polarity of the radical surroundings. In the case of neutral flavin radicals, however, the g_Y component of the $\mathbf{g}$-tensor seems more "responsive" than g_X. This dissimilarity presumably lies in the different symmetries of flavins as compared to nitroxides [37, 39] or para-quinones [40, 41]. The latter have a well-defined symmetry axis with X aligned along the N–O or C=O bonds, respectively, and there are high unpaired electron-spin densities on the oxygens. Flavins, on the other hand, have much lower symmetry with the two carbonyl groups being meta-positioned. Furthermore, the unpaired electron-spin density is mostly localized on C4a and N5 rather than on the carbonyl groups. Hence, hydrogen bonding to H5 is also expected to alter the spin distribution on the isoalloxazine ring. Clearly, additional experimental studies assisted by quantum-chemical calculations of $\mathbf{g}$ principal values are needed to substantiate these preliminary findings further and to rationalize the different behavior of flavins as compared to other organic radicals.

3 ENDOR Investigations of Flavoproteins

ENDOR spectroscopy is nowadays used on a routine basis to determine the geometric and electronic structure of radicals by hyperfine interactions between nuclear magnetic moments and the magnetic moment of the unpaired electron spin. For flavin radicals, these interactions are in most cases too small to be resolved in EPR spectra. Via the hyperfine coupling constant, the electron-spin density at the positions of magnetic nuclei can be evaluated. Several excellent review articles are available which provide detailed descriptions of the basics and the application of this technique for structure determination in paramagnetic proteins and biomolecules [42–44]. In brief, by ENDOR spectroscopy hyperfine couplings of a particular nucleus can be directly determined from pairs of resonance lines that are, according to the condition $\nu_{\mathrm{ENDOR}} = |\nu_{\mathrm{n}} \pm A/2|$, either equally spaced about the magnetic-field-dependent nuclear Larmor frequency, ν_{n}, and separated by the (orientation dependent) hyperfine coupling constant A (for the case $\nu_{\mathrm{n}} > |A/2|$), or centered around $A/2$ and separated by $2\nu_{\mathrm{n}}$ (for $\nu_{\mathrm{n}} < |A/2|$). Traditionally, ENDOR studies on flavoproteins have been performed using the continuous-wave methodology; for reviews see [22, 45, 46]. In recent years, however, pulsed ENDOR techniques (primarily based on the Davies pulse sequence) became increasingly popular [25, 33, 34, 47–53]. In pulsed ENDOR experiments, the signal is obtained by recording an echo intensity as a function of the frequency of a radio-frequency pulse. Changes in the echo intensity occur when the radio frequency is on resonance with an NMR transition, thus generating the ENDOR response [43, 54, 55]. The pulsed methodology offers many advantages over continuous-wave ENDOR, in particular when protein samples in frozen solution with a dilute distribution of paramagnetic centers are to be examined. Pulsed ENDOR generates practically distortion-less line shapes, and is particularly useful when strongly anisotropic hyperfine interactions are to be measured [24, 25]. Furthermore, in the pulsed mode, the ENDOR intensity does not depend on a delicate balance between electron-spin and nuclear-spin relaxation rates and the applied microwave and radio frequency powers as for the continuous-wave technique. Its implementation is therefore much simpler as long as the relaxation times are long enough.

A characteristic pulsed proton ENDOR spectrum of a flavoprotein with the flavin cofactor in its neutral radical form is shown in Fig. 2. In X-band ENDOR, the detected resonances can be grouped into five spectral regions between 1 and about 37 MHz. (1) The central so-called matrix-ENDOR signal extends from about 13 to 16.5 MHz and comprises hyperfine couplings from protons whose nuclear spins interact only very weakly with the unpaired electron spin, e.g., protons from the protein backbone within the cofactor binding pocket, protons of water molecules surrounding the flavin, and also weakly coupled protons directly attached to the 7,8-dimethyl isoalloxazine ring, namely H3, H7α, and H9. (2) Prominent features of axial shape are observed in the flanking 10–12- and 17–19-MHz radio-frequency ranges and arise from the hyperfine couplings of protons of the methyl group attached to C8. Signals of this hyperfine tensor are in general easily detected in proton-ENDOR spectroscopy on flavins [29, 33, 52, 56–59], and are considered to be sensitive probes of the electron-spin density on the outer xylene ring of the flavin isoalloxazine moiety. Furthermore, theory shows that the size of this coupling responds sensitively to polarity changes of the protein surroundings [60]. (3) Flanking the H8α signals at around 9–10 MHz and 19–20 MHz are found the transitions belonging to one of the two β-protons, H1', attached to C1' in the ribityl side chain of the isoalloxazine ring [45, 56]. (4) Signals arising from the hyperfine coupling of the H6 proton occur at

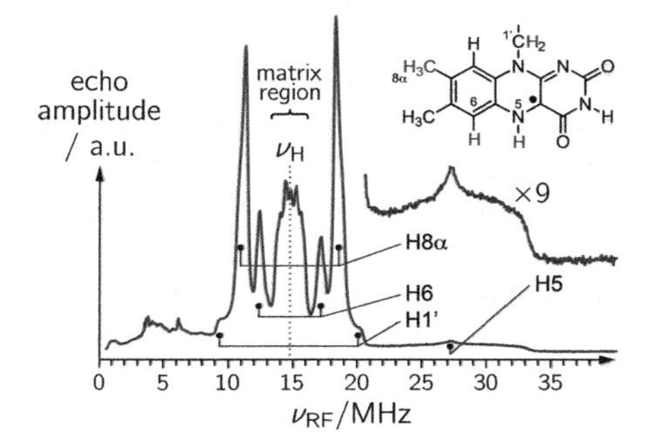

Fig. 2 Pulsed ENDOR on a neutral flavin radical. *E. coli* CPD photolyase was investigated with pulsed Davies ENDOR spectroscopy at $T = 80$ K (for details, see [25]). Detectable protons are marked accordingly

around 12 MHz and 17 MHz. (5) The broad, rhombic ($A_x \neq A_y \neq A_z$) feature extending from 21 to 34 MHz in the pulsed ENDOR spectrum is assigned to the proton bound to N5 [24, 25, 38]. Its contribution to the overall spectrum is easily discriminated from that of other protons in the isoalloxazine moiety due to the exchangeability of H5 with a deuteron upon buffer deuteration. Observation of this very anisotropic hyperfine coupling beautifully demonstrates the advantages of pulsed ENDOR over the conventional continuous-wave methodology. In the latter the first derivative of the signal intensity (with respect to the radio frequency) is recorded, which becomes very small when broad spectral features are to be measured. Hence, such couplings often escape direct detection in continuous-wave ENDOR [25].

A flavin anion radical shows a markedly different proton ENDOR spectrum as compared to that of a neutral radical. The most pronounced differences are, of course, the absence of the signal from H5 and the larger splittings of the signal pairs arising from H8α and H6 in the anion radical case. Hence, in addition to the **g**-tensor, the hyperfine pattern of a flavin radical allows for an unambiguous discrimination of the radical's protonation state [33, 34].

With the commercial availability of pulsed EPR instrumentation, other pulsed methods such as electron-spin echo envelope modulation (ESEEM) or hyperfine sublevel correlation spectroscopy (HYSCORE), which are quite useful to study specific hyperfine and quadrupolar couplings, have also been applied to flavoproteins [59, 61, 62]. These studies have been reviewed recently, e.g., in [46].

3.1 ENDOR Investigations of (6–4) Photolyase

Pulsed ENDOR has been favorably applied to characterize the electronic structure of the $FADH^{\bullet}$ cofactor and its surroundings in (6–4) photolyase. Photolyases are DNA repair enzymes that restore damaged and potentially lethal DNA by splitting UV-generated cyclobutane pyrimidine dimers (CPDs) or (6–4) photoproducts in a light-activated electron-transfer reaction, thereby preventing erratic DNA replication [19, 20]. For CPD-repairing CPD photolyase, the proposed repair mechanism includes a photo-induced single electron-transfer step from the fully reduced FAD cofactor ($FADH^{-}$) to the CPD, resulting in the formation of a CPD anion radical and a neutral $FADH^{\bullet}$ radical [63]. The cyclobutane ring of the putative CPD radical then splits, and subsequently the electron is likely to be transferred back to the $FADH^{\bullet}$ radical, thus restoring the initial redox states [19]. Hence, the entire process represents a true catalytic cycle with net-zero exchanged electrons. In contrast, (6–4) photolyases are unable to restore the original bases from DNA containing (6–4) photoproduct lesions in a single reaction step; rather, following binding of the DNA lesion, the overall repair reaction consists of at least two different steps, one of which could be light-independent while the other must be light-dependent [64–66]. Hitomi and coworkers first proposed a detailed reaction mechanism based on a mutational study, model geometries calculated on the basis of previously published CPD photolyase coordinates [67], and the important finding that the repair rate of (6–4) photolyases strongly depends on the pH [68, 69]. In the initial light-independent step, a 6'-iminium ion intermediate is generated from the (6–4) photoproduct aided by two highly conserved histidines (His354 and His358 in *Xenopus laevis* (6–4) photolyase). The 6'-iminium ion then spontaneously rearranges to an oxetane intermediate by intramolecular nucleophilic attack [66]. The oxetane species was proposed earlier in analogy to the repair mechanism of CPD photolyases, and because it was identified as an intermediate in the formation of (6–4) photoproducts [64, 70]. This putative repair mechanism of (6–4) photolyase requires one histidine to act as a proton acceptor and the other as a proton donor, which implies that the two histidines should have markedly different pK_a values. The subsequent blue-light-driven ($350 < \lambda < 500$ nm) reaction splits the oxetane intermediate presumably via an electron-transfer mechanism similar to that of CPD photolyases.

The unanswered question regarding the detailed repair mechanism of (6–4) photolyases and the involvement of functionally relevant amino acids led to the design of an ENDOR study [56] which will be briefly reviewed here. In general, as the function of a histidine is markedly influenced by its protonation state, it seems likely that the histidines at the solvent-exposed active site cause the unusual pH dependence in the (6–4) photolyase repair activity in vitro [68]. The principal idea was that the protonation of a histidine alters its polarity, which may be indirectly probed by proton-ENDOR spectroscopy using the neutral radical state of the FAD cofactor as the observer molecule. Figure 3 depicts sections of the complete ENDOR spectra (in the radio-frequency region between 17.8 and 21.2 MHz corresponding to hyperfine couplings between about 6 and 13 MHz) of *X. laevis* (6–4) photolyase, where the three H8α and one of the two H1' protons resonate; compare Fig. 2.

It is apparent that the intensity of the H8α ENDOR signal changes significantly as a function of pH. In contrast, the resonances of the other protons are pH independent (data not shown) [56]. For a detailed data analysis it was taken into account that the signal of H8α overlaps with that arising from H1'; by spectral simulation, the individual signal contributions of these protons could be deconvoluted. The overall shapes of the ENDOR spectra of the H358A mutant protein largely resemble those of the wild type at the respective pH conditions. In contrast to the wild type (or H358A), the ENDOR spectra of the H354A protein are markedly different. Thus, replacement of His354 with alanine leads to significant modifications of the cofactor-binding site at the 8α-methyl group and at the linkage of the ribityl side chain. Hence, as a first result, structural information regarding the distance and the location of the two histidines with respect to the flavin observer was obtained: the strong shift of the isotropic hyperfine coupling of H1' in H354A, as compared to the wild type or the H358A protein, observed at all measured pH values, suggested that His354 is close to H1' (it has to be mentioned that experimental data on the structure of a (6–4)-photolyase enzyme was unavailable at the time when the ENDOR experiments were performed). A slight geometrical reorientation due to the histidine-to-alanine replacement results in an altered direction of

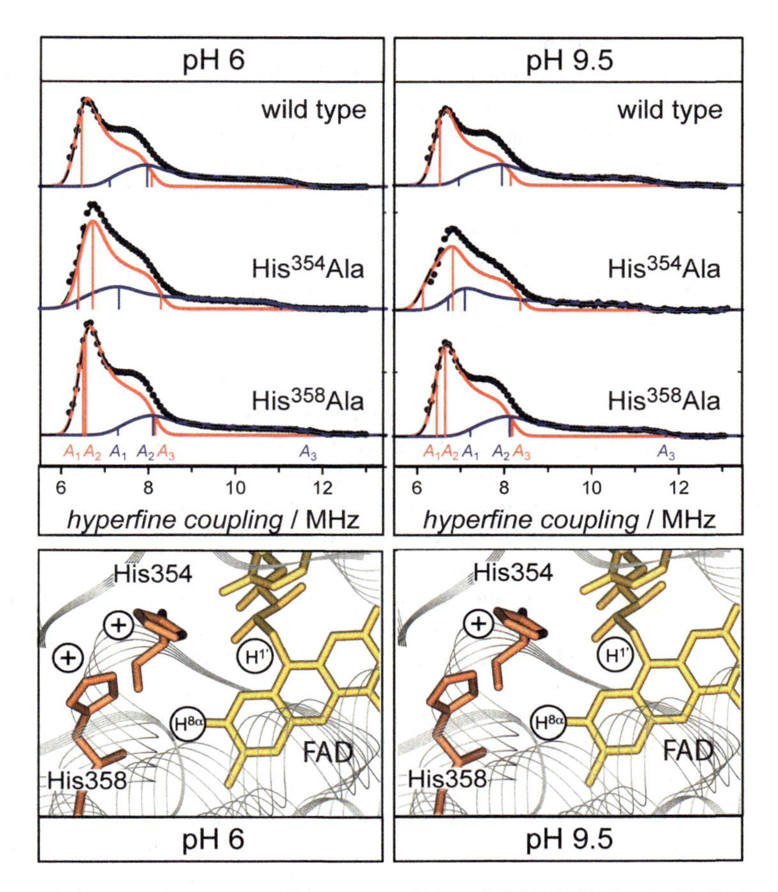

Fig. 3 ENDOR experiments on wild type, H354A and H358A *X. laevis* (6–4) photolyase measured at pH 6.0 and 9.5, respectively. *Upper part*: experimental (*dots*) and simulated (*dashed line*) pulsed ENDOR spectra of different *X. laevis* (6–4) photolyase samples. The *red* and the *blue curves* show the contributions of the H8α and the H1' hyperfine couplings to the overall ENDOR spectrum [56]. *Lower part*: three-dimensional structure of the active site of (6–4) photolyase including selected amino acids [71]. Proposed changes in the protonation state of His354 and His358 upon pH variations are displayed

the C1'–H1' bond with respect to the π-plane of the isoalloxazine ring, thus changing the dihedral angle, and, hence, the strength of the H1' hyperfine coupling. On the other hand, the shift of the isotropic hyperfine coupling of H8α, with respect to the wild type, was larger in the H358A mutant than in H354A. From this finding it was concluded, that His358 is closer than His354 to the H8α protons.

As a second result of the ENDOR study, the H8α ENDOR signal in the H354A mutant was also demonstrated to be strongly pH-dependent, an effect that must originate from a protonation change of His358 when going from pH 9.5 to pH 6 [56]. For steric reasons, it was concluded that at pH 9.5 the (deprotonated) His358 residue should turn towards the smaller Ala354 in the H354A mutant, which affects the axial symmetry of the hyperfine tensor. This implies that His354 does not change its protonation state when going from pH 9.5 to pH 6. Hence, the protonated histidine that is proposed to catalyze intermediate formation must be His354, because His358 is deprotonated at pH 9.5; see Fig. 3. In summary, these findings beautifully demonstrate the potential of ENDOR to study protonation states of amino acids by probing them with a nearby radical, such as the intrinsic flavin radical in the working site of photolyases.

Very recently, the long-awaited crystal structures of *Drosophila melanogaster* (6–4) photolyase in complex with DNA containing a (6–4) photoproduct lesion, and in complex with DNA after in situ repair were presented (see Fig. 3, lower panel) [71]. The overall structure of the (6–4)-photolyase looks surprisingly similar compared to the previously published structures from CPD-photolyases [67, 72], although the binding pocket for the DNA lesion is smaller but deeper than those from the CPD-repairing enzymes, and governed by less hydrophobic amino acids. This change in amino acid composition reflects the altered geometry of the enzyme-bound (6–4) photoproduct and could be an argument for an alternative repair mechanism. The previously discussed two conserved histidines are indeed located in the binding pocket of the substrate, even though only His354 was found to be in direct contact with the (6–4) photoproduct lesion via hydrogen bonding. Based on their structure data, the authors proposed a new mechanism for the repair of the (6–4) photoproduct [71]. However, in contrast to previously suggested reaction schemes, this mechanism does not involve an oxetane intermediate, but electron transfer from the flavin directly to the (6–4) photoproduct. Protonation of the one-electron reduced (6–4) photoproduct's 5-OH group by the nearby histidine then facilitates elimination of a water molecule, which subsequently attacks the acylimine molecule. This intermediate is proposed to split into the two thymines in the DNA lesion and, after back-electron transfer to the flavin, the intact bases are restored. During the last year, a number of studies, from both theoretical and experimental groups, have been published [73–75]. Surprisingly, all these recent studies favor yet other, substantially different, reaction mechanisms for the enzymatic repair of (6–4) photoproducts. Clearly, from structure data alone, the pending question on the reaction mechanism of (6–4) photolyase cannot be solved. Certainly, further EPR/ENDOR spectroscopic studies might be useful to solve this interesting problem.

3.2 ENDOR Investigations of LOV Domains

First discovered as tandem sensor domains in the plant photoreceptor phototropin [76], "light–oxygen–voltage" (LOV) domains have since been found in several plant, fungal, and bacterial proteins [77, 78]. LOV domains are distinguished from other flavin-based blue-light photoreceptors by their characteristic photochemistry. After absorption of a photon in the blue spectral region by the dark-adapted state, the FMN cofactor undergoes within picoseconds efficient intersystem crossing to yield a triplet intermediate state [79]. On a microsecond time scale, a covalent thioether bond between atom C4a of the FMN ring and a conserved nearby cysteine residue is formed [80, 81]. The photoreaction is fully reversible and the signaling state thermally reverts to the dark state. Despite closely similar amino acid sequence and three-dimensional structure [82–84], individual LOV domains differ markedly in the kinetics of their photocycle. For example, the time constants for dark recovery in phototropin LOV domains vary between 10 and 100 s [85], but much longer (or even irreversible) kinetics have also been found [86]. Signal transduction from the location of primary photochemistry to the kinase unit in full-length phototropin is far from being understood, although some key results begin to shed light onto the principal signaling mechanism. Results from NMR spectroscopy show a reordering or even an unfolding of a helix (named Jα-helix) that bridges the LOV domain and the signal transduction domain [87].

The slow dark-state recovery and the remarkable divergence of this reaction are still incompletely understood topics in LOV chemistry. One approach for understanding this phenomenon is to assume that subtle changes in the protein environment lead to specific conversions in protein conformation. These conformational changes in turn could alter the stability of the C4a–sulfur bond, and thus, modulate the reaction rate of the ΔG-driven C–S bond splitting. To find evidence for this notion, an ENDOR study was designed in which parts of the microenvironment in the close vicinity of the FMN cofactor of a LOV domain were altered and, via the determination of hyperfine couplings of the FMN in the radical state, the influence on the protein's reactivity was probed [52]. From these data, the internal energy of the system and, hence, the strength of interaction between the FMN cofactor and its surrounding was estimated.

All mutant *Avena sativa* LOV2 (*As*LOV2) domains examined contained an additional second mutation; the reactive cysteine residue was replaced by an alanine to prevent the light-induced formation of the FMN C4a adduct and to generate a meta-stable neutral FMN radical as an observer instead. Surprisingly, rather unexpected ENDOR spectra were recorded for *As*LOV2; see Fig. 4 (upper panel). Specifically, restricted 8α-methyl-group rotation was observed starting already at rather high temperatures ($T \leq 110$ K). This is quite unusual because it seems generally accepted that methyl group rotation is activated even down to liquid helium temperatures. Apparently, due to steric restrictions, such motion is strongly affected in *As*LOV2. To identify protein-cofactor interactions, a mutagenesis study has been performed: three amino acids, namely Leu496, Phe509, and Asn425, all located near H8α, were examined for their impact on methyl group rotational dynamics (see Fig. 5).

Mutations in these three amino acids clearly showed changed ENDOR spectra, which is consistent with predicted altered sterical interactions (see Fig. 4) [52]. Results obtained from spectral simulations of individual hyperfine couplings of FMN in the *As*LOV2 samples can be divided into two parts. (1) Point mutations of specific amino acid residues that change the temperature dependence of methyl group rotation in *As*LOV2 – proteins containing mutations of the amino acid Leu496 do not seem to have strong impact on the 8α protons as the spectra recorded

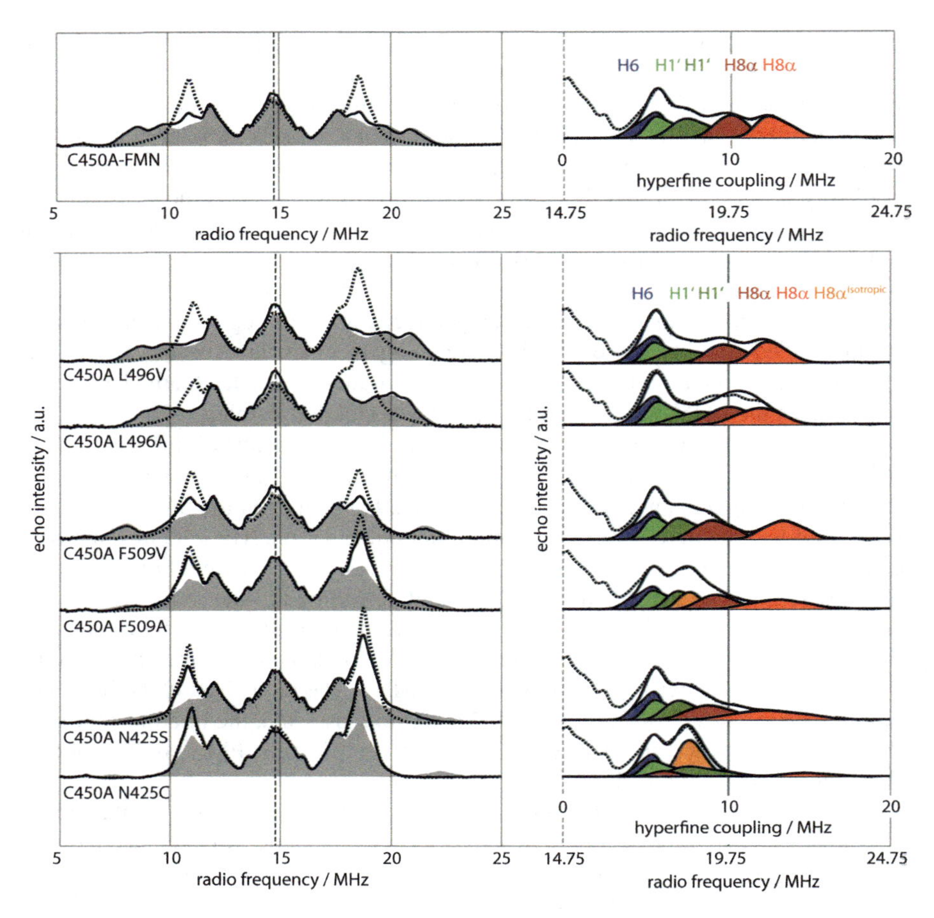

Fig. 4 ENDOR spectra of various *A. sativa* LOV2 domains. *Left panel*: pulsed X-Band Davies-type proton ENDOR spectra of various *As*LOV2 single - and double mutants. Spectra were recorded at 120 K (*dashed lines*), 80 K (*black lines*), and 10 K (*gray shaded*) for all samples. *Right panel*: sections of *As*LOV2 spectra (*dashed lines*) recorded at 10 K with accompanying spectral simulations (spectral contributions of individual hyperfine couplings are shown in color shades of *blue*, *green*, and *red*; the superpositions of all deconvoluted hyperfine contributions are shown as *thick black lines*). Two protein samples, *As*LOV2 C450A/F509A and *As*LOV2 C450A/N425C, require another hyperfine component of axial symmetry for accurate spectral fitting. This feature represents hyperfine coupling from fast rotating 8α-methyl group protons (H8$\alpha^{\text{isotropic}}$) and is shown in *orange*. Adapted from [52]

at 120, 80, and 10 K are highly comparable to those of *As*LOV2 C450A samples. On the other hand, mutations of Phe509 and particularly Asn425 lead to a drastically altered temperature behavior, which is illustrated in the left panel of Fig. 4. Conservative mutations such as F509V cause only subtle changes, whereas F509A, N425S, and especially N425C show an increase in rotational freedom of the methyl group even down to 10 K, which makes spectra of the latter mutation highly

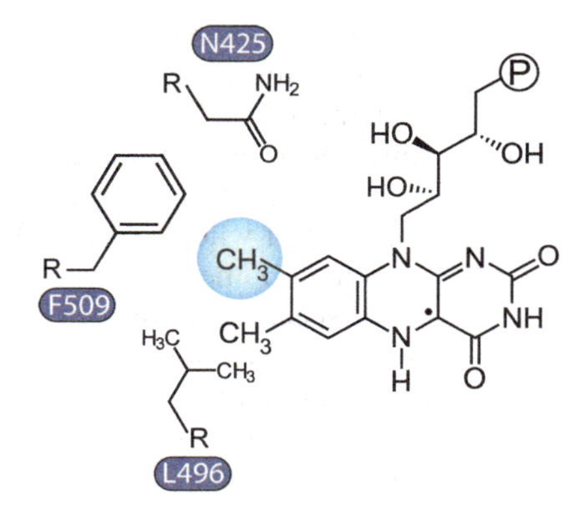

Fig. 5 FMN binding site of *As*LOV2 C450A. Schematic representation of the FMN-binding site of *As*LOV2 with amino acid residues that are in close proximity to the FMN's methyl group at C8α [87]

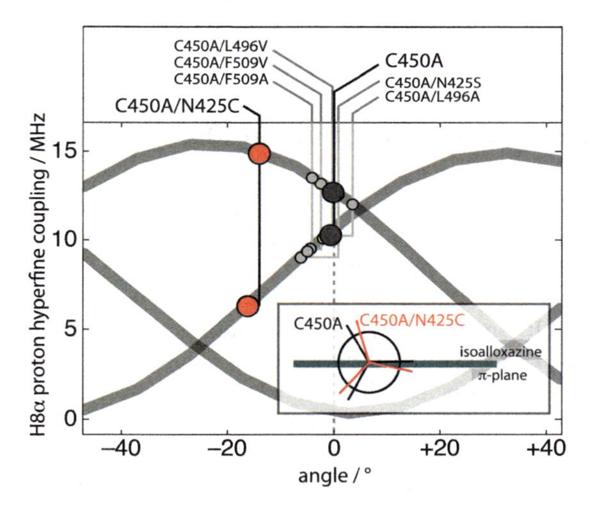

Fig. 6 H8α hyperfine couplings as a function of the rotation angle of the methyl group. Hyperfine coupling values of the three H8α protons as a function of the rotation angle were obtained from density functional theory. In *gray*, the isotropic part of the calculated hyperfine tensor for each of the three hydrogens is shown. The *dots* mark the isotropic hyperfine values of two 8α protons obtained by spectral simulation of the experimental data set. The *inset* shows a graphical view of the angles of the three 8α-methyl-group protons with respect to the isoalloxazine *ring plane* for *As*LOV2 C450A, and for C450A/N425C. Adopted from [52]

comparable to those obtained from other LOV samples, such as LOV1 from *Chlamydomonas reinhardtii* (data not shown) [52]. (2) Point mutations of specific amino acid residues that change the angle of the arrested 8α-methyl protons with respect to the isoalloxazine ring plane – based on results from density functional theory (DFT), dihedral angles could be determined, in which methyl rotation comes to a halt (see Fig. 6). Compared to the C450A samples, mutations of Asn425 change this situation dramatically: whereas the angle in N425S samples is only moderately increased, N425C mutations shifted the dihedral angle by about 10° [52].

It turned out that some of the above-mentioned mutations with strong impact on methyl group rotation also exhibit significantly altered dark-state recovery, i.e., the breakage of the FMN C4a–S bond, when introduced into wild type *As*LOV2. For example, mono-exponential fitting of the back reaction of the N425C mutant results in an adduct-state lifetime of only 7.5 s, which is sixfold decreased as compared to that of wild type *As*LOV2 (45.8 s) [52]. Hence, interaction between Asn425 and the 8α methyl group seems to "coerce" the FMN's isoalloxazine moiety into a specific conformation, thus stabilizing the intrinsically weak FMN C4a–S bond of the cysteinyl-FMN 4a adduct. Replacement of the amino acid leads to an increase in conformational flexibility of the FMN cofactor and, thus, to a destabilization of the S–C4a bond. Hence, careful examinations of hyperfine couplings of a paramagnetic cofactor, and their modifications as a function of the cofactor surroundings, revealed protein sites that have previously escaped attention because they are rather remote from the location of FMN adduct formation.

4 Transient EPR Investigations of Flavoproteins

Short-lived paramagnetic intermediates such as triplet states or radical pairs generated in the course of (photo-)chemical reactions can be favorably studied by measuring the EPR signal intensity as a function of time at a fixed value of the external magnetic field [88, 89]. Typically, the best-possible time response of a conventional EPR spectrometer that is dedicated to continuous-wave fixed-frequency lock-in detection is in the order of about 20 μs. By use of magnetic-field modulation frequencies higher than the 100 kHz usually employed in commercial instruments, the time resolution can be further increased by one order of magnitude, which makes the method well suited for the study of transient free radicals on a microsecond time scale. By completely forgoing the magnetic-field modulation, the time resolution can be pushed into the 10^{-8}- to 10^{-10}-s range [90–92]. A suitably fast data acquisition system comprising a broad-bandwidth microwave frequency mixer read out by a fast transient recorder or a digital oscilloscope is employed to detect directly the transient EPR signal as a function of time at a fixed magnetic field. In transient EPR spectroscopy (TREPR), paramagnetic species are generated on a nanosecond time scale by a short laser flash or radiolysis pulse, which also serves as a trigger to start signal acquisition. A complete TREPR spectrum comprises a series of time profiles recorded at magnetic-field points covering the total spectral width. This yields a two-dimensional variation of the signal intensity with respect to both the magnetic field and the time axis. Spectral information can be extracted from such a plot at any fixed time after the laser pulse as slices parallel to the magnetic-field axis.

In Fig. 7a, the TREPR signal from the photo-generated triplet state of FMN, detected about 1 μs after the laser pulse, is shown as a function of the magnetic field [93, 94]. Due to signal detection in the absence of any effect modulation (e.g., modulation of the magnetic field), the sign of the resonances directly reflects the

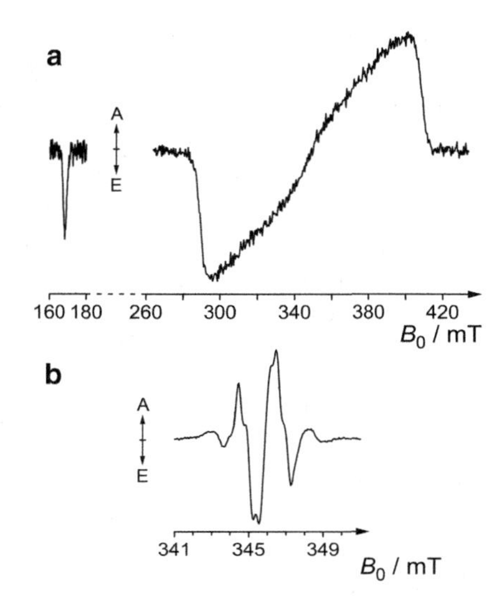

Fig. 7 Triplet and radical-pair TREPR spectra of flavoproteins. (**a**) TREPR data set of the photoexcited triplet state of FMN in frozen aqueous solution measured 1 μs after pulsed laser excitation at 80 K [93]. (**b**) TREPR spectrum of a photogenerated radical pair comprising a flavin and a tryptophan radical in *E. coli* CPD photolyase, measured 1 μs after pulsed laser excitation at 274 K [19]

emissive or enhanced absorptive polarizations of the EPR transitions. Electron-spin polarization arises due to the generation of the electron-spin state with an initial non-equilibrium energy-level population [95–97]. The width of the signal reflects the mutual interaction of the formally two unpaired electron spins in the triplet configuration. Because both spins are localized on the same isoalloxazine moiety, the dipolar spin–spin interactions are quite strong, and hence TREPR spectra of flavin triplet states are fairly broad [93, 94]. The weak transition at low magnetic field represents the "$\Delta M_S = \pm 2$"-transition.

In radical-pair states, the average distance between the two unpaired electron spins is typically much larger than in triplet states. Hence, TREPR spectra of photogenerated (and electron-spin polarized) radical pairs are narrower due to the reduced mutual dipolar and exchange interactions as compared to flavin triplets. This is shown in Fig. 7b, where the TREPR signal of a flavin-based radical pair in photolyase is depicted [19]. Analysis of the spectral shapes of TREPR signals yields information on the chemical nature of the individual radicals of the radical-pair state, and the interaction of the radicals with each other and with their immediate surroundings.

4.1 TREPR Studies of Reactive Paramagnetic Intermediates in Cryptochrome

Redox reactions have been proposed to play a key role in light-responsive activities of cryptochromes [98, 99], blue-light photoreceptors in plants, animals, and bacteria with widespread functions ranging from the regulation of circadian rhythms of plants and animals [13] to the sensing of magnetic fields in a number of species

[17]. Both in vitro and in vivo experiments suggest that the FAD redox state is changed from fully oxidized (FAD^{ox}) to the radical form when it adopts the signaling state [50, 51, 98]. The results agree with the redox activity of photolyases, which share (to a high degree) amino acid sequence and three-dimensional structure with cryptochromes [20]. Despite this similarity, cryptochromes generally do not show DNA repair activity. In photolyases, when starting from FAD^{ox}, photoinduced intraprotein electron transfer produces a radical pair, comprising an FAD radical and either a tyrosine or tryptophan radical, which is directly observable by time-resolved EPR [19, 100]. The specific amino acid involved in photoreduction of FAD in *Escherichia coli* CPD photolyase was first identified by a comprehensive point-mutational study in which each individual tryptophan of the enzyme was replaced by phenylalanine [101]. Only the W306F mutation abolished photoreduction of FAD^{ox} or $FADH^{\bullet}$. Trp306 is situated at the enzyme surface at a distance of approximately 20 Å to FAD [67]. However, this distance is too large for a rapid *direct* electron transfer between Trp306 and the FAD, which is completed within 30 ps, as has been determined recently by time-resolved optical spectroscopy [102]. Hence, a chain of tryptophans comprising Trp359 and Trp382 was postulated early on upon elucidation of the three-dimensional structure [67] of the enzyme to provide an efficient multistep electron-transfer pathway through well-defined intermediates between Trp306 and the FAD [103]. This chain of tryptophans has been conserved throughout all structurally characterized photolyases to date and is also found in cryptochromes. While the relevance of this intraprotein electron transfer for photolyase function is still under debate [104], the cascade is believed to be critical for cryptochrome signaling. For example, it has been shown recently that substitutions of the surface-exposed tryptophan or the tryptophan proximal to FAD reduce in-vivo photoreceptor function of *Arabidopsis* cryptochrome-1 [105].

Radical pairs generated along the tryptophan chain by light-induced electron transfer to FAD^{ox} in cryptochromes have been proposed to function as essential ingredients of a biological compass to collect information on the geomagnetic orientation in a large and taxonomically diverse group of organisms [106]. In principle, a compass based on radical-pair photochemistry requires (1) generation of a spin-correlated radical pair with coherent interconversion of its singlet and triplet states in combination with a spin-selective reaction, such as further "forward" reactions that compete with charge recombination (which regenerates the ground-state reactants) – the latter is only allowed for the singlet radical-pair but not the triplet radical-pair configuration, (2) modulation of the singlet-to-triplet interconversion by Zeeman magnetic interactions of the unpaired electron spins with the magnetic field, and (3) sufficiently small inter-radical exchange and dipolar interactions such that they do not attenuate the radical pair's singlet-to-triplet interconversion [17]. Hence, understanding the suitability and potential of cryptochromes for magnetoreception requires identification of radical-pair states and examination of their origin, and the detailed characterization of magnetic interaction parameters and kinetics. TREPR with a time resolution down to 10 ns allows real-time observation of such spin states generated by pulsed laser excitation. Cryptochromes of the DASH type are ideal paradigm systems for such

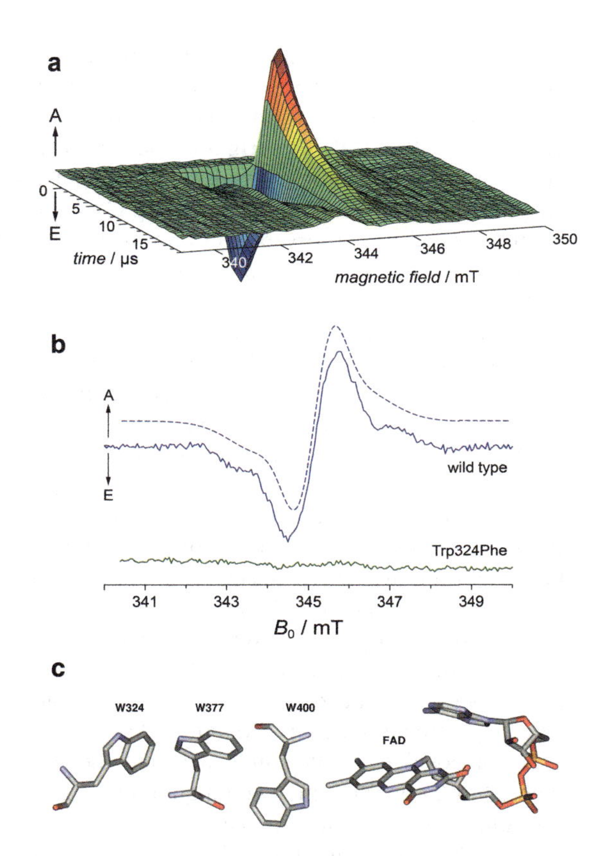

Fig. 8 TREPR spectrum of a photo-generated radical pair in *X. laevis* cryptochrome-DASH. (**a**) Complete TREPR data set of *X. laevis* cryptochrome-DASH measured at 274 K [109]. (**b**) TREPR spectrum of wild type (*solid blue curve*) and W324F (*solid green curve*) *X. laevis* cryptochrome-DASH recorded 500 ns after pulsed laser excitation. The *dashed curve* shows a spectral simulation of the EPR data of the wild-type protein using parameters described in the text and in [109]. (**c**) The conserved tryptophan triad of *X. laevis* cryptochrome-DASH. Figure taken from [125]

studies, because these proteins can be expressed from diverse species, their three-dimensional structure is known [107, 108], and they are stable and available in the amounts required for spectroscopic studies.

Recently, a TREPR study of light-induced paramagnetic intermediates from wild-type cryptochrome DASH of the frog *X. laevis* was presented and the results were compared with those from a mutant protein (W324F) lacking the terminal tryptophan residue of the conserved putative electron-transfer chain [109]; see Fig. 8.

The TREPR signal of wild-type cryptochrome recorded at physiologically relevant temperature is depicted in three dimensions as a function of the magnetic field and the time after pulsed laser excitation. Positive and negative signals indicate enhanced absorptive and emissive electron-spin polarization of the EPR transitions, respectively. The signal is assigned to a radical pair based on its spectral shape and narrow width. A spin-polarized flavin triplet state detected under comparable experimental conditions would span more than 150 mT due to the large spin–spin interactions between the two unpaired electrons (see above). The time evolution reveals that the radical-pair state exists for at least 6 µs; a more precise determination is not possible because the exponential signal decay is affected by spin relaxation of the electron-spin polarization to the Boltzmann equilibrium population. The spectrum of *X. laevis* cryptochrome resembles those obtained earlier on light-induced short-lived radical-pair species in FAD photoreduction of photolyases [19, 100]. The origin of the radical-pair signal in cryptochrome could be unraveled by examination of a single-point mutant, W324F, which lacks the enzyme-surface exposed tryptophan Trp324 (equivalent to Trp306 in *E. coli* CPD photolyase) of the conserved electron-transfer cascade. Under identical experimental conditions, the mutant protein did not exhibit any TREPR signal. The conclusion that Trp324 is the terminal electron donor in the light-induced electron-transfer reaction to the flavin chromophore in *X. laevis* cryptochrome is supported by spectral simulations, which have been performed on the basis of the correlated coupled radical-pair model and assuming the fixed orientations of the Trp324$^•$ radical and the flavin radical given by the three-dimensional structure of the protein [109]. The strength of the dipolar interaction between the two radicals was estimated based on the point-dipole approximation, which yielded $d = -0.36$ mT assuming an inter-radical distance of about 2.0 nm between Trp324$^•$ and FADH$^•$. Principal values for the **g**-tensors of both radicals were taken from high-magnetic-field/high-microwave-frequency examinations (see above). However, a satisfactory simulation of the TREPR signal of the Trp324$^•$FAD$^•$ radical pair was only obtained if a non-zero and positive exchange interaction parameter $J = +73$ μT was taken into account. Together with recent findings from optical spectroscopy on the FAD^{ox} photoreaction of the related *E. coli* CPD photolyase from which an electronic singlet precursor state of Trp306$^•$···FAD$^•$ radical pair formation was confirmed [110], a positive J value indicates that the triplet radical-pair configuration is favored by $2J$ over its singlet configuration. Both exchange and dipolar interactions are rather large as compared to the strength of the geomagnetic field, which is on the order of about 50 μT in Europe. Hence, the rather strong radical–radical interactions may inhibit the magnetic-field dependence of singlet-to-triplet interconversion of radical-pair states, and hence, make radical pairs of the type of Trp324$^•$···FAD$^•$ in cryptochromes unsuitable as sensors for the Earth's magnetic field unless cancellation effects are in effect, as has recently been suggested by Efimova and Hore [111].

Examination of the radical-pair species generated by pulsed laser light by TREPR at two different magnetic-field ranges and microwave frequencies, namely at X-band and at Q-band, revealed that the overall electron spin polarization (i.e., the integral of the TREPR signal amplitude over the magnetic field range of the entire radical-pair resonances) is zero at both EPR working points [112]: one half of the spectrum is in enhanced absorption, and the other half is in emission. This can only be rationalized if the electronic precursor multiplicity of radical-pair formation is a pure singlet state because a singlet has an equal number of spins with "spin-down" and "spin-up" projection, while the triplet levels $|T_+\rangle$, $|T_0\rangle$, and $|T_-\rangle$ are not populated. This situation applies at all ranges of the magnetic field, e.g., at the B_0-values applied in X-band as well as in Q-band EPR. However, when the radical-pair precursor is a spin-polarized triplet state, the initial TREPR spectrum typically exhibits either net absorption or net emission. This is understood in terms of the three triplet wave functions, $|T_+\rangle$, $|T_0\rangle$, and $|T_-\rangle$, which can now be populated. Similar to the singlet state, $|S\rangle$, the $|T_0\rangle$-state has an equal number of electrons with "spin-down" and "spin-up" projection. The $|T_+\rangle$- and $|T_-\rangle$-states, on the other hand, may contain different populations because of second-order effects created by the size of the zero-field splitting parameters (D and E) relative to the strength of the external magnetic field B_0. If $|T_-\rangle$ is in excess, then net enhanced absorption occurs because $|T_-\rangle$ contains two electrons with "spin-down" projection. Accordingly, if $|T_+\rangle$ is in excess, net emission is observed (because of the two electrons with "spin-up" projection). The stronger B_0 is relative to the size of zero-field splitting of the triplet zero-field sublevels $|T_X\rangle$, $|T_Y\rangle$, and $|T_Z\rangle$ (X, Y and Z are the principal axes of the dipolar coupling tensor in the triplet), the smaller the population difference between the high-field triplet levels $|T_-\rangle$ and $|T_+\rangle$ despite the non-Boltzmann population of the zero-field triplet levels [113]. In this so-called "high-magnetic-field limit," where the energies of $|T_-\rangle$ and $|T_+\rangle$ decrease and increase, respectively, in a linear fashion with the external magnetic field B_0, one expects increasingly symmetric TREPR spectra [112]. In other words, the higher the magnetic field, the more efficiently the enhanced absorptive and emissive spectral contributions will cancel out, and the more the integral of the TREPR signal intensity over the magnetic field will approach zero. Given typical zero-field splitting parameters ($|D| \approx 70$ mT, $|E| \approx 20$ mT) and zero-field populations ($\rho_X = 2/3$, $\rho_Y = 1/3$, and $\rho_Z = 0$) of a flavin triplet state at physiological pH [93], a net emissive TREPR spectrum is expected from a potentially triplet-born cryptochrome radical pair at X-band, whereas at Q-band the spectrum will become more symmetric with vanishing net electron-spin polarization. This is, however, not observed [112]. Similar considerations can be applied to the potential case of a triplet precursor state that is not spin-polarized but thermally relaxed prior to radical-pair generation. In this situation TREPR spectra become more asymmetric, i.e., net electron-spin polarization shifts towards enhanced absorption, with increasing magnetic field B_0. At X-band EPR, fairly symmetric spectra are expected, whereas at Q-band net-absorption will be observed [112]. This nicely demonstrates that a two-microwave-frequency TREPR approach can be used to draw conclusions on the nature of the precursor electronic state in light-induced spin-correlated radical-pair formation.

The TREPR results clearly show that cryptochromes (exemplified for the DASH-type) readily form radical-pair species upon photoexcitation. Spin correlation of such radical-pair states (singlet vs triplet), which is a necessary condition for magnetoselectivity of radical-pair reactions, manifests itself as electron-spin polarization of EPR transitions, which can be directly detected by TREPR in real time. Such observations support the conservation of photo-induced radical-pair reactions and their relevance among proteins of the photolyase/cryptochrome family. The results are of high relevance for studies of magnetosensors based on radical-pair (photo-)chemistry in general [114], and for the assessment of the suitability of cryptochrome radical pairs in animal magnetoreception in particular [17, 115].

4.2 TREPR Studies of BLUF Proteins

Blue-light sensitive photoreceptor BLUF ("blue light using FAD") domains are also flavoproteins, which regulate various, mostly stress-related processes in bacteria and eukaryotes. BLUF domains were found in various proteins, e.g., in the N-terminus of the AppA protein of the purple bacterium *Rhodobacter sphaeroides*. In this organism, AppA acts as a transcriptional antirepressor and interacts with the photosynthesis repressor protein PpsR to form a stable AppA-$(PpsR)_2$ complex in the dark and under low-light conditions [116]. The BLUF photoreceptor class was also discovered in a series of gene products derived from proteobacteria, cyanobacteria, and some eukaryotes [117]. Three-dimensional single-crystal and NMR structures for a number of BLUF proteins reveal a novel FAD binding fold (see, e.g., [118–122]). Photo-excitation of BLUF domains induces a small but characteristic red-shift of the FAD absorption of about 10–15 nm in the UV/vis region. This was rationalized with a reaction mechanism, in which a very short-lived biradical species comprising an FAD radical and a tyrosine radical are generated upon optical excitation of the FAD [123]. The proposed radical-pair species are, however, too short-lived (<1 ns) to be directly detected by TREPR at physiological temperatures, first because the temporal resolution of TREPR rarely reaches values below 10 ns, and second because the initial electron-spin polarization is quickly abolished by overall protein motion in liquid solution. In frozen aqueous solution, however, paramagnetic species could be detected by TREPR on a time scale of up to 100 μs [124]; see Fig. 9.

Triplet states were observed for various dark-adapted BLUF domains from different organisms. Based on the zero-field splitting parameters, the spin-polarized TREPR signals could be assigned to flavin triplet states. Radical-pair signatures were also detected, although, despite high conservation of amino acid sequence and three-dimensional structure, only from BLUF domains of *Synechocystis* sp. PCC

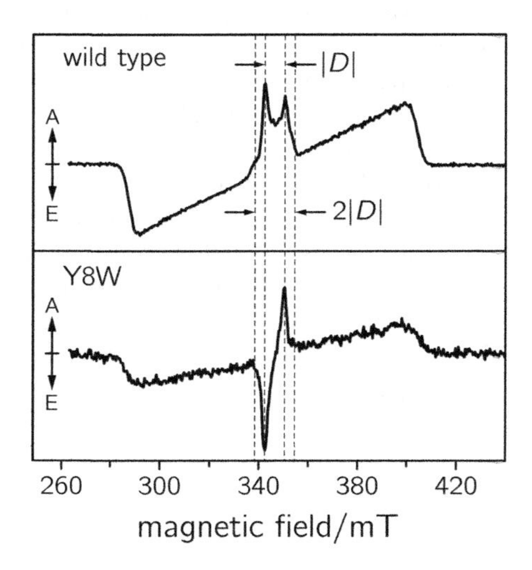

Fig. 9 TREPR spectra of wild-type and mutant Slr-BLUF domains. *Upper panel*: dark-adapted wild-type Slr-BLUF TREPR spectrum recorded 2 μs after pulsed laser excitation and at a temperature of 80 K. *Lower panel*: dark-adapted Y8W Slr-BLUF sample. Figure adapted from [124]

6803 (Slr-BLUF). The radical-pair signal of the wild type exhibits strong net-enhanced absorptive spin polarization and is superimposed by a similarly strongly spin-polarized flavin triplet spectrum. The narrow feature near g $\approx$ 2 was assigned to a strongly coupled radical pair comprising the flavin chromophore and an electron-donating amino acid, both in the radical form. Two aromatic amino acid residues are reasonable close to the isoalloxazine ring of the FAD to transfer efficiently an electron upon photoexcitation: Tyr8 and Trp91 in Slr-BLUF are located within 0.5–0.6 nm to C4a of FAD, respectively (measured from the points of highest unpaired electron-spin density of the respective residue). From the magnetic-field position of extreme and inflection points, the zero-field splitting parameters of the dipole–dipole coupling tensor can be extracted. Under the assumption of perfect axial symmetry (i.e., $E = 0$) of the strongly-coupled radical pair, the inflection points of the outer wings are separated by $2|D|$, and the inner features by $|D|$; see Fig. 9. Examination of the radical-pair feature yielded $|D| = (10.3 \pm 0.3)$ mT. Using the point-dipole approximation, a distance of about 0.7 nm was calculated as the average distance between the two unpaired electron spins of the radical pair, which is close to the distances found between Tyr8 and FAD, or Trp91 and FAD. Hence, based on the dipolar coupling alone, the precise origin of the radical pair TREPR signal cannot be unambiguously determined. To assign the amino acid involved in electron transfer to the FAD, the radical-pair signal of the wild type Slr-BLUF domain was compared to that of a mutant, where Tyr8 was replaced by a tryptophan, Y8W. Removal of Tyr8 completely changed the radical-pair feature, which now showed emissively and enhanced absorptively spin-polarized signal contributions. These results show that Tyr8 is directly involved in light-induced electron transfer to the flavin. However, the precise origin of the radical pair signal from Y8W Slr-BLUF remains unclear. The electron-donating residue could be Trp8, Trp91, or another as yet unidentified aromatic amino acid. This issue needs to be addressed by a more comprehensive inspection of aromatic amino acids and their involvement in the photocycle using TREPR preferably performed at higher magnetic fields than in X-band EPR to resolve better the radical pair signals by means of the different g-values of the coupled radicals.

5 Concluding Remarks

In recent years, a wealth of information on photolyase-mediated DNA repair and cryptochrome- and phototropin-mediated blue-light responses was obtained through the combined efforts of biologists, chemists, and physicists, from both experimental and theoretical studies. Here we have chosen recent examples of experimental work from our group to demonstrate the potential of modern EPR methods to answer mechanism-related questions and to study reactive intermediates in photo-induced electron transfer. Nevertheless, some key questions of these reactions remain to be solved. Important questions are as follows. What are the precise differences between CPD photolyase and (6–4) photolyase regarding substrate binding and DNA repair? Why are the cryptochromes, despite their high protein-sequence homology to (6–4) photolyases, incapable of repairing UV-induced DNA lesions? Are cryptochromes capable of sensing and transducing magnetic-field information, and if so, how is this task achieved in detail? Solving these questions will be a challenge for the next decade(s). We are confident that application of modern EPR will make an important contribution to this.

Acknowledgements We thank our colleagues, collaborators, and coworkers, who contributed substantially over the past 10 years to the work that is reviewed here: Adelbert Bacher (Technical University Munich), Alfred Batschauer (University of Marburg), Till Biskup (Oxford University), Robert Bittl (Free University Berlin), Richard Brosi (Free University Berlin), Lars-Oliver Essen (University of Marburg), Markus Fischer (University of Hamburg), Martin Fuchs (SLS, Paul-Scherrer-Institute Villigen), Elizabeth D. Getzoff (Scripps Research Institute, La Jolla), Peter Hegemann (Humboldt-University of Berlin), Kenichi Hitomi (Scripps Research Institute, La Jolla), Boris Illarionov (University of Hamburg), Sylwia Kacprzak (University of Freiburg), Chris Kay (University College London), Radoslaw Kowalczyk (University of Surrey, UK), Anthony R. Marino (University of Chicago), Tilo Mathes (Humbolt-University of Berlin), Klaus Möbius (Fee University Berlin), James R. Norris (University of Chicago), Asako Okafuji, Gerald Richter (Cardiff University), Alexander Schnegg (Free University Berlin), and Takeshi Todo (Osaka University). This work was supported by the Deutsche Forschungsgemeinschaft (DFG We 2376/3-1).

References

1. Frey PA, Hegeman AD, Reed GH (2006) Chem Rev 106:3302
2. Stubbe J, Nocera DG, Yee CS, Chang MCY (2003) Chem Rev 103:2167
3. Stubbe J, van der Donk WA (1998) Chem Rev 98:705
4. Frey PA (2001) Annu Rev Biochem 70:121
5. Himo F, Siegbahn PEM (2003) Chem Rev 103:2421
6. Fetzner S, Steiner RA (2010) Appl Microbiol Biotechnol 86:791
7. Massey V (2000) Biochem Soc Trans 28:283
8. Fraaije MW, Mattevi A (2000) Trends Biochem Sci 25:126
9. Joosten V, van Berkel WJH (2007) Curr Opin Chem Biol 11:195
10. Massey V, Palmer G (1966) Biochemistry 5:3181
11. Massey V (1994) J Biol Chem 269:22459
12. Losi A (2007) Photochem Photobiol 83:1283
13. Lin C, Todo T (2005) Genome Biol 6:220
14. Demarsy E, Fankhauser C (2009) Curr Opin Plant Biol 12:69
15. Liu B, Liu H, Zhong D, Lin C (2010) Curr Opin Plant Biol 13:578
16. Losi A, Gärtner W (2011) Photochem Photobiol 87:491
17. Rodgers CT, Hore PJ (2009) Proc Natl Acad Sci USA 106:353
18. Ritz T, Ahmad M, Mouritsen H, Wiltschko R, Wiltschko W (2010) J R Soc Interface 7:S135
19. Weber S (2005) Biochim Biophys Acta 1707:1
20. Sancar A (2003) Chem Rev 103:2203
21. Müller M, Carell T (2009) Curr Opin Struct Biol 19:277
22. Edmondson DE (1985) Biochem Soc Trans 13:593
23. Kay CWM, Weber S (2002) In: Gilbert BC, Davies MJ, Murphy DM (eds) Electron paramagnetic resonance (specialist periodical reports), vol 18. Royal Society of Chemistry, Cambridge, p 222

24. Schleicher E, Wenzel R, Okafuji A, Ahmad M, Batschauer A, Essen L-O, Hitomi K, Getzoff ED, Bittl R, Weber S (2010) Appl Magn Reson 37:339
25. Weber S, Kay CWM, Bacher A, Richter G, Bittl R (2005) Chemphyschem 6:292
26. García JI, Medina M, Sancho J, Alonso PJ, Gómez-Moreno C, Mayoral JA, Martínez JI (2002) J Phys Chem A 106:4729
27. Acocella A, Jones GA, Zerbetto F (2010) J Phys Chem B 114:4101
28. Fuchs M, Schleicher E, Schnegg A, Kay CWM, Törring JT, Bittl R, Bacher A, Richter G, Möbius K, Weber S (2002) J Phys Chem B 106:8885
29. Barquera B, Morgan JE, Lukoyanov D, Scholes CP, Gennis RB, Nilges MJ (2003) J Am Chem Soc 125:265
30. Kay CWM, Bittl R, Bacher A, Richter G, Weber S (2005) J Am Chem Soc 127:10780
31. Schnegg A, Kay CWM, Schleicher E, Hitomi K, Todo T, Möbius K, Weber S (2006) Mol Phys 104:1627
32. Schnegg A, Okafuji A, Bacher A, Bittl R, Fischer M, Fuchs MR, Hegemann P, Joshi M, Kay CWM, Richter G, Schleicher E, Weber S (2006) Appl Magn Reson 30:345
33. Okafuji A, Schnegg A, Schleicher E, Möbius K, Weber S (2008) J Phys Chem B 112:3568
34. Kay CWM, El Mkami H, Molla G, Pollegioni L, Ramsay RR (2007) J Am Chem Soc 129:16091
35. Pauwels E, Declerck R, Verstraelen T, De Sterck B, Kay CWM, Van Speybroeck V, Waroquier M (2010) J Phys Chem B 114:16655
36. Möbius K (2000) Chem Soc Rev 29:129
37. Steinhoff H-J, Savitsky A, Wegener C, Pfeiffer M, Plato M, Möbius K (2000) Biochim Biophys Acta 1457:253
38. Kay CWM, Schleicher E, Hitomi K, Todo T, Bittl R, Weber S (2005) Magn Reson Chem 43: S96
39. Gullá AF, Budil DE (2001) J Phys Chem B 105:8056
40. Sinnecker S, Reijerse E, Neese F, Lubitz W (2004) J Am Chem Soc 126:3280
41. Teutloff C, Hofbauer W, Zech SG, Stein M, Bittl R, Lubitz W (2001) Appl Magn Reson 21:363
42. Murphy DM, Farley RD (2006) Chem Soc Rev 35:249
43. van Doorslaer S, Vinck E (2007) Phys Chem Chem Phys 9:4620
44. Kulik L, Lubitz W (2009) Photosynth Res 102:391
45. Kay CWM, Feicht R, Schulz K, Sadewater P, Sancar A, Bacher A, Möbius K, Richter G, Weber S (1999) Biochemistry 38:16740
46. Medina M, Cammack R (2007) Appl Magn Reson 31:457
47. Bittl R, Kay CWM, Weber S, Hegemann P (2003) Biochemistry 42:8506
48. Barquera B, Ramirez-Silva L, Morgan JE, Nilges MJ (2006) J Biol Chem 281:36482
49. Nagai H, Fukushima Y, Okajima K, Ikeuchi M, Mino H (2008) Biochemistry 47:12574
50. Banerjee R, Schleicher E, Meier S, Muñoz Viana R, Pokorny R, Ahmad M, Bittl R, Batschauer A (2007) J Biol Chem 282:14916
51. Hoang N, Schleicher E, Kacprzak S, Bouly JP, Picot M, Wu W, Berndt A, Wolf E, Bittl R, Ahmad M (2008) PLoS Biol 6:e160.1559
52. Brosi R, Illarionov B, Mathes T, Fischer M, Joshi M, Bacher A, Hegemann P, Bittl R, Weber S, Schleicher E (2010) J Am Chem Soc 132:8935
53. Hamdane D, Guérineau V, Un S, Golinelli-Pimpaneau B (2011) Biochemistry 50:5208
54. Goldfarb D, Arieli D (2004) Annu Rev Biophys Biomol Struct 33:441
55. Goldfarb D (2006) Phys Chem Chem Phys 8:2325
56. Schleicher E, Hitomi K, Kay CWM, Getzoff ED, Todo T, Weber S (2007) J Biol Chem 282:4738
57. Kurreck H, Bock M, Bretz N, Elsner M, Kraus H, Lubitz W, Müller F, Geissler J, Kroneck PMH (1984) J Am Chem Soc 106:737
58. Çinkaya I, Buckel W, Medina M, Gómez-Moreno C, Cammack R (1997) Biol Chem 378:843

59. Medina M, Lostao A, Sancho J, Gómez-Moreno C, Cammack R, Alonso PJ, Martínez JI (1999) Biophys J 77:1712
60. Weber S, Richter G, Schleicher E, Bacher A, Möbius K, Kay CWM (2001) Biophys J 81:1195
61. Martínez JI, Alonso PJ, Gómez-Moreno C, Medina M (1997) Biochemistry 36:15526
62. Medina M, Vrielink A, Cammack R (1997) FEBS Lett 400:247
63. Jordan SP, Jorns MS (1988) Biochemistry 27:8915
64. Kim S-T, Malhotra K, Smith CA, Taylor J-S, Sancar A (1994) J Biol Chem 269:8535
65. Zhao X, Liu J, Hsu DS, Zhao S, Taylor J-S, Sancar A (1997) J Biol Chem 272:32580
66. Hitomi K, Kim S-T, Iwai S, Harima N, Otoshi E, Ikenaga M, Todo T (1997) J Biol Chem 272:32591
67. Park H-W, Kim S-T, Sancar A, Deisenhofer J (1995) Science 268:1866
68. Hitomi K, Nakamura H, Kim S-T, Mizukoshi T, Ishikawa T, Iwai S, Todo T (2001) J Biol Chem 276:10103
69. Lv XY, Qiao DR, Xiong Y, Xu H, You FF, Cao Y, He X, Cao Y (2008) FEMS Microbiol Lett 283:42
70. Varghese AJ, Wang SY (1968) Biochem Biophys Res Commun 33:102
71. Maul MJ, Barends TRM, Glas AF, Cryle MJ, Domratcheva T, Schneider S, Schlichting I, Carell T (2008) Angew Chem Int Ed 47:10076
72. Mees A, Klar T, Gnau P, Hennecke U, Eker APM, Carell T, Essen L-O (2004) Science 306:1789
73. Sadeghian K, Bocola M, Merz T, Schütz M (2010) J Am Chem Soc 132:16285
74. Domratcheva T, Schlichting I (2009) J Am Chem Soc 131:17793
75. Li J, Liu Z, Tan C, Guo X, Wang L, Sancar A, Zhong D (2010) Nature (London) 466:887
76. Christie JM, Salomon M, Nozue K, Wada M, Briggs WR (1999) Proc Natl Acad Sci USA 96:8779
77. Christie JM (2007) Annu Rev Plant Biol 58:21
78. Briggs WR (2007) J Biomed Sci 14:499
79. Swartz TE, Corchnoy SB, Christie JM, Lewis JW, Szundi I, Briggs WR, Bogomolni RA (2001) J Biol Chem 276:36493
80. Salomon M, Christie JM, Knieb E, Lempert U, Briggs WR (2000) Biochemistry 39:9401
81. Salomon M, Eisenreich W, Dürr H, Schleicher E, Knieb E, Massey V, Rüdiger W, Müller F, Bacher A, Richter G (2001) Proc Natl Acad Sci USA 98:12357
82. Crosson S, Moffat K (2001) Proc Natl Acad Sci USA 98:2995
83. Crosson S, Moffat K (2002) Plant Cell 14:1067
84. Fedorov R, Schlichting I, Hartmann E, Domratcheva T, Fuhrmann M, Hegemann P (2003) Biophys J 84:2474
85. Kasahara M, Swartz TE, Olney MA, Onodera A, Mochizuki N, Fukuzawa H, Asamizu E, Tabata S, Kanegae H, Takano M, Christie JM, Nagatani A, Briggs WR (2002) Plant Physiol 129:762
86. Swartz TE, Tseng T-S, Frederickson MA, Paris G, Comerci DJ, Rajashekara G, Kim J-G, Mudgett MB, Splitter GA, Ugalde RA, Goldbaum FA, Briggs WR, Bogomolni RA (2007) Science 317:1090
87. Harper SM, Neil LC, Gardner KH (2003) Science 301:1541
88. Stehlik D, Möbius K (1997) Annu Rev Phys Chem 48:745
89. Bittl R, Weber S (2005) Biochim Biophys Acta 1707:117
90. Furrer R, Thurnauer MC (1981) Chem Phys Lett 79:28
91. van Tol J, Brunel LC, Angerhofer A (2001) Appl Magn Reson 21:335
92. van Tol J, Brunel LC, Wylde RJ (2005) Rev Sci Instrum 76:074101
93. Kowalczyk RM, Schleicher E, Bittl R, Weber S (2004) J Am Chem Soc 126:11393
94. Schleicher E, Kowalczyk RM, Kay CWM, Hegemann P, Bacher A, Fischer M, Bittl R, Richter G, Weber S (2004) J Am Chem Soc 126:11067
95. Turro NJ, Kleinman MH, Karatekin E (2000) Angew Chem Int Ed 39:4436

96. Woodward JR (2002) Prog React Kinet Mech 27:165
97. Hirota N, Yamauchi S (2003) J Photochem Photobiol C Photochem Rev 4:109
98. Merrow M, Roenneberg T (2001) Cell 106:141
99. Froy O, Chang DC, Reppert SM (2002) Curr Biol 12:147
100. Weber S, Kay CWM, Mögling H, Möbius K, Hitomi K, Todo T (2002) Proc Natl Acad Sci USA 99:1319
101. Li YF, Heelis PF, Sancar A (1991) Biochemistry 30:6322
102. Lukacs A, Eker APM, Byrdin M, Brettel K, Vos MH (2008) J Am Chem Soc 130:14394
103. Byrdin M, Sartor V, Eker APM, Vos MH, Aubert C, Brettel K, Mathis P (2004) Biochim Biophys Acta 1655:64
104. Kavakli IH, Sancar A (2004) Biochemistry 43:15103
105. Zeugner A, Byrdin M, Bouly J-P, Bakrim N, Giovani B, Brettel K, Ahmad M (2005) J Biol Chem 280:19437
106. Ritz T, Adem S, Schulten K (2000) Biophys J 78:707
107. Brudler R, Hitomi K, Daiyasu H, Toh H, Kucho K-i, Ishiura M, Kanehisa M, Roberts VA, Todo T, Tainer JA, Getzoff ED (2003) Mol Cell 11:59
108. Klar T, Pokorny R, Moldt J, Batschauer A, Essen L-O (2007) J Mol Biol 366:954
109. Biskup T, Schleicher E, Okafuji A, Link G, Hitomi K, Getzoff ED, Weber S (2009) Angew Chem Int Ed 48:404
110. Henbest KB, Maeda K, Hore PJ, Joshi M, Bacher A, Bittl R, Weber S, Timmel CR, Schleicher E (2008) Proc Natl Acad Sci USA 105:14395
111. Efimova O, Hore PJ (2008) Biophys J 94:1565
112. Weber S, Biskup T, Okafuji A, Marino AR, Berthold T, Link G, Hitomi K, Getzoff ED, Schleicher E, Norris JR (2010) J Phys Chem B 114:14745
113. Mi Q, Ratner MA, Wasielewski MR (2010) J Phys Chem A 114:162
114. Maeda K, Henbest KB, Cintolesi F, Kuprov I, Rodgers CT, Liddell PA, Gust D, Timmel CR, Hore PJ (2008) Nature (London) 453:387
115. Ritz T, Wiltschko R, Hore PJ, Rodgers CT, Stapput K, Thalau P, Timmel CR, Wiltschko W (2009) Biophys J 96:3451
116. Masuda S, Bauer CE (2002) Cell 110:613
117. Gomelsky M, Klug G (2002) Trends Biochem Sci 27:497
118. Anderson S, Dragnea V, Masuda S, Ybe J, Moffat K, Bauer C (2005) Biochemistry 44:7998
119. Kita A, Okajima K, Morimoto Y, Ikeuchi M, Miki K (2005) J Mol Biol 349:1
120. Yuan H, Anderson S, Masuda S, Dragnea V, Moffat K, Bauer C (2006) Biochemistry 45:12687
121. Wu Q, Gardner KH (2009) Biochemistry 48:2620
122. Grinstead JS, Hsu S-TD, Laan W, Bonvin AMJJ, Hellingwerf KJ, Boelens R, Kaptein R (2006) Chembiochem 7:187
123. Gauden M, van Stokkum IHM, Key JM, Lührs DC, van Grondelle R, Hegemann P, Kennis JTM (2006) Proc Natl Acad Sci USA 103:10895
124. Weber S, Schroeder C, Kacprzak S, Mathes T, Kowalczyk RM, Essen L-O, Hegemann P, Schleicher E, Bittl R (2011) Photochem Photobiol 87:574
125. Schleicher E, Bittl R, Weber S (2009) FEBS J 276:4290

Top Curr Chem (2012) 321: 67–90
DOI: 10.1007/128_2011_236

Published online: 28 September 2011

EPR Spectroscopy in Polymer Science

Dariush Hinderberger

Abstract Synthetic polymers belong to the vast realm of soft matter and are one of the key types of materials to address societal needs at the beginning of the twenty-first century. Polymer science progressively addresses questions that deal with tuning mesoscopic and macroscopic structures and functions of polymers by understanding the effects that govern these systems on the nanoscopic level. EPR spectroscopy as a local, sensitive, and extremely specific magnetic resonance technique in many cases shows sensitivity on well-suited length- (0–10 nm) and time scales (μs–ps) and can deliver unique information on structure, dynamics, and in particular function of polymeric systems. A short review of recent literature is given and the power of simple EPR methods, especially CW EPR performed on a low-cost benchtop spectrometer, to elucidate complex polymeric materials is shown with specific examples from thermoresponsive polymer systems. These bear great potential in molecular transport and biomedical applications (e.g., drug delivery) and insights into interactions between carrier and small molecule are fundamental for designing and tuning these materials.

Keywords ESR/EPR spectroscopy · Non-covalent interactions · Polymers · Responsive polymers · Soft matter

Contents

D. Hinderberger
Max-Planck-Institut für Polymerforschung, Ackermannweg 10, 55128 Mainz, Germany
e-mail: dariush.hinderberger@mpip-mainz.mpg.de

1 Introduction

Synthetic polymers were one of the major driving forces of industrial and scientific development in the second half of the last century. A large portion of research on polymers initially dealt with understanding macroscopic properties, in particular their mechanical features. Nowadays, research in the polymer or macromolecular sciences has evolved far beyond these questions and specifically aims at understanding, manipulating, and designing the molecular level of polymers. Furthermore, not only classical "structural" polymers like polyethylene or polystyrene are studied but also the focus has shifted towards functional and smart polymeric materials.

In a recent article, leading researchers in the field have summarized challenges and opportunities for research in macromolecular science [1]. Five societal fields were identified, in which polymers can play a decisive role in the coming years and that researchers in polymer science should be aware of. These fields are energy, sustainability, health, and security, informatics/defense/protection. This extremely broad collection of fields that certainly shape our near future already implies that polymers (or even more generally, synthetic soft matter systems) are extremely versatile in applications. This huge range of possible applications is a direct consequence of the extreme variability of physical properties that can be achieved by chemical fine tuning on the molecular level. Modern polymer science is highly interdisciplinary, covering fields from chemical synthesis via processing technology and physical characterization all the way into biology and theoretical chemistry/physics. In the above-mentioned summary of current challenges and opportunities in polymer science, key research fields are named, in which advances are urgently needed to achieve progress for beneficial application of polymers in the mentioned societal issues: (1) new synthetic routes, (2) complex polymer systems formed through weak, non-covalent interactions, (3) characterization and properties, (4) theory and simulations, and (5) macromolecular processing and assembly.

Of these five key fields, electron paramagnetic resonance (EPR) spectroscopy as a local, sensitive, and highly selective technique can "naturally" add important information to fields (2) and (3), i.e., formation of complex polymer systems through non-covalent interactions and characterization and properties thereof, respectively. Polymeric systems have been under investigation with EPR spectroscopy for more than three decades. A fundamental introduction of EPR spectroscopic methods, examples of past work, and to a large extent future possibilities for EPR in polymer science were excellently reviewed in 2006 in the definitive book "Advanced ESR Methods in Polymer Research" edited by Shulamith Schlick [2].

It is in the nature of review articles that the choice of presented work is subjective and to a certain degree reflects the interests of the author(s). Here I will only briefly summarize a narrow selection of the most recent EPR-related research in polymer science and specifically put it in the context of the above-mentioned, identified needs and roles of polymers for the next decade. I will first use a few examples to put the spotlight on the specific question of how weak non-covalent interactions shape the structure and properties of responsive polymeric systems.

Using these examples, which are nowhere near complete, I aim at highlighting in a conceptual manner which valuable insights EPR spectroscopy, mainly through spin labeling and spin probing approaches, can add to modern polymer science in the second decade of the twenty-first century.

1.1 Non-covalent Interactions Shape the Molecular Scale of Complex Polymeric Systems

Polymers are one major part of the very broad field of "soft matter." This term is a classification that summarizes a wide variety of systems, from liquids via liquid crystals, micelles, synthetic macromolecules, membranes, colloidal suspensions, foams, and gels, right into the realm of biology with biomacromolecules (proteins, enzymes, DNA, RNA), viruses, and even whole cells [3–10]. All these systems, man-made or biological, have in common that their structure and function are governed by physical processes and interactions that are comparable to the thermal energy at room temperature ($k_B T = 4.1 \times 10^{-21}$ J). Some of these interactions are schematically depicted in Fig. 1. This differentiates them from "hard matter," in which typical energies are several orders of magnitude larger.

One of the most fascinating albeit complicating features of soft matter in general and complex polymers in particular is that usually many (weak), often counteracting, interactions *on the molecular scale* control their unique mesoscopic and/or macroscopic structure and function. In addition, the contribution of the chemical composition of the individual interacting species and entropic contributions play a substantial role. One aspect that is of utmost importance in particular for function in biological polymers is the existence of non-equilibrium processes on the molecular scale [11–13]. To achieve such *molecularly controlled non-equilibrium* states it is in fact advantageous to utilize weak interactions. With interactions that are too strong, living systems would be much better structured but their dynamics and hence their function would simply disappear.

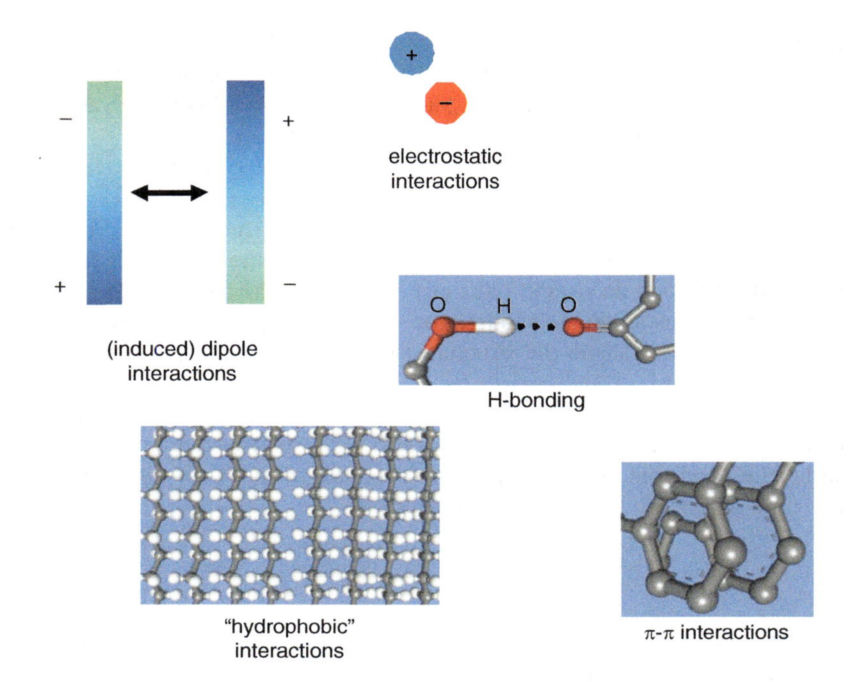

Fig. 1 A selection of weak interactions shaping polymers and soft matter in general on the molecular scale. These interactions, often in combination with each other, govern polymer structure and function

Ever since De Gennes and contemporaries shaped this vast field of condensed matter research, many experimental and theoretical studies focused on their meso- and macroscopic scale properties. Due to the weakness of the individual forces, the complexity of the many contributing interactions and the varying degrees of order, fewer studies have dealt with the molecular scale origin of soft matter and polymer properties [6, 9, 14–16].

The relevant length scales in polymers span a range of at least four orders of magnitude, from Ångstroms (see the forces in Fig. 1) up to at least a few micrometers. In soft matter and polymer theory the *multiscale modeling* approach, which attempts to simulate such systems in total including all relevant interactions, has recently attracted a lot of attention [17]. In experimental polymer research an analogous approach would be the combination of results from many techniques [18].

The central point of this chapter is to show that methods of EPR spectroscopy, in particular simple CW EPR, can unravel *structural, dynamic,* and *functional* properties of complex, synthetic polymers on the *molecular scale*. It will be shown that although EPR spectroscopy is a local method that has greatest versatility in the range of several nanometers, it is possible to obtain insight into fundamental interactions on the short length scales and into macroscopically observable functions on much longer length scales.

The work presented in Sect. 3 shows that with simple CW EPR on spin probes that are directed to the site of interest solely through non-covalent interactions it is possible to unravel and characterize structural and functional inhomogeneities on the nanoscale in thermoresponsive polymeric systems. As no chemical attachment was needed, this could be called a *site-directed spin probing* approach (see also, e.g., [19] or Chap. 7 of [2]). The molecular-scale characterization in these cases delivered evidence that the macroscopically observable thermal response has its origin in these inhomogeneities on the nanoscale.

Section 4 then gives a very short summary of other polymer-oriented studies of the last 5 years, again with no claim of completeness.

It should be noted that EPR spectroscopy in fields like bioinorganic chemistry or protein research is nowadays often combined with other experimental and simulation methods. In contrast, for rather randomly structured materials such as thermoresponsive polymer systems presented here, one can obtain meaningful insight with simple CW EPR methods. Before the specific recent examples are presented, the necessary basics of CW EPR spectroscopy are explained in the next section.

2 Radicals and CW EPR for the Study of Polymeric Systems on the Molecular Scale

EPR today is a spectroscopic method for determining the structure, dynamics, and spatial distribution of paramagnetic species [20]. For the first 25 years after its discovery in Kazan in the winter of 1944, EPR was a playground for scientists studying the fundamental physical properties of condensed matter. There are several reasons why this has changed markedly today and why EPR nowadays has found its firm albeit specialized place in materials science and biophysics: (1) the seminal work by Freed and others [21–24] starting in the late 1960s on the detailed analysis of molecular dynamics from EPR spectra, (2) the advent of several pulse EPR techniques [25, 26] and the first commercially available pulse EPR spectrometer in the 1980s, and (3) the breakthrough of making possible distance measurements in the range between 1 and ~8 nm in the late 1990s [27–29].

EPR applications are nonetheless limited due to lack of naturally occurring paramagnetic systems. In many cases one has to introduce stable free radicals artificially. These are called spin labels if they are covalently linked to the investigated system and spin probes if their interaction is of non-covalent nature [19, 30–32]. This apparent disadvantage, though, can often be of great value, as these spin probes can be detected very specifically and with higher sensitivity as compared to NMR on, e.g., ^{1}H-nuclei [33].

This section gives a concise introduction to CW EPR spectroscopy and only highlights the points needed for understanding the examples from polymer science in the following section. For a more detailed introduction to EPR spectroscopic techniques the reader is referred to [19, 34].

2.1 Radicals, EPR Spectra, and Their Information Content

The starting point for every kind of EPR study is the choice of the radical that is best suited to deliver valuable information on the systems that are investigated. In this chapter only three types of radicals are used for this purpose. Nitroxide radicals such as TEMPO (2,2,6,6-tetramethylpiperidine-1-oxyl; Fig. 2), its derivatives, and its five-ring analogs (Fig. 2) are the by far most widely used spin probes and spin labels and they are the spin probes of choice for the majority of EPR spectroscopic applications in naturally diamagnetic systems [2, 19].

For completeness one should note that the second very important type of paramagnetic species that are studied intensively are paramagnetic transition metals, mainly in the form of complexes. Whether transition metal ions are paramagnetic or not depends on the oxidation state and, without giving a complete list, examples often found in the literature are Cu^{2+}, Cr^{3+}, Mn^{2+}, Fe^{3+}, Co^{2+}, and $Ni^{1/3+}$ [35].

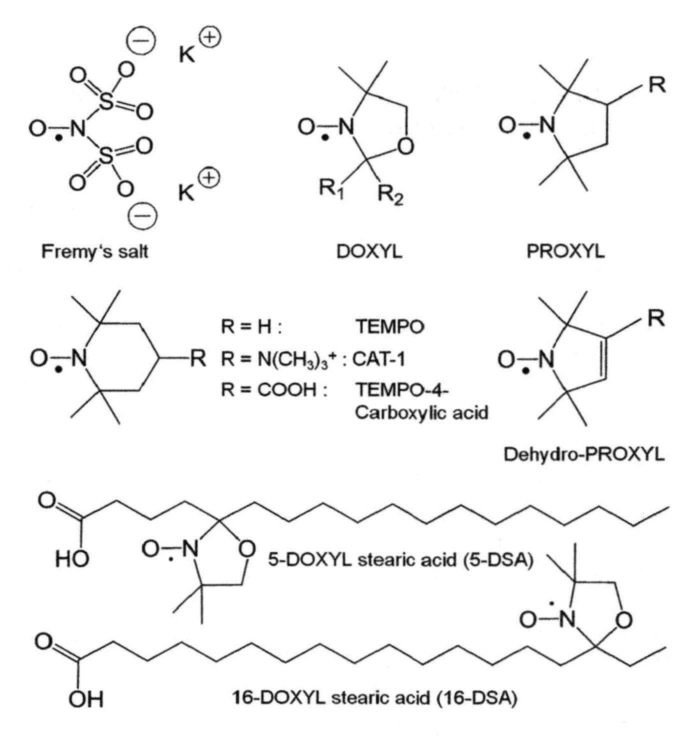

Fig. 2 Chemical structures of common nitroxide spin probes: *Fremy's salt* (potassium nitrosodisulfonate); *TEMPO* and derivatives (2,2,6,6-tetramethylpiperidine-1-oxyl), *DOXYL* (4,4-dimethyloxazolidine-1-oxyl); PROXYL (2,2,5,5-tetramethylpyrrolidine-1-oxyl); Dehydro-PROXYL (2,2,5,5-tetramethylpyrroline-1-oxyl); *5-DSA* (5-DOXYL stearic acid); *16-DSA* (16-DOXYL stearic acid)

Summaries of the information content of EPR spectroscopic methods (in particular on nitroxide radicals) and the length scales of interest are given in Fig. 3. Focusing on one radical ("observer spin"), the standard method *continuous wave (CW) EPR* at any temperature and echo-detected (ED) EPR at low temperatures give valuable information on the fingerprint of the radical. This is mainly the electronic but can also be the geometric structure of the radical center. From CW EPR spectral analysis and/or simulations, rotational motion on the time scale 10 ps – 1 μs can be characterized qualitatively and quantitatively. Furthermore, in CW EPR, radicals also intrinsically report on their immediate (usually up to a few solvation layers, maximum up to ~2 nm) chemical environment (e.g., polarity, proticity, etc.).

There are also pulse EPR methods that probe the chemical or rather magnetic environment. These are pulse electron nuclear double resonance (ENDOR) and hyperfine sublevel correlation (HYSCORE) spectroscopy, which allow measuring hyperfine couplings from the unpaired electron spin to surrounding magnetically active nuclei ([20]; in Fig. 3 this is a ^{31}P nucleus). As these experiments are performed in frozen solution (e.g., in all examples of this chapter) or in solids, from the anisotropy and orientation dependence of the hyperfine coupling one can obtain valuable information on the structure up to ~1 nm.

Finally, in recent years one pulse EPR method that measures the dipole–dipole couplings between *two* ("remote" and "observer") unpaired electron spins has attracted a lot of attention, in particular in the field of biophysical structure determination [27–29, 36–41]. This method, double electron–electron resonance

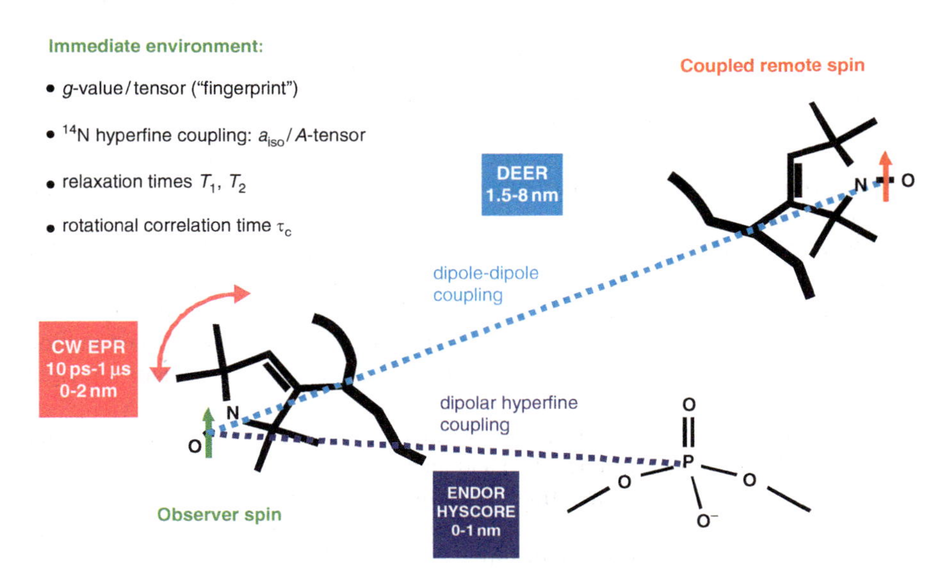

Fig. 3 Graphical summary of the information content of EPR spectroscopic methods on nitroxide radicals and a summary of the length and time scales of interest

(DEER), allows distance measurements through the $\omega_{dd} \propto r^{-3}$ dependence (ω_{dd} = dipolar coupling frequency; r = interspin distance). Due to the frequency range of EPR interactions, the accessible distances are in the extremely interesting range of 1.5–8 nm. Particularly for biopolymers and soft matter in general, this closes a gap between distance information from NMR (<1 nm, [33]) and from fluorescence-based methods, which on the other hand can address single molecule (>3 nm, [42–45]).

2.2 CW EPR: A Measure of Dynamics and of the Chemical Environment

For nitroxides in dilute liquid solution, the generally anisotropic spin Hamilton operator is simplified tremendously and, if unresolved proton hyperfine couplings are treated as line broadening, only the electron-Zeeman interaction and the hyperfine coupling to the magnetic ^{14}N nucleus ($I = 1$) remain [20]. The g- and hyperfine (A-) tensors are averaged to isotropic values due to fast motion of the spin probe and the resonance condition for the irradiated microwave becomes

$$\Delta E_{\text{nit}} = \hbar\omega_S = g\beta_e B_0 + a_{\text{iso}} m_I \tag{1}$$

which explains the three-line pattern of the nitroxide CW EPR spectra shown in Fig. 4. Due to the interaction with the nitrogen nucleus ($I = 1$) there are three transitions allowed that lead to the three-line nitroxide spectrum.

Equation (1) also nicely shows that the major advantage of performing high-field/high-frequency EPR, e.g., going to W-band (~94 GHz) frequencies, is the improved g-resolution, while the hyperfine resolution remains unaltered. It thus becomes possible to separate contributions to the spectra from electron-Zeeman and hyperfine anisotropies.

The effect of rotational motion and chemical environment on X-band CW EPR spectra is also summarized in Fig. 4 [19]. Rotational diffusion of spin probe molecules in CW EPR (Fig. 4a) can be roughly characterized by the rotational correlation time τ_c as belonging to one of the following regimes: (1) fast motion ($\tau_c \leq 100$ ps), (2) intermediate motion (100 ns $> \tau_c >$ 1 ns), and finally (3) rigid limit ($\tau_c \geq 1$ μs).

Rotational motion can be isotropic or anisotropic (e.g., when spin labels are attached to larger polymer molecules) and analysis of CW EPR spectra most often is quantified by spectral simulations assuming a rotational model of some sort (e.g., isotropic Brownian or uni-axial motion or more complicated models like "microscopic order, macroscopic disorder," or MOMD; see [19, 21]).

The effect of the local environment on the nitroxide spectra is illustrated in Fig. 4b. The electronic structure of a nitroxide is slightly altered depending on the

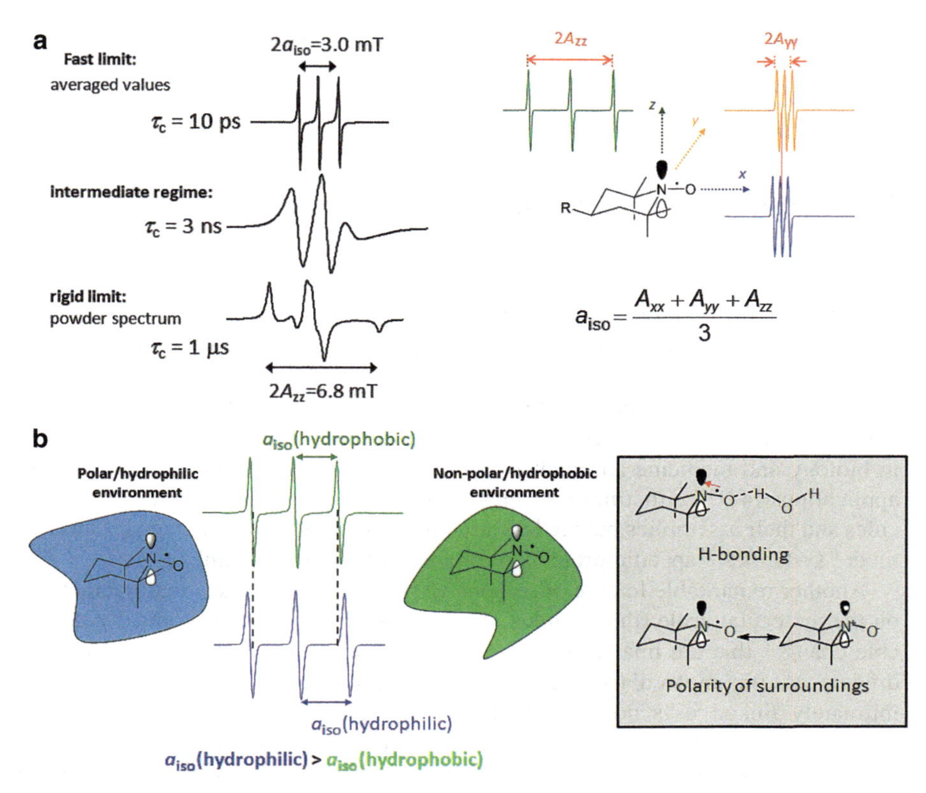

Fig. 4 Information from nitroxide CW EPR spectra. (**a**) *Right*: principal axis system of electron-Zeeman and hyperfine tensors (collinear). *Left*: the effect of rotational dynamics on the CW EPR spectra. Fast rotation (i.e., faster than a typical rotational correlation time of $\tau_c \sim 10$ ps) leads to the averaged spectrum. The isotropic g-value g_{iso} determines the center of the central line and spacing between the lines that is dominated by a_{iso}. In the intermediate motion regime 100 ns $> \tau_c >$ 1 ns and the rigid limit is reached at $\tau_c \sim 1$ μs [19]. (**b**) Influence of the chemical environment on CW EPR spectra. As both, hydrophilic and polar environments lead to an increased electron spin density at the nitroxide ^{14}N nucleus (see *gray inset*), a_{iso} and hence the line splitting in the spectra in hydrophilic and polar surroundings is larger than in non-polar and hydrophobic environments

interactions with molecules in its surrounding. In systems of different polarity and/or proticity (pH), the *same* spin probe features slightly different hyperfine couplings:

$$a_{iso}(\text{hydrophilic/polar environments}) > a_{iso}(\text{hydrophobic/non} - \text{polar})$$

The changes in the line splitting are small but add up at the high-field line. Hence, spin probes in different nanoscopic environments in inhomogeneous samples can be distinguished and analyzed separately. In fact, this ability to distinguish regions of different environments is used to reveal nano-inhomogeneities in the thermoresponsive polymeric systems, which are described in the next section.

When the much higher resolution of high-field EPR is available, one can even obtain a more detailed picture and distinguish polarity- and proticity-based effects [46, 47].

3 Nano-inhomogeneities in Thermoresponsive Polymeric Systems

For a multitude of reasons, synthetic polymers that exhibit environmentally responsive behavior are a very interesting class of polymeric materials [8]. Most prominently, living cells are regulated by macromolecules that either create or respond to environmental triggers, and they can be considered as biomimetic. Hence, their development is an integral part of the emerging field of smart applications in biology and medicine and makes them a valuable class of polymers for health applications [48]. Furthermore, as they to a certain degree mimic biomacromolecules and their assemblies but are less structurally complex, one may regard them as model systems for specific proteins or biomacromolecular assemblies.

Another remarkable feature of responsive polymeric systems is that interactions on the molecular scale (the stimulus of some sort) lead to macroscopically detectable changes that are finally employed for the function (e.g., directed delivery of drugs). As the molecular-scale interactions and macroscopic function are so intimately linked it is noteworthy that rather few studies have dealt with the nanoscopic level of these materials. This may be due to the fact that many conventional methods of physical polymer characterization may simply not be able to resolve the many different, often counteracting interactions [18, 49, 50]. In processes like a response of any kind, solvent-polymer, solvent–solvent, and polymer–polymer interactions all play a crucial role. Better understanding of the structure and interactions on the nanoscale is not only of value in itself but it may also shed light on similar processes in biomacromolecules and may aid the design and control of responsive polymers with respect to their applications [8, 48, 49]. These applications can be counted to the above-mentioned societal need of *health*, as responsive polymers are hot candidates for, e.g., drug or nucleic acid delivery purposes.

In this section, the EPR spectroscopic characterization of thermoresponsive polymeric systems is presented. The polymeric systems are water-swollen at lower temperatures and upon temperature increase the incorporated water is driven out and the system undergoes a reversible phase separation. Simple CW EPR spectroscopy (see above), carried out on a low-cost, easy-to-use benchtop spectrometer, is used here to reveal and characterize inhomogeneities on a scale of several nanometers during the thermal collapse. Further, neither any physical model of analysis nor chemical synthesis to introduce radicals had to be utilized. Adding amphiphilic TEMPO spin probes as guest molecules to the polymeric systems leads to self-assembly of these tracer molecules in hydrophilic and hydrophobic regions of the systems. These probes in different environments can be discerned and one can directly analyze the nanoscopic inhomogeneities from their CW EPR spectra. In three-dimensional hydrogels based on poly(*N*-isopropylacrylamide), PNIPAAM, as the thermoresponsive polymer we find static inhomogeneities not only of structure but most remarkably also of function (Sect. 3.1 [51]). In contrast, in thermoresponsive dendronized polymers, dynamic inhomogeneities over a broad temperature range are accompanied by one of the sharpest macroscopic thermal responses ever recorded (Sect. 3.2 [52–54]).

3.1 Static Nano-inhomogeneities of Structure and Function in Poly(NIPAAM) Hydrogels

A well-studied representative of thermoresponsive polymeric materials is poly (*N*-isopropylacrylamide) (PNIPAAM) in various modifications, which in water exhibits a lower critical solution temperature (LCST) around 32 °C [8, 55, 56]. Though responsive crosslinked hydrogels, especially PNIPAAM, have been studied extensively in the literature, the bulk of the scientific research dealt with the dependence of the macroscopic swelling behavior on the chemical structure. Despite the fact that microstructure and inhomogeneities on the sub-micrometer scale significantly influence the swelling and collapse behavior, considerably less publications dealt with the structural morphology of hydrogels explicitly [57, 58].

In this section, it is shown that EPR spectroscopy of amphiphilic nitroxide spin probes offers an interesting way to study the release of small molecules from PNIPAAM hydrogels. EPR spectroscopy on these reporter molecules shows high selectivity and site-specificity and delivers a large variety of information on local guest–host interaction, the distribution of guest molecules (on the nanometer scale), and accessibility by solvents. A picture of the collapse process at a molecular level can be drawn that is specifically based on variations of the chemical environment and rotational dynamics [32, 59–61].

In PNIPAAM hydrogels, even at temperatures well below the LCST, CW EPR signals from two types of TEMPO species are found. One of the species features a significantly reduced hyperfine splitting constant a_{iso} of 43.2 MHz (similar to values in organic solvents like chloroform), while the other species still has an a_{iso} of 47.3 MHz, close to that in bulk water. The signal of the species with lower a_{iso} gradually increases with increasing temperature; see Fig. 5a. This suggests that with increasing amounts of expelled water more and more polymer chains remain in the direct surrounding of the spin probe, which provide a much more hydrophobic environment for the low a_{iso} species than water. Note that this is in good agreement with nuclear magnetic resonance (NMR) based results obtained by Kariyo et al. [58]. On the other hand, small molecules are continuously released from the polymeric network when the temperature is increased; see Fig. 5a, b, although macroscopically the polymeric network collapses rather sharply at the LCST (at around 29 °C) of the polymer (Fig. 5b). Note that the maximum deflection of the EPR-derived curve coincides with this LCST [62].

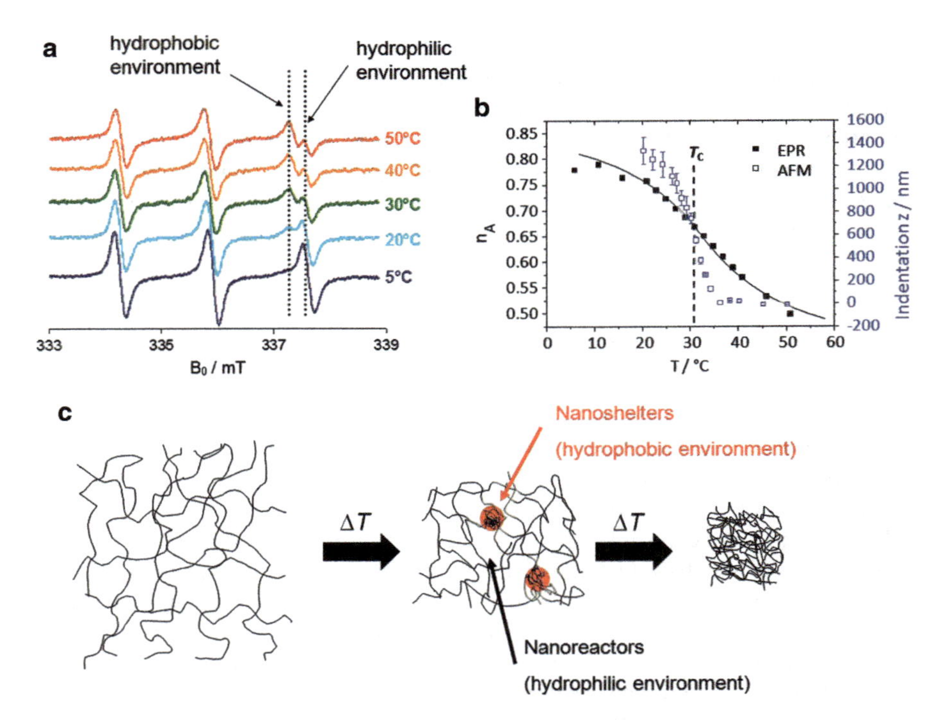

Fig. 5 (**a**) Temperature-dependent CW EPR spectra of TEMPO in an aqueous solution with PNIPAAM-based hydrogels and (**b**) comparison of the fraction of spin probes in hydrophilic environment n_A, as found from combining spectral simulations of both species (according to $S_{combined} = n_A \cdot S_A + (1 - n_A)S_B$) and a macroscopic (AFM-based) observation of hydrogel collapse. (**c**) Model of network collapse as seen by EPR spin probes: individual pockets continuously collapse before the macroscopic collapse happens

These findings have led to the conclusion that the polymer network collapse is a continuous, nano-inhomogeneous process, in which individual polymeric "pockets" are in a collapsed state even at temperatures significantly below the LCST and that the macroscopic collapse takes place only when a certain number and/or volume of collapsed "pockets" is reached.

This is schematically depicted in Fig. 5c. These nano-inhomogeneities can furthermore be considered as static on the EPR-time scale, as they neither change with time nor with temperature. With temperature, only the number or volume fraction of collapsed regions increases but there is no dynamic opening/closing of the collapsed regions. Once they are formed and once they have trapped TEMPO molecules, the only way to "open" the collapsed pockets for the aqueous phase is to reduce temperature again. It is worthwhile to note that, if the cationic and more hydrophilic radical Cat1 is used instead of TEMPO, no distinct second species can be observed; Cat1 does not sample hydrophobic regions.

Remarkably, the hydrophilic regions further form *nanoreactors*, which strongly accelerate acid-catalyzed disproportionation reactions if acidic protons are present,

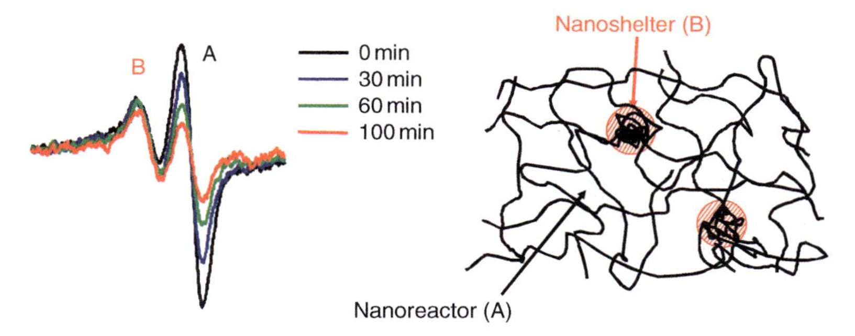

Fig. 6 The high-field peak region of TEMPO CW EPR spectra in methacrylic-acid containing PNIPAAM copolymer networks (*left*). "A" denotes the peak from TEMPO in hydrophilic environment, while "B" denotes the peak from TEMPO in the collapsed regions. The spectra were recorded in the course of 120 min and in this timeframe the "B" type species is only minimally reduced, while the hydrophilic "A" species is significantly reduced due to acid-catalyzed disproportionation of TEMPO to diamagnetic compounds. The coexistence of regions, in which this chemical reaction is facilitated ("nanoreactors") and simultaneously of regions, in which TEMPO molecules are protected from the reaction ("nanoreactors") is schematically depicted on the *right*

e.g., from methacrylic acid co-monomers in the hydrogel, while simultaneously the hydrophobic regions act as *nanoshelters*, in which enclosed spin probes are protected from decay (see Fig. 6). The results show that the system consisting of a statistical binary or tertiary copolymer displays remarkably complex behavior that mimics spatial and chemical inhomogeneities observed in functional biopolymers such as enzymes.

Note that in a subsequent study a similar result of micro-/nanophase separation has been observed in block copolymers of PNIPAAM and *N*-isopropyl*meth*acrylamide (PNIPMAM). This small-angle neutron scattering (SANS) study used a scattering analysis with a new form factor model taking into account a nanophase separated internal morphology [50].

The simple and cheap EPR method presented here not only *directly* mirrors the nanophase separation in the CW EPR spectra without the use of any model but even allows the characterization of the nano-inhomogeneity on the functional level of a chemical reaction.

3.2 Dynamic Nano-inhomogeneities During the Thermal Transition of Dendronized Polymers

Ever since Wu's discovery of the coil–globule transition of *single* PNIPAAM chains near the LCST [63–66], the collapse mechanism including the formation of stable mesoglobules has been a topic of intense research [8, 48, 50, 67–73]. Despite these efforts, a molecular scale picture of what happens when thermoresponsive polymers start to dehydrate at a certain temperature, subsequently

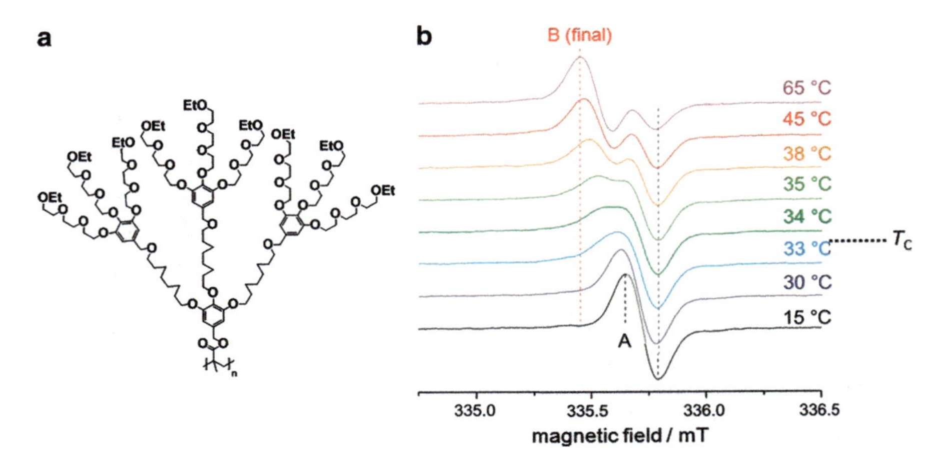

Fig. 7 (**a**) Chemical structure of the thermoresponsive dendronized polymer PG2(ET), denoting a second generation dendron that is capped with ethoxy groups. (**b**) The high-field peak region of TEMPO CW EPR spectra in PG2(ET). In contrast to the spectra in PNIPAAM (Figs. 5 and 6) there is a temperature dependent shift of the hyperfine splitting. This indicates dynamic inhomogeneities on the nanoscale, as explained in the text

collapsing and assembling into mesoglobules, does not exist. This severely hampers rational materials design.

In studies aiming at detection of unusual properties of dendronized polymers [74–78], Schlüter, Zhang, and coworkers recently discovered that such systems based on oligoethyleneglycol (OEG) units exhibit fast and fully reversible phase transitions with a sharpness that is amongst the most extreme ever observed (see Fig. 7a) [79–81].

These dendronized polymers with terminal ethoxy groups are soluble in water and their LCST is found in a physiologically interesting temperature regime between 30 and 36 °C. It is also interesting to note that the LCST of these OEG dendronized polymers is as low as is known for poly(ethylene oxide) and long chain ethylene oxide oligomers. For the latter the influence of hydrophobic end groups on the LCST has been thoroughly investigated both experimentally and theoretically [82]. Given this extraordinary behavior, these polymers are particularly suited to gain a deeper fundamental understanding of the processes involved. As described in the previous section, there are clear indications that thermal responses proceed via the formation of structural inhomogeneities of variable lifetimes on the nanometer scale that are still poorly understood. In this section, the focus is on a clearer understanding of the formation, structure, and lifetimes of these local inhomogeneities, the effect of the individual chemical structures on the physical processes, and the influence of the local heterogeneities on the aspired function (e.g., drug delivery).

The remarkable macroscopic behavior of such materials results from the systems being far from classical macroscopic equilibrium. It can be viewed as an example of "molecularly controlled non-equilibrium." Such macromolecule-based processes far from equilibrium are extensively found in nature, e.g., in DNA replication, to obtain high specificity in the noisy environment of a cell. Investigations into similar concepts in synthetic macromolecular systems are still rare [11–13].

To obtain insights into the molecular processes associated with the thermal transition again the amphiphilic radical TEMPO was used. While – as for TEMPO in PNIPAAM, see Fig. 5. – the low field and center peaks remain almost unaffected, the high-field line, most sensitive to structural and dynamic effects, changes considerably and is displayed in Fig. 7b for various temperatures. The apparent splitting of this line at elevated temperatures again originates from two nitroxide species A and B that are placed in local environments with different polarities. This gives rise to considerable differences of the isotropic hyperfine coupling constants a_{iso} (and the g-values g_{iso}).

The spectral parameters for component A again coincide with those of TEMPO in pure water ($a_{\mathrm{A}} \sim 48.3$ MHz), i.e., this spin probe is located in a strongly hydrated, hydrophilic environment. The observed decrease of a_{iso} by 3.7 MHz for species B ("final" at 65 °C) is indicative of much more hydrophobic and less hydrated surroundings for these spin probes (comparable to chloroform or *tert*-butyl alcohol [83]). At temperatures below the collapse temperature T_{C}, only the hydrophilic spectral component A is observed since all dendritic units are water-swollen. Above the critical temperature of 33 °C an increasing fraction of hydrophobic species B is observed with increasing temperature. The dehydration of the dendritic units thus leads to a local phase separation with the formation of hydrophobic cavities. Unlike in PNIPAAM hydrogels (Sect. 3.1), here hydrophobic regions are not observed below the macroscopic collapse temperature.

Strikingly, the peak position of the spectral component B is not fixed but approaches its final value only at temperatures well above T_{C}. This indicates a dynamic exchange of the spin probes between hydrophilic and hydrophobic regions. This exchange leads to an intermediate hyperfine coupling constant that is an effective weighted average between the two extreme values of the hydrophilic and the (static on the EPR time scale) hydrophobic regions (at 65 °C). Thus, the inhomogeneities formed upon the phase separation are not static, but dynamic and strongly influence the EPR spectral shape. This is a marked discrepancy again when compared to the case of PNIPAAM where only static hydrophobic regions are found.

The exchange detected by the spin probes can be caused by two effects: hopping of the spin probe between collapsed and hydrated polymer aggregate regions, or fluctuations of the aggregates themselves. The latter can be viewed as fast opening and closing of hydrophobic cavities or a fast swelling and de-swelling of regions surrounding the spin probe. The size of the inhomogeneities can be estimated by the translational displacement of TEMPO in the polymer matrix, given by $\langle x^2 \rangle = 6D_{\mathrm{T}}\tau_{\mathrm{T}}$. At $T = 34$ °C, a maximum translational displacement $\langle x^2 \rangle^{1/2} \leq 5.1$ nm of the spin probes due to diffusion is obtained, which is assisted by fluctuations of the polymer undergoing the thermal transition [34, 84].

Hence, slightly above T_C the few hydrophobic cavities formed are still small, i.e., in the range of a few nanometers. Then spin probe movement and/or local polymer fluctuations lead to an exchange of the probe molecules on the EPR time scale between the hydrophobic and large hydrophilic regions. The latter are still overwhelmingly more abundant. The spin probes thus mainly sample the interface between the two fundamentally different regions. Note that a few local dynamic heterogeneities on a nanometer scale are sufficient to induce a macroscopically observable (by turbidity measurements) transition in the sample. This macroscopic transition is detected at the same temperature as the change in the EPR spectra occurs. This suggests that the small hydrophobic regions detected by EPR might be visualized as cross-links affecting the organization of the dendronized macromolecules on much larger scales. The sharp macroscopic transition can then be viewed as the onset of a complex de-swelling process that is broad on the molecular scale rather than a sharp transition.

When increasing the temperature, not only the fraction of the hydrophobic regions but also their size grows and exchange of probe molecules between hydrophobic and hydrophilic sites becomes unlikely. The spin probes now sample the bulk hydrophobic (and remaining hydrophilic) regions rather than their interface. Together with the increase in size, the dynamics of the polymer fluctuations slow down, as both effects are coupled. In combination, a final state of distinct hydrophobic and hydrophilic regions is observed that are "static" on the EPR time scale (denoted "B (final)" in Fig. 7b). The complex collapse transition is schematically illustrated in Fig. 8.

The aggregation and the collapse can further be characterized by analyzing the effective hyperfine coupling constants of those TEMPO molecules in hydrophobic environments a_B and the fraction of TEMPO in hydrophilic environments y_A as a function of temperature. By plotting these parameters vs the reduced temperature $(T - T_C)/T_C$ it was possible to check whether the collapse results from a well-behaved phase transition. Both parameters do not follow one straight line, expected

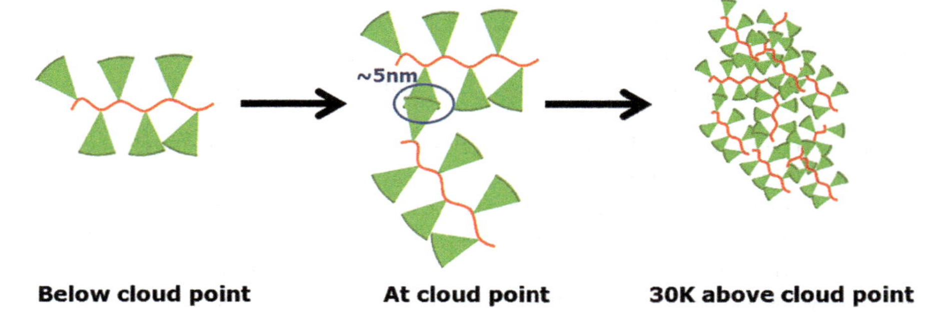

Fig. 8 Model of the collapse of thermoresponsive dendronized polymers as seen by EPR spin probes: few individual hydrophobic and dynamic patches of ~5 nm are sufficient to achieve a macroscopic collapse at the cloud point. Only at temperatures well above T_c a static state (on EPR time scales) is reached

for a simple phase transition, but instead strongly deviate from linearity [85]. Thus, in this wide temperature range, a complex dehydration takes place which cannot be described in the picture of a classical phase transition based on a single de-swelling process.

Furthermore, the effect of the chemical structure in the core of the dendrons on the thermoresponsive behavior can be tested. It is found that the initial dehydration and aggregation process at T_c turns out to be most effective when the dehydration is supported by a hydrophobic core. It deteriorates when the core contains oxyethylene groups, which can trap more water [53].

Altogether, a collapse transition as sketched in Fig. 8 is found: below T_c, no aggregation of dendrons is observed. At and above T_c, there is a growing number of uncorrelated hydrophobic regions up to a concentration and/or a volume fraction that is similar to that of the remaining hydrophilic regions. This could be an indication that the growth of hydrophobic regions reaches a threshold that could be interpreted as a percolation point. When the fractions of species A and B become equal, the likelihood of two hydrophobic regions (which up to that point can be largely uncorrelated) becoming neighbors increases immensely and the role of the interface becomes less important. Hence, at temperatures well above T_c (i.e., at least 30 °C), a "static" situation is achieved where TEMPO molecules are either trapped in hydrophilic or hydrophobic regions. There is no exchange on relevant EPR time scales between these two regions any more.

4 EPR on Polymers: A Short Survey

Several groups have in recent years employed similar principles of self-assembly as shown in the examples from the previous section.

Ottaviani and various coworkers have in the past often used a spin probing/ spin labeling approach to study self-assembled soft matter systems. In a recent paper they describe the complexes formed between cationic surfactants and hydrophobically modified anionic polymers. Analyzing the changes in dynamics as well as environment (hydrophobicity) of the long-chain fatty acid spin probe 5-DSA (see Fig. 2), they were able to extract indirectly information of the structure and dynamics of the complex formed [86] in dependence of the surfactant concentration and the pH. Such complexes of hydrophobically modified polyelectrolytes or ionic polymers with surfactant have unusual properties (in particular rheological) and are therefore widely used in, e.g., cosmetics or food applications.

In a series of papers, Jeschke and coworkers [87–90] as well as Schlick and coworkers [91, 92] studied self-assembled nanocomposite materials made from polymers and natural clays or artificial silicates. These nanocomposite materials have superior mechanical and also heat resistant properties and are hence interesting for applications in the fields of *defense* and *protection*. Such polymer–inorganic hybrid materials form complex structures and EPR spectroscopy has to be combined with other physical techniques such as NMR spectroscopy, small and wide angle X-ray diffraction, FT infrared spectroscopy, or differential scanning calorimetry.

Jeschke et al. studied the impact of surfactants on the nanocomposites in particular and in a combined CW EPR and pulse (ESEEM-based) approach demonstrated that there are two types of surfactants present in the composites. They could distinguish surfactants that adhere strongly to the surface of the nanoplatelets from those that are intercalated and interact most strongly with the added polymer (in this case poly(styrene)) [87].

The group of Schlick has shown how spin-labeled polymers like poly(methyl acrylate) or PEO can be used to study the self assembly of these polymers with clays and artificial silicates [91, 92]. As an example, through temperature-dependent CW EPR measurements they could very well distinguish polymer segments that strongly interact with the inorganic platelets from those that have a significantly increased segmental mobility.

The group of Goldfarb and coworkers have in recent years explored how (spin-labeled) thermoresponsive triblock copolymers of the Pluronic©-type (PEO-PPO-PEO, poly(ethylene oxide)-poly(propyleneoxide)-poly(ethyleneoxide)) can be used to build templates, e.g., for the formation of mesoporous frameworks [93, 94]. These structures bear great potential as carrier materials for catalysts and hence could aid societal needs in energy and sustainability.

This group could also show how such spin-labeled amphiphilic polymers self-assemble and decorate carbon nanotubes in aqueous dispersion and can help to solubilize them by attaching their hydrophobic PPO cores to the hydrophobic nanotubes while expanding their PEO blocks into water [95].

Similarly, Wasserman and coworkers have studied a wide selection of polymeric materials in aqueous solution that are associative of some kind, i.e., that form some sort of self-assembly through non-covalent interactions [96]. Their study mainly deals with hydrogels of hydrophobically modified polymers, aqueous solutions of polymeric micelles created by block copolymers, and hydrogels based on poly (acrylic acid) and macrodiisocyanates. The spin probes of choice were hydrophobic, such as 5- and 16-DSA (see Fig. 2) or even spin labeled polymers. It was, e.g., possible to screen for the effect of chemical structure on the gel formation by recording and understanding the local mobility of the hydrophobic, long chain spin probes as a function of temperature.

As pioneered by Schlick and coworkers, EPR imaging (EPRI/ESRI) combined with a spin trapping approach can also be used in polymer science, mainly when studying the progress of polymer degradation [97]. By introducing hindered amine stabilizers that convert short lived radicals produced during degradation into longer lived nitroxide radicals, one can follow the degradation with one- and two-dimensional EPRI. This is particularly relevant as such stabilizers are routinely added to commercial polymers to suppress radical chain degradation processes. Information on the photo-, mechanical, or thermal stability of polymer-based devices is immensely important and EPRI can reveal in a destruction-free manner that the polymer decay in the presence of stabilizers is inhomogeneous on a millimeter length scale.

Almost naturally, CW EPR spectroscopy can also contribute to understanding the electronic properties of polymers that are envisioned for application in the fields of *polymer electronics* (e.g., [98, 99]) and *photovoltaics* (e.g., [100–102]). Unlike in the studies highlighted before, in these cases all materials are EPR active and paramagnetic probes do not have to be added. Going beyond the conventional use of simple CW EPR spectroscopy to study electronic defects, Van Doorslaer, Goovaerts, Groenen, and coworkers have used multifrequency (X-, Q-, and W-band) EPR techniques such as HYSCORE and pulse ENDOR (see Sect. 2.1) to elucidate the extension of polarons in films of electro-active polymers [103].

Finally, the use of nanoscale distance measurements using DEER (see Sect. 2.1) in polymer science has been nowhere near as widespread as it already is in biophysics [108, 109]. Jeschke and coworkers very convincingly applied DEER to study the shape persistence of nominally rigid structures made from para-phenylene, ethynylene, and butadiynylene building blocks [104, 105]. Fitting two types of physical models, a worm-like chain and a newly developed harmonic segmented chain model, to DEER data obtained from doubly spin-labeled oligomers, they could quantify the backbone flexibility [105]. In combination with molecular dynamics simulations, they could even predict values for the persistence lengths of (13.8 ± 1.5) nm for poly(*p*-phenyleneethynylene)s and of (11.8 ± 1.5) nm for poly(*p*-phenylenebutadiynylene)s. A simple and more general description of how the conformational flexibility can be assessed in oligomers or larger molecules in general from DEER data has been developed by Prisner and coworkers [106].

Finally, a DEER study on models for molecular wires made from butadiyne-linked zinc porphyrin oligomers, end-labeled with nitroxide radicals, was performed by Lovett, Anderson, and coworkers [107]. Unlike in [104–106], one can control the conformations of these metalloporphyrin-based structures by self-assembly with multidentate amine ligands, which bend the rigid oligomeric structure. The experimentally found end-to-end distances in these complexes match the predictions from molecular dynamics calculations. This study thus presents a proof-of-principle that DEER spectroscopy is also well suited for understanding more complex supramolecular structures.

5 Conclusions and Outlook

Modern polymer science is highly interdisciplinary, covering fields from chemical synthesis via processing technology and physical characterization all the way into biology and theoretical chemistry/physics. Increasingly complex systems are in the focus nowadays which are often structured on the nanoscopic scale by non-covalent interactions. The studies presented here generally have in common that the great disadvantage of the probe molecule-based technique EPR can be turned to great advantage: the use of radicals as spin probes that are allowed to self assemble in largely disordered and complex systems results in excellent selectivity.

The interesting structural and functional properties of thermoresponsive polymers can be determined on the molecular or nanometer scale. Here it is shown that very simple and "cheap" CW EPR on small amphiphilic spin probes can characterize how thermal responses proceed in two very different thermoresponsive polymer systems. In these disordered systems the complexity is due to their statistical nature and the many counteracting forces, and the thermal collapse is intricately linked to static (Sect. 3.1) and dynamic (Sect. 3.2) inhomogeneities on a scale of several nanometers. Considering that these studies were performed on a low-cost, easy-to-use benchtop CW EPR spectrometer, one could arguably imagine that CW EPR on physically added, suitable (small) spin probes could in fact be used as a standard tool for the characterization of responsive, self-assembling functional polymers. In particular the desired use as transporting vehicles (drug, DNA/RNA delivery, etc.) makes understanding of non-covalent interactions between carrier systems and transported molecules indispensable, and this simple CW EPR-*site directed spin probing* approach could deliver valuable insights.

Acknowledgments I gratefully acknowledge my coworkers in my polymer-related EPR research, Matthias J.N. Junk and Dennis Kurzbach for their work as well as Hans W. Spiess and Gerhard Wegner for helpful discussions. I am grateful to my cooperation partners A. Dieter Schlüter, Afang Zhang, and Ulrich Jonas, for some of the thermoresponsive materials and helpful discussions.

References

1. Ober CK, Cheng SZD, Hammond PT, Muthukumar M, Reichmanis E, Wooley KL, Lodge TP (2009) Macromolecules 42:465–471
2. Schlick S (2006) Advanced ESR methods in polymer research. Wiley-Interscience, Hoboken, NJ, USA
3. de Gennes PG (1992) Rev Mod Phys 64:645–648
4. Kamien RD (2002) Rev Mod Phys 74:953–971
5. Jain S, Bates FS (2003) Science 300:460–464
6. Whitesides GM, Lipomi DJ (2009) Faraday Trans 143:373–384
7. George M, Weiss RG (2006) Acc Chem Res 39:489–497
8. Schild HG (1992) Prog Polym Sci 17:163–249
9. MacKintosh FC, Schmidt CF (2010) Curr Opin Cell Biol 22:29–35
10. Janmey PA, Hvidt S, Kas J, Lerche D, Maggs A, Sackmann E, Schliwa M, Stossel TP (1994) J Biol Chem 269:32503–32513
11. Cady F, Qian H (2009) Phys Biol 6:036011
12. Brutlag D, Kornberg A (1972) J Biol Chem 247:241–248
13. Drobny GP, Long JR, Karlsson T, Shaw W, Popham J, Oyler N, Bower P, Stringer J, Gregory D, Mehta M, Stayton PS (2003) Annu Rev Phys Chem 54:531–571
14. Israelachvili JN (1991) Intermolecular and surface forces, 2nd edn. Academic Press, London
15. Monkenbusch M, Richter D (2007) C R Physique 8
16. Messina R (2009) J Phys Condens Matter 21:113102
17. Praprotnik M, Delle Site L, Kremer K (2008) Annu Rev Phys Chem 59:545–571
18. Spiess HW (2010) Macromolecules 43:5479–5491

19. Hinderberger D, Jeschke G (2006) Site-specific characterization of structure and dynamics of complex materials by EPR spin probes. Modern Magnetic Resonance 3:1509–1517
20. Schweiger A, Jeschke G (2001) Principles of pulse electron paramagnetic resonance. Oxford University Press, Oxford
21. Schneider DJ, Freed JH (1989) In: Berliner LJ, Reuben J (eds) Biological magnetic resonance vol 8: Spin labeling-theory and applications. Plenum Press, New York
22. Goldman SA, Bruno GV, Polnaszek CF, Freed JH (1972) J Chem Phys 56:716
23. Hwang JS, Mason RP, Hwang LP, Freed JH (1975) J Phys Chem 79:489
24. Eastman MP, Bruno GV, Freed JH (1970) J Chem Phys 52:2511
25. Höfer P, Grupp A, Nebenführ H, Mehring M (1986) Chem Phys Lett 132:279–282
26. Milov AD, Ponomarev AB, Tsvetkov YuD (1984) Chem Phys Lett 110:67
27. Martin RE, Pannier M, Diederich F, Gramlich V, Hubrich M, Spiess HW (1998) Angew Chem 110:2994–2998; Angew Chem Int Ed 1998, 37:2834–2837
28. Pannier M, Veit S, Godt A, Jeschke G, Spiess HW (2000) J Magn Reson 142:331–340
29. Jeschke G (2002) Macromol Rapid Commun 23:227–246
30. Hubbell WL, Cafiso DS, Altenbach C (2000) Nat Struct Biol 7:735
31. Kocherginsky N, Swartz HM (1995) Nitroxide spin labels – reactions in biology and chemistry. CRC Press, Boca Raton
32. Hinderberger D, Spiess HW, Jeschke G (2010) Appl Magn Reson 37:657–683
33. Schmidt-Rohr K, Spiess HW (1996) Multidimensional solid-state NMR and polymers. Academic Press, London
34. Atherton NM (1993) Principles of electron spin resonance. Ellis Horwood, New York
35. Calle C, Sreekanth A, Fedin MV, Forrer J, Garcia-Rubio I, Gromov IA, Hinderberger D, Kasumaj B, Léger P, Mancosu B, Mitrikas G, Santangelo MG, Stoll S, Schweiger A, Tschaggelar R, Harmer J (2006) Helv Chim Acta 89:2495–2521
36. Milov AD, Salikhov KM, Shirov MD (1981) Fizika Tverdogo Tela 23:975–982
37. Schiemann O, Prisner TF (2007) Q Rev Biophys 40:1–53
38. Dockter C, Volkov A, Bauer C, Polyhach Y, Joly-Lopez Z, Jeschke G, Paulsen H (2009) Proc Natl Acad Sci U S A 106:18485–18490
39. Hilger D, Jung H, Padan E, Wegener C, Vogel KP, Steinhoff H-J, Jeschke G (2005) Biophys J 89:1328–1338
40. Schiemann O, Piton N, Plackmeyer J, Bode BE, Prisner TF, Engels JW (2007) Nat Protoc 2:904–923
41. Schiemann O, Cekan P, Margraf D, Prisner TF, Sigurdsson ST (2009) Angew Chem 121:3342–3345; Angew Chem Int Ed 2009, 48:3292–3295
42. Lipman EA, Schuler B, Bakajin O, Eaton WA (2003) Science 301:1233–1235
43. Lacoste TD, Michalet X, Pinaud F, Chemla DS, Alivisatos AP, Weiss S (2000) Proc Nat Acad Sci U S A 97:9461–9466
44. Basché T, Mörner WE, Orrit M, Talon H (1992) Phys Rev Lett 69:1516–1519
45. Gensch T, Hofkens J, Heirmann A, Tsuda K, Verheijen W, Vosch T, Christ T, Basché T, Müllen K, De Schryver FC (1999) Angew Chem 111:3970–3974; Angew Chem Int Ed 1999, 38:3752–3756
46. Steinhoff HJ, Savitsky A, Wegener C, Pfeiffer M, Plato M, Mobius K (2000) Biochim Biophys Acta 1457:253–262
47. Akdogan Y, Heller J, Zimmermann H, Hinderberger D (2010) Phys Chem Chem Phys 12:7874–7882
48. de las Heras Alarcón C, Pennadam S, Alexander C (2005) Chem Soc Rev 34:276–285
49. Dreher MR, Simnick AJ, Fischer K, Smith RJ, Patel A, Schmidt M, Chilkoti A (2008) J Am Chem Soc 130:687–694
50. Keerl M, Pedersen JS, Richtering W (2009) J Am Chem Soc 131:3093–3097
51. Junk MJN, Jonas U, Hinderberger D (2008) Small 4:1485–1493
52. Junk MJN, Li W, Schlüter AD, Wegner G, Spiess HW, Zhang A, Hinderberger D (2010) Angew Chem 122:5818–5823; Angew Chem Int Ed 2010, 49:5683–5687

53. Junk MJN, Li W, Schlüter AD, Wegner G, Spiess HW, Zhang A, Hinderberger D (2011) Macromol Chem Phys 212:1229–1235
54. Junk MJN, Li W, Schlüter AD, Wegner G, Spiess HW, Zhang A, Hinderberger D (2011) J Am Chem Soc 133:10832–10838
55. Hirotsu S, Hirokawa Y, Tanaka T (1987) J Chem Phys 87:1392–1395
56. Yu H, Grainger DW (1993) J Appl Polym Sci 49:1553–1563
57. Ikkai F, Shibayama M (2005) J Polym Sci B Polym Phys 43:617–628
58. Kariyo S, Küppers M, Badiger MV, Prabhakar A, Jagadeesh B, Stapf S, Blümich B (2005) Magn Reson Imaging 23:249–253
59. Hinderberger D, Schmelz O, Rehahn M, Jeschke G (2004) Angew Chem 2004, 116:4716–4721; Angew Chem Int Ed 2004, 43:4616–4621
60. Harvey RD, Schlick S (1989) Polymer 30:11–16
61. Rex GC, Schlick S (1987) Polymer 28:2134–2138
62. Beines PW, Klosterkamp I, Menges B, Jonas U, Knoll W (2007) Langmuir 23:2231–2238
63. Wu C, Zhou SQ (1995) Macromolecules 28:5388–5390
64. Wu C, Zhou SQ (1995) Macromolecules 28:8381–8387
65. Wu C, Zhou SQ (1996) Phys Rev Lett 77:3053–3055
66. Wang X, Qiu X, Wu C (1998) Macromolecules 31:2972–2976
67. Van Durme K, Verbrugghe S, Du Prez FE, Van Mele B (2004) Macromolecules 37:1054–1061
68. Luo S, Xu J, Zhu Z, Wu C, Liu S (2006) J Phys Chem B 110:9132–9138
69. Ono Y, Shikata T (2006) J Am Chem Soc 128:10030–10031
70. Cheng H, Shen L, Wu C (2006) Macromolecules 39:2325–2329
71. Van Durme K, Van Assche G, Aseyev V, Raula J, Tenhu H, Van Mele B (2007) Macromolecules 40:3765–3772
72. Ono Y, Shikata T (2007) J Phys Chem B 111:1511–1513
73. Keerl M, Smirnovas V, Winter R, Richtering W (2008) Angew Chem 120:344–347; Angew Chem Int Ed 2008, 47:338–341
74. Schlüter AD, Rabe JP (2000) Angew Chem 112:860–880; Angew Chem Int Ed 2000, 39:864–883
75. Zhang A, Shu L, Bo Z, Schlüter AD (2003) Macromol Chem Phys 204:328–339
76. Schlüter AD (2005) Top Curr Chem 245:151–191
77. Frauenrath H (2005) Prog Polym Sci 30:325–384
78. Rosen BM, Wilson CJ, Wilson DA, Peterca M, Imam MR, Percec V (2009) Chem Rev 109:6275–6540
79. Li W, Zhang A, Feldman K, Walde P, Schlüter AD (2008) Macromolecules 41:3659–3667
80. Li W, Zhang A, Schlüter AD (2008) Chem Commun 5523–5525
81. Li W, Wu D, Schlüter AD, Zhang A (2009) J Polym Sci A Polym Chem 47:6630–6640
82. Dormidontova EE (2004) Macromolecules 37:7747–7761
83. Knauer BR, Napier JJ (1976) J Am Chem Soc 98:4395
84. Kovarskii AL, Wasserman AM, Buchachenko AL (1972) J Magn Reson 7:225–237
85. Chaikin PM, Lubensky TC (1995) Principles of condensed matter physics. Cambridge University Press, Cambridge
86. Deo P, Deo N, Somasundaran P, Moscatelli A, Jockusch S, Turro NJ, Ananthapadmanabhan KP, Ottaviani MF (2007) Langmuir 23:5906–5913
87. Schleidt S, Spiess HW, Jeschke G (2006) Colloid Polym Sci 284:1211–1219
88. Panek G, Schleidt S, Mao Q, Wolkenhauer M, Spiess HW, Jeschke G (2006) Macromolecules 39:2191–2200
89. Mao Q, Schleidt S, Zimmermann H, Jeschke G (2007) Macromol Chem Phys 208:2145–2160
90. Mao Q, Schleidt S, Zimmermann H, Jeschke G (2008) Phys Chem Chem Phys 10:1156–1167
91. Miwa Y, Drews AR, Schlick S (2006) Macromolecules 39:3304–3311
92. Miwa Y, Drews AR, Schlick S (2008) Macromolecules 41:4701–4708

93. Ruthstein S, Schmidt J, Kesselman E, Popovitz-Biro R, Omer L, Frydman V, Talmon Y, Goldfarb D (2008) Chem Mater 20:2779–2792
94. Ruthstein S, Schmidt J, Kesselman E, Talmon Y, Goldfarb D (2006) J Am Chem Soc 128:3366–3374
95. Shvartzman-Cohen R, Florent M, Goldfarb D, Szleifer I, Yerushalmi-Rozen R (2008) Langmuir 24:4625–4632
96. Wasserman AM, Yasina LL, Motyakin MV, Aliev II, Churochkina NA, Rogovina LZ, Lysenko EA, Baranovsky VYu (2008) Spectrochim Acta A 69:1344–1353
97. Motyakin MV, Schlick S (2006) Polym Degrad Stab 91:1462–1470
98. Gomar-Nadal E, Mugica L, Vidal-Gancedo J, Casado J, Lopez Navarrete JT, Veciana J, Rovira C, Amabilino DB (2007) Macromolecules 40:7521–7531
99. Adhikari AR, Huang M, Bakhru H, Chipara M, Ryu CY, Ajaya PM (2006) Nanotechnology 17:5947–5953
100. Deepa M, Bhandari S, Arora M, Kant R (2008) Macromol Chem Phys 209:137–149
101. Colladet K, Fourier S, Cleij TJ, Lutsen L, Gelan J, Vanderzande D, Nguyen LH, Neugebauer H, Sariciftci S, Aguirre A, Janssen G, Goovaerts E (2007) Macromolecules 40:65–72
102. Berridge R, Skabara PJ, Pozo-Gonzalo C, Kanibolotsky A, Lohr J, McDouall JJW, McInnes EJL, Wolowska J, Winder C, Sariciftci NS, Harrington RW, Clegg W (2006) J Phys Chem B 110:3140–3152
103. Aguirre A, Gast P, Orlinskii S, Akimoto I, Groenen EJJ, El Mkami H, Goovaerts E, Van Doorslaer S (2008) Phys Chem Chem Phys 10:7129–7138
104. Godt A, Schulte M, Zimmermann H, Jeschke G (2006) Angew Chem Int Ed 45:7560–7564
105. Jeschke G, Sajid M, Schulte M, Ramezanian N, Volkov A, Zimmermann H, Godt A (2010) J Am Chem Soc 132:10107–10117
106. Margraf D, Bode BE, Marko A, Schiemann O, Prisner TF (2007) Mol Phys 105:2153–2160
107. Lovett JE, Hoffmann M, Cnossen A, Shutter ATJ, Hogben HJ, Warren JE, Pascu SI, Kay CWM, Timmel CR, Anderson HL (2009) J Am Chem Soc 131:13852–13859
108. Drescher M (2011) Top Curr Chem doi:10.1007/128_2011_235
109. Bordignon E (2011) Top Curr Chem doi:10.1007/128_2011_243

Top Curr Chem (2012) 321: 91–120
DOI: 10.1007/128_2011_235

Published online: 9 August 2011

EPR in Protein Science

Intrinsically Disordered Proteins

Malte Drescher

Abstract Intrinsically disordered proteins (IDPs) form a unique protein category characterized by the absence of a well-defined structure and by remarkable conformational flexibility. Electron Paramagnetic Resonance (EPR) spectroscopy combined with site-directed spin labeling (SDSL) is amongst the most suitable methods to unravel their structure and dynamics. This review summarizes the tremendous methodological developments in the area of SDSL EPR and its applications in protein research. Recent results on the intrinsically disordered Parkinson's disease protein α-synuclein illustrate that the method has gained increasing attention in IDP research. SDSL EPR has now reached a level where broad application in this rapidly advancing field is feasible.

Keywords α-Synuclein · DEER/PELDOR · EPR · Intrinsically disordered proteins · Site-directed spin labeling

Contents

M. Drescher
Department of Chemistry, Zukunftskolleg, and Konstanz Research School Chemical Biology, University of Konstanz, 78457 Konstanz, Germany
e-mail: malte.drescher@uni-konstanz.de

The main purpose of this chapter is to summarize the state of the art in Electron Paramagnetic Resonance (EPR) spectroscopy combined with spin-labeling approaches as a tool for studying structure and dynamics of proteins. It should give the non-specialist reader an overview of the tremendous methodological developments and applications which have a huge impact on the field of biophysics, biology, and biochemistry. In particular, contributions of site-directed spin labeling EPR (SDSL EPR) to the rapidly advancing field of Intrinsically Disordered Proteins (IDPs) are described.

SDSL EPR as pioneered by W. L. Hubbel and co-workers has become a powerful tool for studying structure and dynamics of macromolecules, in particular biological macromolecules as proteins, which do not necessarily contain endogenous paramagnetic centers [1–4]. While SDSL EPR is applied to many biomacromolecules, this chapter provides a rather selective insight into the field of SDSL EPR of proteins and is organized as follows.

First a basic introduction into approaches of SDSL of proteins is given, followed by a summary of the important approaches for mobility measurements, accessibility studies, and distance determination. Finally, we will address IDPs. Since structure and dynamics of IDPs drastically depend on the environment and corresponding details are notoriously difficult to unravel by NMR or X-ray structure determination, SDSL EPR can significantly contribute in the investigation of those systems. To showcase the use of SDSL EPR in this field, recent results on α-synuclein being a canonical model among the IDPs are reviewed.

1 Site-Directed Spin-Labeling of Proteins

Usually, nitroxides are used as spin-labels in SDSL EPR [5, 6]. However, with increasing spread of high-field EPR the relevance of other paramagnetic labels, e.g., Gd^{3+}, could gain increased relevance [7, 8]. Nitroxides are stable free radicals of the general form $\cdot O\text{-}NR^1R^2$. The unpaired electron required for EPR detection is (de-) localized on the N–O bond, about 40% of the spin density at the nitrogen atom and 60% at the oxygen atom. Since many biological macromolecules are diamagnetic, the nitroxide resonance is most often the only signal in the EPR spectrum. Nitroxide radicals are very stable, e.g., they can be stored for months under ordinary conditions. This stability is mainly due to steric protection of the N–O bond. Other parts of the molecule can be modified in order to attach the radical covalently to larger molecules [9].

Nitroxides can be used either as spin probes or as spin labels. Spin probes are subject to non-covalent interactions with the system under study. In many cases, spin probes very similar to one component of the system, e.g., spin-labeled lipids, are introduced. In contrast, spin labels are covalently linked to a complex structure, in many cases to a specific site, e.g., of a protein (SDSL) [10]. An alternative approach includes spin labeling of ligands interacting with the protein under study [11, 12].

The most common spin labeling strategy for proteins uses cysteine substitution mutagenesis followed by modification of the unique sulfhydryl group with specific

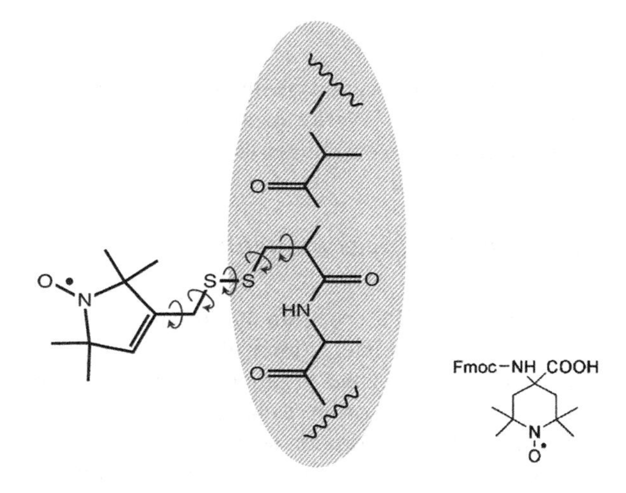

Fig. 1 *Left*: Structure of MTSSL bound to a cysteine residue within an amino acid sequence (*shaded*). The rotational degree of freedom caused by participating single bonds is indicated. *Right*: The unnatural spin-labeled amino acid TOAC

labeling agents, e.g., MTSSL (1-oxyl-2,2,5,5-tetra-methylpyrroline-3-methyl)-methanthiosulfonate (Fig. 1, left) [13]. The modified side chain resulting from the reaction is often designated as R1. For a variety of reasons, side chain R1 has become the spin label of choice in SDSL studies. On one hand, R1 is tolerated surprisingly well at the vast majority of sites at which it has been introduced in many different proteins. On the other hand, its EPR spectra are exquisitely sensitive to features of the local environment and provide a fingerprint for virtually every site [14].

Other possible spin labels specific for sulfhydryl groups are nitroxides containing an iodoacetamide group [15]. Spin labels featuring different specificity are available [16, 17]. After the labeling procedure, excess spin label should be removed, for instance using a size exclusion spin column. For labeling procedures, see [148].

Nitroxide spin labels are small and have been shown to have minimal effects on protein structures [18, 19]. However, for SDSL, as for all labeling techniques, control experiments comparing wild type protein and labeled mutants are essential to exclude distortion of protein conformation and function due to the label.

An alternative strategy for the introduction of spin labels can be used in protein and peptide chemistry by spin label building blocks, e.g., 4-amino-1-oxyl-2,2,6,6-tetramethyl-piperidine-4-carboxylic acid (TOAC, Fig. 1, right) which are directly incorporated into the peptide during chemical synthesis [20–25].

2 SDSL EPR Methods

From EPR experiments with spin labels, four primary parameters are obtained: (1) side chain mobility, (2) distances to other paramagnetic centers, e.g., a second spin label or a metal ion within the very same or another molecule, (3) solvent and oxygen accessibility, and (4) a measure for polarity of the environment of the spin-label.

In the following sections methods for obtaining the first three types of information are explained. Polarity measurements are particularly useful for membrane proteins and are reviewed in [148].

2.1 Mobility Measurements

The typical CW EPR spectrum of nitroxide labels in X band consists of three lines due to the hyperfine interaction with the ^{14}N nucleus featuring a nuclear spin $I = 1$. As a consequence of the common experimental procedure which includes signal detection using a lock-in technique, EPR spectra are usually shown as first derivatives of the absorption spectrum (Fig. 2). In a homogeneous magnetic field the EPR signal does not depend on the spatial position of the label, and therefore motion and translational diffusion cannot be detected unless applying magnetic field gradients [27, 28]. However, since the Zeeman interaction and, in particular, the hyperfine interaction of nitroxides are anisotropic, the EPR signal is sensitive to the molecular orientation of the label with respect to the external magnetic field. Thus, rotational diffusion can generally be detected by EPR.

Rotation of the label with correlation times of the order of the inverse width of the nitroxide spectrum (5.5 ns) partially averages the anisotropy, resulting in spectral changes (Fig. 2). Slight narrowing of the spectrum is detectable up to rotational correlation times of 1 μs and anisotropy-related line broadening is detectable down to rotational correlation times of 10 ps. Therefore, SDSL EPR is sensitive to dynamics on the picosecond to microsecond timescales, covering a variety of the important biological molecular mechanisms such as the dynamics of proteins in solution [29–31].

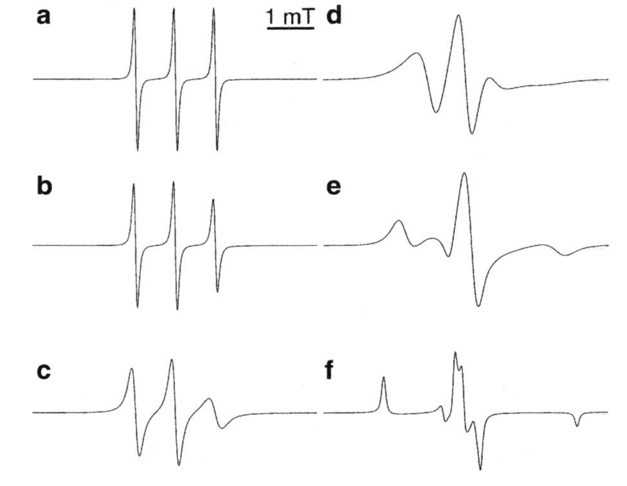

Fig. 2 Simulated EPR spectra for nitroxides in X band (9.7 GHz) assuming different (isotropic) rotational mobilities. (**a**) Very fast rotational mobility corresponding to the isotropic limit, rotational mobilities corresponding to rotational correlation times of (**b**) 100 ps, (**c**) 1 ns, (**d**) 3 ns, (**e**) 10 ns, (**f**) very slow rotational mobility corresponding to rigid limit. Simulated using EasySpin functions Garlic (**a–c**), Chili (**d**, **e**), and Pepper (**f**) [26]

The spectra do not directly report on the dynamics of the labeled macromolecule as a whole but contain information on three types of motion: (1) internal motion of the nitroxide about the chemical bonds of the linker (cf. Fig. 1, left), (2) motion of the site of attachment relative to the rest of the macromolecule (conformational flexibility), and (3) motion of the macromolecule as a whole. The internal motion of the label may be restricted by the environment, depending on the extent to which the molecular environment engulfs the label. These three dynamic components significantly complicate the spectral analysis. However, a spectrum can often be approximated by a simple motional model to provide information on the properties of the macromolecule [32]. Temperature dependent experiments can adjust the contributions of the different types of motion to the motional properties reported by the EPR spectra [33].

The accessible timescale depends on the experimental frequency, e.g., the slow overall and collective motion will show up best at lower operating frequencies and fast motion will show up best at higher operating frequencies. In particular, high field EPR can be used to analyze anisotropic motional dynamics.

One EPR spectrum measured at one single frequency does not allow complete description of the spin label motion. Therefore, multifrequency EPR studies are preferable to separate various motional modes in a protein according to their timescales [34, 35].

In order to analyze the rotational mobility of the spin label quantitatively, spectral simulations are required. Simulation programs for CW EPR spectra are available [26, 36–38]. For the case of fast isotropic motion, approximate values of the rotational correlation time can be calculated from the line height ratios [39].

A semiquantitative measure for nitroxide mobility is the inverse central line width [18], another measure is the inverse second moment of the entire spectrum. Plotting the inverse central line width vs the inverse second moment allows for distinguishing different topological regions. So, different categories, namely sites in loops or unfolded regions, sites on the surface of ordered structures, e.g., helices, or sites that are buried inside the core of a protein can be identified [18, 40–43]. The periodic dependence of mobility along a sequence can be used to identify secondary structure elements and protein topography [44].

While quantitative labeling is often checked by mass spectrometry, free labels and labels attached to a macromolecule can be distinguished by EPR mobility measurements (Fig. 3). For comparison, the correlation time of the unbound, free label in aqueous solution at room temperature is ~0.05 ns; in the example shown in Fig. 3 its mobility is reduced to a correlation time of 0.8 ns upon attachment to a protein fragment. Labels immobilized in well folded proteins feature typical correlation times in the order of several nanoseconds [46].

2.2 Distance Measurements

Exploiting different experimental approaches, EPR spectroscopy can access distances between paramagnetic centers in the range between 1 and 8 nm [1, 13, 47–54].

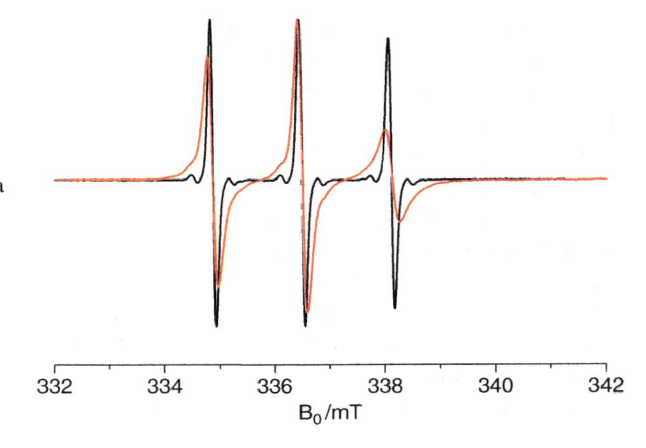

Fig. 3 Experimental CW EPR spectra of free label (*black*) and the same label attached to a soluble protein fragment (TonB) in aqueous solution at room temperature in X band. (Experimental data taken from [45])

By proper data analysis, distance *distributions* can be obtained [55–61] allowing for analyzing flexible structures or coexisting conformations. Such techniques have been successfully applied on soluble proteins [62–64].

While the complementary technique Förster resonant energy transfer (FRET), which is widely used for studying distances in proteins requires two different, relatively large chromophores, which must be chosen according to the expected distance, EPR distance measurements can be performed using two identical much smaller nitroxide labels and are precise over a broad range of distances [51, 65, 66].

Depending on the labeling strategy, inter- and intramolecular distances are accessible. While for measuring intramolecular distance constraints doubly spin labeled molecules can be used, intermolecular distances can be determined between singly labeled molecules (Fig. 4).

Distance measurements by EPR rely on the dipole–dipole coupling between spins, which is inversely proportional to the cube of the distance [67]. Additionally, the dipole–dipole coupling also depends on the angle between the spin–spin vector and the magnetic field (Fig. 5a). For oriented samples this angular dependency can be observed (Fig. 5b). Fast reorientation of the spin–spin vector, e.g., fast rotational diffusion of the doubly labeled protein under investigation, results in averaging over all possible orientations and cancels the dipole–dipole interaction to zero. Therefore, most often distance measurements are performed in a frozen state upon shock freezing in glass forming solution, e.g., aqueous buffer solution mixed with glycerol, resulting in an isotropic orientation distribution. Hence, the dipolar spectrum of such a macroscopically isotropic sample consists of a superposition of dipolar spectra of all possible orientations of the spin–spin-vector resulting in a classic Pake pattern (Fig. 5c).

Folding kinetics of proteins can be determined by a combination of rapid freeze-quench experiments and SDSL EPR distance measurement or stopped-flow EPR [68, 69].

In EPR distance measurements, two cases have to be distinguished. For distances between nitroxide spin labels below 2 nm the dipole–dipole coupling

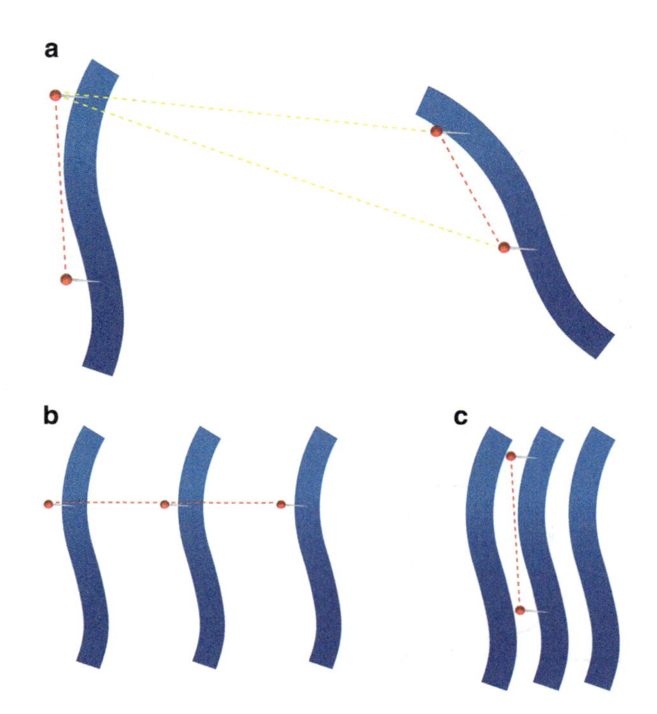

Fig. 4 Cartoon representation of EPR distance measurements. (**a**) Doubly labeled monomeric proteins give rise to intra- and intermolecular spin–spin interactions. In order to determine intramolecular distances, experimental data has to be corrected for intermolecular contributions. (**b**) Intermolecular distance measurements using singly labeled proteins in protein oligomers or aggregates. Multiples of the distance are also expected. This may be even more complicated for different types of aggregation and can be analyzed by studying a series of samples with increasing content of non-labeled molecules (diamagnetic dilution). (**c**) To measure intramolecular distances within oligomers/aggregates, a mixture of doubly labeled and non-labeled proteins can be used

exceeds the inhomogeneous line width of the EPR spectrum caused by unresolved hyperfine couplings and g-anisotropy. In this case, distances can be derived by CW EPR. For distances between nitroxide spin labels larger than approximately 2 nm, the dipole–dipole coupling is much smaller than competing interactions of the spin Hamiltonian. Thus the dipole–dipole coupling has to be separated from those larger interactions, which is usually done by pulsed EPR approaches, among these the four-pulse double electron electron resonance (DEER) [48, 51, 70, 71]. By using multiple techniques a more complete picture is obtained than can be found by a single technique [72]. In the borderline region of applicability of CW EPR and DEER (1.7–2.0 nm) both methods have to be combined in order to obtain accurate inter-nitroxide distances [73]. The lower limit of precise distance measurements is determined by the exchange coupling between two spins. With respect to the dipole–dipole interaction it can usually be neglected for distances larger than about 1.2 nm [74, 75]. On the other hand, for much shorter distances characteristic

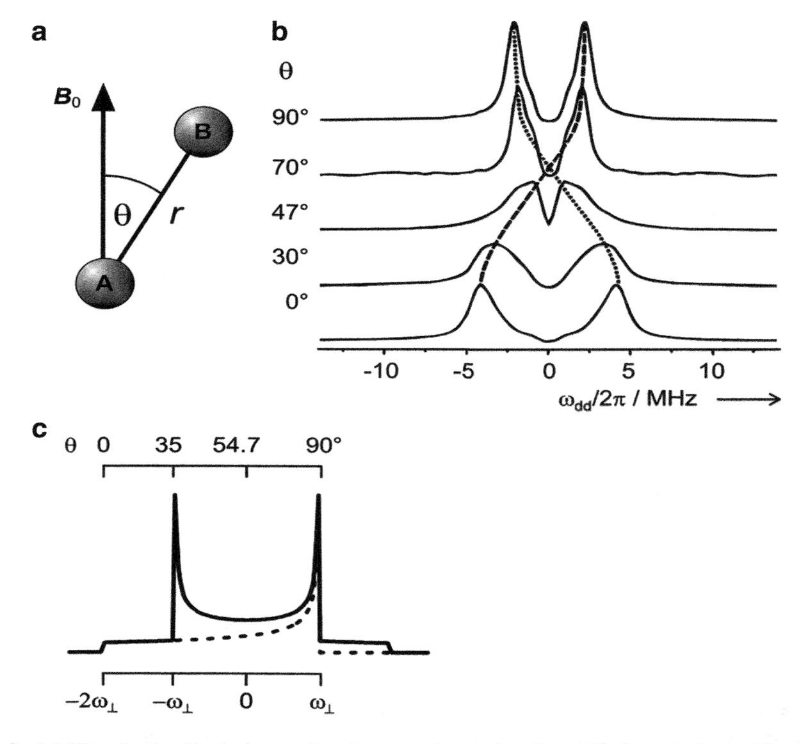

Fig. 5 (**a**) The dipole–dipole interaction between two spins A and B depends inter alia on the angle θ between the spin–spin-vector and the external magnetic field B_0. (**b**) This angular dependency can be observed for oriented samples as shown with this experimental data for a biradical in a liquid crystal under different orientations. (**c**) Simulated dipolar spectrum for a macroscopically isotropic sample (*Pake pattern*). Adapted from [66], copyright Wiley-VCH Verlag GmbH and Co. KGaA. Reproduced with permission

exchange narrowed single line EPR spectra indicate orbital overlap between multiple spin labels in close contact [76–78].

2.2.1 CW EPR Experiments

CW EPR experiments for distance determination can be performed on standard spectrometers, most commonly in X band which are quite generally accessible, and the measurements are technically not very demanding. Typical sample volumes are 50 μL at concentrations of about 50 μM.

Distances in solution at physiological temperatures can at least be estimated under conditions where the reorientation rate of the spin–spin-vector is reduced by other mechanisms, e.g., embedding the proteins in membranes or upon addition of viscosity agents [79]. In this case the dipole–dipole interaction is partly averaged out, making accurate distance measurement difficult. Quantitative

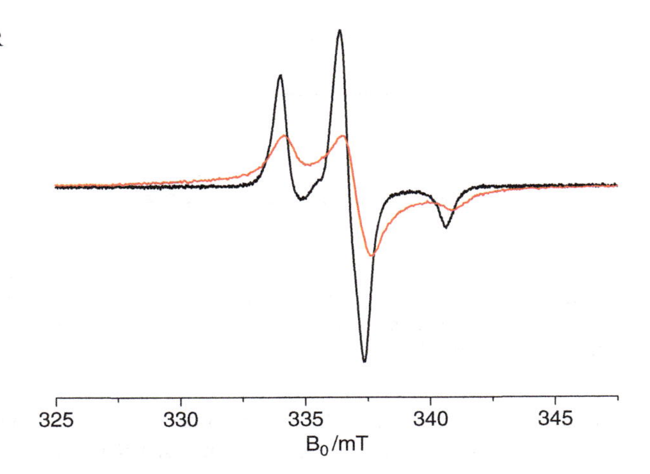

Fig. 6 Spin-normalized EPR spectra of aggregated peptides showing dipolar broadening at $T = 120$ K. The broadened spectrum originates from a sample containing 100% singly labeled peptides (*red*), while the reference sample (*black*) contains 90% unlabeled and 10% singly labeled peptides. This diamagnetic dilution suppresses dipolar broadening. (Experimental data taken from [80])

distance measurement is performed in frozen solution, the optimum temperature being $T = 120$ K.

Dipolar interaction can lead to an EPR line broadening (Fig. 6). The spectrum of the interacting spins can be treated as the convolution of the non-interacting powder pattern spectrum with a dipolar broadening function which is known as Pake pattern in randomly oriented samples.

When EPR spectra are normalized to the same number of contributing spins (as in Fig. 6), dipolar broadening is apparent by a decrease of the signal amplitude, which can be recognized more easily.

The non-interacting powder pattern which is required for distance analysis is experimentally accessible by measuring EPR spectra of samples containing either singly labeled proteins or, in the case of intermolecular distances, containing both labeled and unlabeled proteins ("diamagnetic dilution") to avoid interspin distances below 2.0 nm.

Software for extraction of distances from CW spectra, e.g., by analysis of spectral lineshapes by simulation or lineshape deconvolution [50, 81], is available and reviewed in [148].

2.2.2 Pulsed Methods

Pulsed methods [82] increase the range of distance sensitivity. They can be used to separate the dipole–dipole interaction from other contributions of the spin Hamiltonian. At very large available microwave power, distances can be measured well by double quantum coherence (DQC) that uses a single frequency. With the power available on commercial spectrometers, double electron electron resonance [DEER, an acronym which is synonymously used with PELDOR (Pulsed Electron Double Resonance)] is the more sensitive technique and is thus most widely applied in the

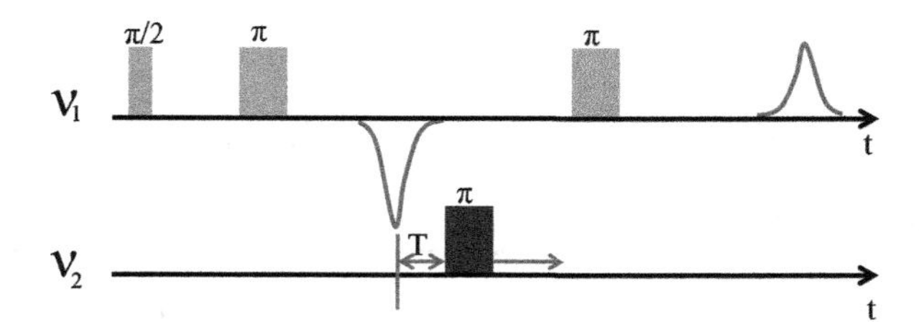

Fig. 7 Pulse sequence of the dead-time free four-pulse DEER experiment. The pulse sequence at frequency ν_1 (refocused spin echo) addresses the A spins, only, while the pulse at ν_2 flips the B spins. Applying the pump pulse at variable time T results in a modulation of the refocused echo intensity V (cf. Fig. 8a)

field. It requires a pulsed EPR spectrometer equipped with a two-frequency setup, which are commercially available. The DEER pulse sequence is depicted in Fig. 7 and is described below. Measurements are typically conducted using the dead-time free four-pulse DEER sequence and require typically 12 h of signal averaging for biological samples in X band [70]. Note in this context that one DEER measurement most often enables one to derive *one* distance constraint, only.

The Four-Pulse DEER Experiment

Originally, DEER was introduced as a three-pulse experiment [83, 84]. The dead time inherent with this pulse sequence prevents one from recording the important first data points of the DEER curve. Therefore, a dead-time free four-pulse variant of DEER was introduced [70] and is now used extensively.

The pulse sequence of the four-pulse DEER experiment is shown in Fig. 7. We consider a system of electron spins A (observer spins) and B (pumped spins) possessing a resonant microwave frequency ν_1 and ν_2, respectively. Spins A and B are usually chemically identical nitroxide labels. Since the nitroxide spectrum features a width of approximately 180 MHz, it is possible to apply pulses at two different frequencies with non-overlapping excitation profiles and subsequently divide the nitroxides in the sample into A- and B-spins, respectively. Accordingly, the pulse sequence at ν_1 addresses the A spins only, while the pulse at ν_2 flips the B spins. At frequency ν_1 a two-pulse Hahn-echo sequence is first applied resulting in an echo depicted in Fig. 7 with negative phase, which is followed by a third pulse at this so-called observer frequency which leads to a refocused echo of the observer spins A. The dipolar interaction can be studied by insertion of an additional π-pulse at the second frequency ν_2 between the two π-pulses at ν_1. This pulse affects the B spins only. Their inversion leads to a change of the local magnetic field at the A spins. Applying the pump pulse at variable time T results in a modulation of the refocused echo intensity V with the frequency of the dipolar coupling between A and B spins. Plotting V vs T yields the typical DEER curve (Fig. 8a).

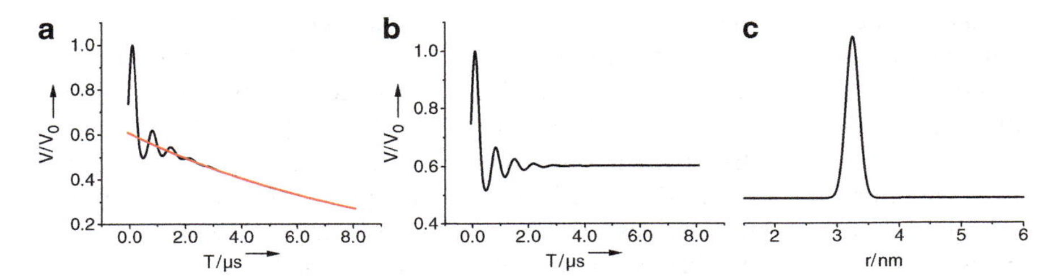

Fig. 8 (**a**) Simulated DEER data for a doubly labeled model system. The intensity V of the refocused observer echo (cf. Fig. 7) is plotted vs the delay T of the pump pulse (*black*). The DEER curve can be corrected for a background signal (*red*) originating from intermolecular interactions (cf. Fig. 4a). (**b**) Dipolar evolution (form factor) derived from DEER data in (**a**) by correcting for the intermolecular background. (**c**) Corresponding distance distribution. Simulations were performed using DEERAnalysis [61]

Measuring in frozen solution is desired in order to avoid the averaging out of the dipole–dipole interactions and, in particular, the strong decrease in transverse relaxation time T_2 that is induced even by moderate spin label dynamics. Additionally, the proper choice of temperature is important in pulsed EPR to optimize relaxation rates. $T = 50$ K is ideal for DEER at nitroxides in aqueous solution, so liquid helium cooling is advantageous [85].

A dramatic increase in sensitivity can be obtained by lengthening the transverse relaxation time by choosing the right solvent. At low temperatures the transverse relaxation time is significantly longer in a fully deuterated matrix than in a protonated one. Deuteration of the underlying protein, as well as the solvent, extends the transverse relaxation time to a considerable degree and gives enhanced sensitivity and an extended accessible distance range [86].

Typical sample volumes for X-band measurements are in the order of some 10 μL at minimum concentration of some 10 μM. There is an optimum concentration depending on the required maximum accessible distance. For distances of up to 2.5 nm, concentrations up to 4 mM can be used; for measuring distances up to 8 nm, the concentration should not exceed 0.35 mM [66].

While most experiments reported in the literature were performed in X band, Q band DEER gains increasing attention owing to its superior sensitivity revealing higher-quality distance data as well as significantly increased sample throughput [87, 88]. DEER in W band gives access to the relative orientation of spin labels due to orientation selection at high fields. More precisely, selective excitation by microwave pulses may unravel if the orientation of the spin–spin vector is correlated to the orientations of the molecular frame of the two nitroxides [89, 90]. However, due to conformational freedom of the labels, such correlation is often not very strong.

In most cases, EPR distance measurements are performed to determine a distance within a nanoobject, e.g., the spin–spin distance in a doubly labeled protein. It is desirable to consider an isolated pair of spin labels; therefore dipolar interactions to spins of neighboring objects, e.g., intermolecular interactions, should be suppressed. This can be achieved by diamagnetic dilution, e.g., mixing with non-labeled wild type protein or protein labeled with a diamagnetic label analog. In the case of studying intramolecular distances in protein oligomers, diamagnetic dilution is of particular importance (cf. Fig. 8) [79, 91].

In any case, the DEER signal has to be corrected for the background originating from couplings to spins outside of the interesting nanoobject (Fig. 8a) [85]. Experimental background functions can be derived from singly labeled samples; they can be used for correcting the background in corresponding doubly labeled samples. If experimental background functions from singly labeled molecules are not available, theoretical functions taking homogeneous distributions of nanoobjects into account can be applied for correction. In many samples the distribution is homogeneous in $d = 3$ dimensions. Proteins bound to a membrane surface may be confined to $d = 2$ dimensions.

The data after background correction is often referred to as form factor (Fig. 8b).

The assumption of well separated spin pairs may not always be valid, e.g., in singly labeled trimers. Those cases lead to signal contributions from sum and difference combinations of dipolar frequencies which are not easy to analyze in terms of distances [92].

However, a parameter being rather easily determined is the number of spins per nanoobject, e.g., the number of proteins in an oligomer [57, 93, 94]. It is directly related to the modulation depth of the DEER curve after background correction, which additionally depends strongly on the excitation position, length, and flip angle of the pump pulse. Uncertainty in the degree of spin labeling affects interpretation of the oligomerization state [93–95]. In turn, reduced modulation depth for intramolecular distance measurements can indicate a fraction of de facto singly labeled molecules.

In analogy, from the background of the DEER signal reflecting homogeneously distributed spins, *local* spin concentrations up to 20 mM can be measured [96].

For evaluation of experimental DEER data several software packages are available [59, 61]. They cater either for data analysis based on a model of the distance distribution [97–99] or for model-free methods, e.g., Tikhonov regularization [57, 59]. For the model-free approach, the underlying mathematical problem is (moderately) ill-posed, i.e., quality of the analyzed data is very crucial. Incomplete labeling of double mutants results in (1) lower signal to noise of the primary data with increasing number of completely unlabelled molecules and (2) reduced modulation depth with decreasing number of doubly labeled molecules.

It is important to note that the distance between the spin density on the nitroxides differs from the corresponding distances of the protein backbone, since distance measurements utilize spin-labels as MTSSL which possess a number of single bonds in their linker allowing for different rotamers (Fig. 1 left) and thus are not conformationally unambiguous [100]. This introduces an uncertainty of the backbone–spin distance and complicates the interpretation of the spin–spin distances in terms of the protein backbone [64], although the uncertainty becomes less important for longer distances between the labeled sites [101].

This uncertainty can be reduced by molecular modeling of the spin label behavior. Several approaches were made to overcome this problem [64, 102, 103]. For instance, the program package MMM describes spin labels by a set of alternative conformations, rotamers, which can be attached without serious clashes with atoms of other residues or cofactors. The individual rotamers are assigned Boltzmann populations corresponding to an estimate of the sum of their internal energy and interaction energy with the protein. All simulations of experiments on spin labels are then based on the population weighted average over the ensemble of rotamers [104].

Experimental data for DEER experiments can be predicted for a modeled structure and favorable attachment sites can be predicted by scanning the whole protein [105].

2.3 Accessibility Measurements

Secondary structure can be obtained by studying the accessibility of the nitroxide label to paramagnetic colliders. The collision rate (more precisely the Heisenberg exchange frequency) with the spin label influences the relaxation time of the latter which can be measured and be used to estimate the local concentration of a paramagnetic quencher near a nitroxide spin label.

Water-soluble quenchers are transition metal complexes such as chelated nickel [nickel(II)-ethylenediamine-*N*,*N*′-diacetic acid (NiEDDA)]. A ubiquitous paramagnetic quencher is triplet oxygen, which is much more soluble in apolar environments, such as lipid bilayers, than in polar environments [3, 106]. As a consequence, by measuring the respective local concentration, membrane-exposed sites can be distinguished from solvent-exposed sites.

At very high local concentrations such quenchers cause line broadening. At low local concentration exchange broadening is insignificant and the transverse relaxation time T_2 is the same in the presence and absence of the quencher. In this case, the influence of the quenchers on the longitudinal relaxation time T_1 can be quantified by measuring saturation curves. For this, the peak-to-peak amplitude of the first derivative central line of the nitroxide spectrum is measured as a function of microwave power. Measuring on a reference substance such as dilute diphenylpicrylhydrazyl powder in KCl and defining a dimensionless accessibility parameter Π enables one to eliminate the dependency of those saturation curves on T_2 and the conversion efficiency of the microwave resonator [107–110].

3 Applications to Intrinsically Disordered Proteins

The most powerful techniques for protein structure determination in general are X-ray crystallography and nuclear magnetic resonance (NMR) spectroscopy. Very limited structural information is available if these techniques are not applicable.

This holds true for many membrane proteins which are difficult to crystallize or concentrate; therefore the determination of their structures is one of the most challenging fields in structural biology. Furthermore, structure determination of membrane proteins is an important application of SDSL EPR which is reviewed [33, 111–113, 148].

X-ray crystallography and NMR spectroscopy are also less successful in determination of structure and dynamics of IDPs [114]. IDPs have been recognized as a unique protein class as well, justified by the clear structural and functional separation which they have in common, and again, SDSL EPR can significantly contribute to their characterization as illustrated in the following.

IDPs comprise a large fraction of eukaryotic proteins (>30%). They lack a well-defined three-dimensional fold and display remarkable conformational flexibility. This property potentially enables them to be promiscuous in their interactions and to adapt their structure according to the needed function.

Since structure and dynamics of IDPs drastically depend on the environment, corresponding details are notoriously difficult to unravel. Because of their inherent flexibility, IDPs often fail to crystallize in their free form. When crystallization is successful, it only leads to a snapshot of a single conformation not representing the whole conformational ensemble [115]. The most common goal of structural studies, the determination of unique high-resolution structures, is not attainable for IDPs due to the absence of a well-defined structure.

Upon interaction with other cellular components, IDPs adopt more highly ordered conformations. These are subject to high-resolution structures in some cases; however, at least in some cases, the bound states of IDPs remain highly non-compact and retain substantial mobility [116, 117].

SDSL EPR offers a powerful tool to study IDPs. Nitroxide spin labels introduce a minimal perturbation of the system, and probe the very local environment of the label [3, 4, 38, 118]. The higher sensitivity of EPR compared to NMR allows for much lower concentrations of protein samples.

Several IDPs have already been subjected to SDSL EPR investigations. So EPR data showed decreased flexibility in a region of residual helical structure in the disordered C-terminal domain of the measles virus nucleoprotein, and demonstrated the further ordering of this region upon interaction with a binding partner [119, 120].

The amyloid β peptide Aβ was the subject of several studies [91, 121–123]; structural constraints on $A\beta_{1-40}$ fibrils were obtained from measurements of CW EPR spectra and determination of spin–spin couplings in a series of spin-labeled cysteine mutant samples. Conclusions about molecular structure and supramolecular organization were drawn from these data. The observation of co-fibrillization of $A\beta_{1-40}$ and $A\beta_{1-42}$ suggested the absence of large structural differences between $A\beta_{1-40}$ and $A\beta_{1-42}$ fibrils.

Further examples of IDPs studied by SDSL EPR are the prion protein H1 [124], ubiquitin [125], or serum albumin [12]. In the following, recent results concerning α-synuclein are reviewed to illustrate the potential of SDSL EPR in the field of IDPs.

3.1 α-Synuclein

Intrinsic disorder is highly abundant among proteins associated with neurodegenerative diseases. The canonical model among the IDPs is α-synuclein (ASYN), a 140-residue protein that is abundantly present in the Lewy bodies characteristic of Parkinson's disease (PD). PD is the most common age-related movement disorder and the second most common neurodegenerative disorder. ASYN with its high propensity to aggregate and its conformational flexibility is an ideal model system for IDPs and for understanding PD and related disorders.

Depending on the environment, it adopts a variety of structurally distinct conformations including the intrinsically unfolded state, an amyloidogenic partially folded conformation, and different α-helical and β-sheet conformations. This conformational flexibility led to the term "protein-chameleon."

The exact physiological role of ASYN has yet to be determined, but membrane binding seems to be important for its function. As a consequence, the membrane bound form has received considerable attention in the last several years. Preferentially, ASYN binds to negatively charged lipid surfaces.

Human ASYN does not contain any cysteine residues. Singly and doubly labeled ASYN derivatives can be generated by site-directed mutagenesis introducing cysteines and subsequent spin labeling with MTSSL. Spin-labeled ASYN in solution at room temperature gives rise to sharp EPR line shapes characteristic for loop or unfolded regions indicating that ASYN is intrinsically disordered, e.g., largely unfolded in solution (Fig. 9a, b).

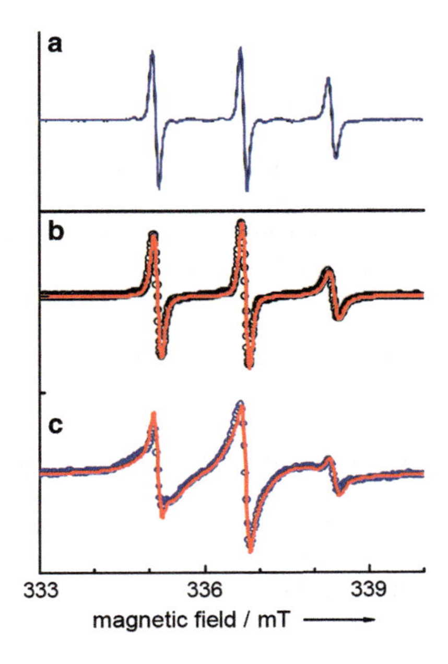

Fig. 9 CW EPR spectra at room temperature of (**a**) ASYN labeled at residue 140 in solution (*black*) and upon vesicle binding (*blue*), and ASYN labeled at residue 90 (**b**) in solution and (**c**) upon vesicle binding including corresponding spectral simulations (*red*). Taken from [126]

3.1.1 Fibrils

The aggregation of proteins into amyloid fibrils is associated with several neurodegenerative diseases. It is believed that the aggregation of ASYN from monomers by intermediates into amyloid fibrils is the toxic disease-causative mechanism of PD.

A large set of singly labeled ASYN derivatives were used in order to investigate the structural features of ASYN fibrils. Fibrils grown from spin-labeled ASYN featured a fibril morphology being very similar to fibrils taken from wild type ASYN as verified by electron microscopy. Additionally, co-mixing of wild type and labeled ASYN indicated that both species are able to adopt similar structures within the fibril, confirming that the introduction of a spin label is tolerated remarkably well in amyloid fibrils [91].

Analyzing the intermolecular spin–spin interaction within the fibrils in terms of dipolar broadening depending on diamagnetic dilution, it was shown that similar sites from different molecules come into close proximity. While the accuracy of this analysis was not sufficient to distinguish fully whether parallelism occurred between strands or sheets (corresponding to distances of 4.7 or 10 Å) [127], a highly ordered and specifically folded central core region of about 70 amino acids was identified. The N-terminus is structurally more heterogeneous; the C-terminus consisting of 40 amino acids is completely unfolded [128].

The latter encouraged Chen et al. [127] to employ a C-terminal truncation mutant of ASYN containing residues 1–115. This allowed for optimizing spectral quality and minimizing components from non-fibrillized protein or other background labeling possibly due to codon mistranslation [129]. Single-line, exchange narrowed EPR spectra were observed for the majority of all sites within the core region of ASYN fibrils. Such exchange narrowing requires the orbital overlap between multiple spin labels in close contact and therewith confirmed that the core region of ASYN fibrils is arranged in a parallel, in-register structure wherein similar residues from different molecules are stacked on top of each other. This core region extends from residue 36 to residue 98 and is tightly packed. Accessibility measurements suggested the location of potential β-sheet regions within the fibril. Furthermore, the data provide structural constraints for generating three-dimensional models.

3.1.2 Membrane Binding

Not only misfolding and fibril formation of ASYN but also membrane binding are of particular interest, especially for unraveling its physiological role. The N-terminus of ASYN contains 7 repeats, each of which is made up of 11 amino acids. Sequence analysis suggested that this part is likely to mediate lipid interactions [130, 131]. NMR studies are limited by the size of the complex under investigation. Hence, the structural information available concerns NMR studies of ASYN on micelles [132–135].

Micelles, however, differ in important aspects from biological membranes. Micelles have typical diameters of 5 nm and therefore may be too small to mimic organellic membranes. In order to understand the conformational changes that occur upon membrane binding of monomeric ASYN, SDSL EPR was performed with ASYN bound to phospholipid vesicles, e.g., small or large unilamellar vesicles (SUVs or LUVs, respectively).

To characterize the structural changes induced by membrane binding, the EPR spectra of 47 singly spin labeled ASYN derivatives were recorded in solution and upon binding to small unilamellar vesicles [42, 128].

As already mentioned, spin-labeled ASYN in solution gives rise to sharp EPR line shapes characteristic of intrinsic disorder (Fig. 9a, b). Upon membrane binding, spectral changes were observed for ASYN derivatives labeled within the repeats region (Fig. 9c). In contrast, little or no changes were detected for labels at positions within the last 40 amino acids, confirming that conformational changes upon membrane binding do not occur within the C-terminal portion (Fig. 9a).

The spectra upon membrane binding still contained residual sharp spectral components allocated to unbound ASYN. The spectra can be corrected for this component by subtraction of the spectrum of the free label. Analyzing the resulting spectra originating exclusively from membrane bound ASYN labeled within the repeat region exhibited line shapes indicating lipid- or solvent-exposed helix surface sites.

Additionally, O_2 and NiEDDA accessibilities (ΠO_2 and ΠNiEDDA, respectively) were determined for the labels in the repeat regions. Nonpolar O_2 preferentially partitions into the membrane whereas the more polar NiEDDA preferentially partitions into the solvent. As a consequence, membrane-exposed sites show enhanced accessibility to O_2, whereas solvent-exposed sites are preferentially accessible to NiEDDA. In agreement with the formation of a helical structure, ΠO_2 and ΠNiEDDA exhibit continuous periodic oscillations. The accessibility data for both colliders can be conveniently summarized by the depth parameter Φ [$\Phi = \ln(\Pi O_2/\Pi \mathrm{NiEDDA})$] (Fig. 10), which increases linearly with increasing immersion depth [136].

The consecutive scan of the residues with respect to mobility and accessibilities demonstrates the formation of a single, elongated helix, wherein each 11 amino acid repeat takes up 3 helical turns.

Without subtracting the residual sharp spectral components of spin-labeled ASYN in the presence of SUVs, a multi-component spectral simulation strategy is required in order to describe the experimental data (Fig. 9b, c). Three different contributions featuring different isotropic rotational mobilities can be allocated to free spin labels, labeled residues not bound to SUVs, and residues bound to SUVs by the following approach. The spectra of ASYN in the absence of liposomes are well described by a superposition of two components, S_1 and S_2, where S_1 corresponds to the spectrum of the free spin label MTSSL measured independently. In the presence of SUVs, an additional component S_3 is needed, corresponding to the broadened part of the spectra. The shape of component S_3 and the prefactors

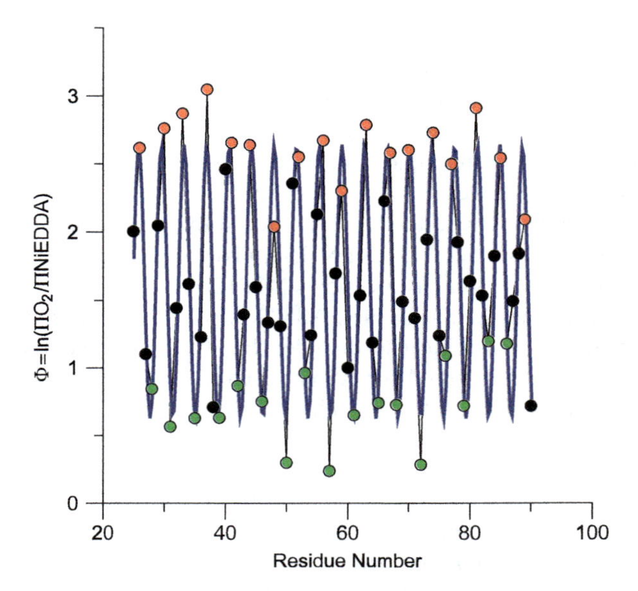

Fig. 10 Solvent accessibility analysis of singly labeled ASYN derivatives. The ratios of the accessibilities to O_2 and NiEDDA (ΠO_2 and Π NiEDDA, respectively) for residues 25–90 summarized by the depth parameter $\Phi = \ln(\Pi O_2/\Pi \text{ NiEDDA})$, with increasing Φ values indicating deeper membrane immersion depth. The *blue line* indicates the best fit to a cosine function and the resulting periodicity corresponds to the theoretically predicted periodicity of 3.67 amino acids per turn. Copyright 2005 National Academy of Sciences, USA, reproduced from [136]

a, b, and c are determined by least square fits to the data according to $S = aS_1 + bS_2 + cS_3$ [126].

Hence, the *local* binding affinity can be determined. Using this approach a systematic study varying the charge density of the membrane allowed for a locally resolved analysis of the protein–membrane binding affinity. The results showed that binding of ASYN to artificial phospholipid membranes is initiated by the N-terminus (Fig. 11) [126].

3.1.3 Conformation of Membrane Bound α-Synuclein

The NMR structure of ASYN bound to SDS micelles, commonly used for membrane mimicking, revealed a break in the helix, resulting in two antiparallel alpha-helices [132]. This model was confirmed by distance measurements exploiting SDSL EPR utilizing 13 different ASYN double mutants each containing 2 spin-labeled cysteines (horseshoe model, Fig. 12) [137].

In this study, one mutant includes a pair of cysteines placed within a single helix to provide an internal distance control. Distance distributions were obtained by DEER measurements and Tikhonov regularization. Studying ASYN bound to detergent and lysophospholipid micelles, it has been shown that the inter-helical

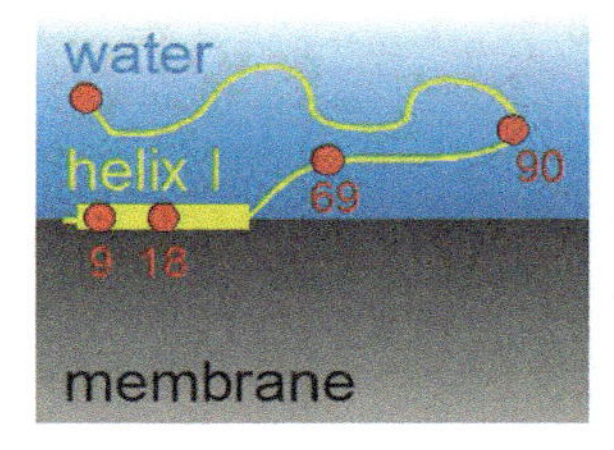

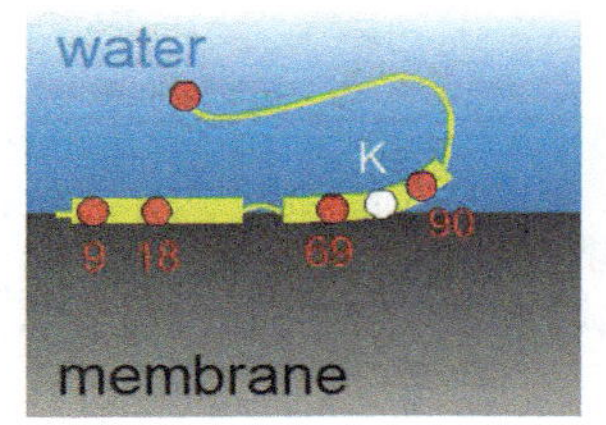

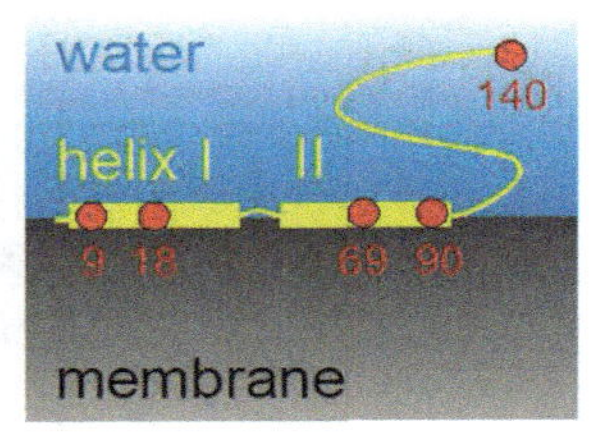

Fig. 11 Spin-label EPR revealed that ASYN membrane binding is triggered by its N-terminus. Schematic representation of ASYN at the membrane–water interface. Positions of spin labels used in this study are depicted as *red circles*, and the number of the labeled residue is given. Representing electrostatic interactions, the cationic residue K80 is shown as a *white circle*. Adapted from [126]

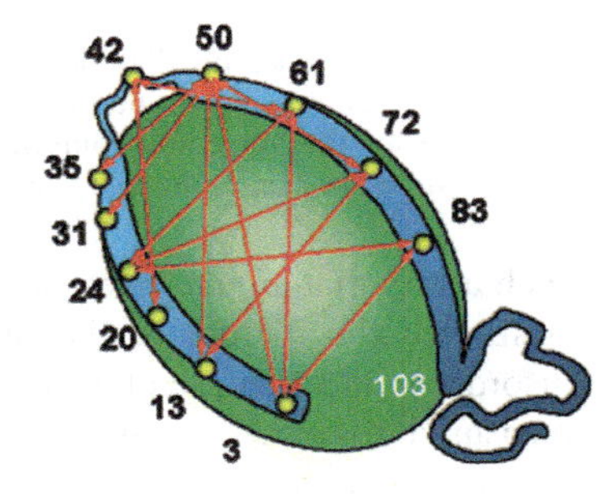

Fig. 12 Cartoon representation of the two helices and linker region of ASYN bound to an ellipsoidal micelle, illustrating the different distances measured using pulsed EPR (taken from [137])

separation between the two helices formed upon binding to micelles is dependent on micelle composition, with micelles formed from longer acyl chains leading to an increased splaying of the two helices. The distance constraints were in accord with the NMR data. The data suggested that the topology of ASYN is not strongly constrained by the linker region between the two helices and instead depends on the geometry of the surface to which the protein is bound.

The geometry of micelles, however, differs significantly from those of biological membranes. Micelles have typical diameters of 5 nm and therefore may be too small to accommodate ASYN in the extended conformation (around 15 nm for an extended helix of 100 residues). Therefore, it had been postulated that the small size of the micelles may have artificially constrained the protein into a horseshoe structure.

A subsequent study considered two selected possible conformations for ASYN bound to SUVs, namely an extended helix and the horseshoe structure. Theoretically expected spin–spin distance distributions for doubly labeled ASYN taking the possible rotamers of the spin labels into account were calculated. This enabled one to identify label positions in the crucial location close to the potential linker region between the two horseshoe helices which would allow distinguishing between these conformations by CW EPR distance measurements. CW EPR spectra of correspondingly labeled ASYN bound to POPC SUVs were measured and, using the theoretical distance distributions, calculated. The authors interpreted their results in

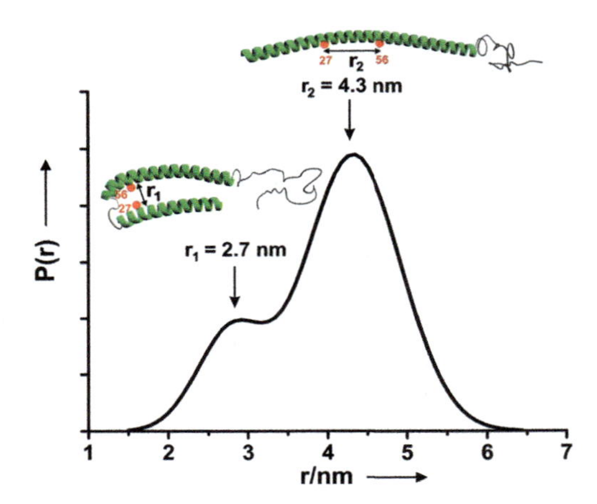

Fig. 13 The distance distribution for ASYN bound to LUV and labeled at residue 27 and 56 clearly consists of two contributions. The shorter distance agrees well with the expected distance of 2.7 nm for the horseshoe conformation derived from the NMR structure (pdb access code 1XQ8) while the longer distance is consistent with an extended alpha-helix. Taken from [140]

such a way that an unbroken helical structure around residue 40 was ruled out and confirmed the picture of the interhelix region characterized by conformational disorder [138]. Later, a close inspection of the data resulted in suggesting that the measured distances may be more consistent with an extended helix form than with the horseshoe model [139].

ASYN bound to vesicles, bicelles, and rodlike micelles was also studied by DEER allowing for measuring longer distances. Jao et al. reported results suggesting an extended helix conformation being significantly different from that of ASYN in the presence of SDS micelles. Their DEER study showed that for several double mutants the average distance per residue was $\pm 1\%$ of that for an alpha-helix, which argues strongly for an extended helix [137].

Already in this study, a number of samples have yielded somewhat bimodal distance distributions, suggesting distinct conformations of the protein. Actually, this was confirmed by a further DEER study [140], which used the ability of DEER to measure distance *distributions* for direct evidence of coexisting horseshoe and extended helix conformations of membrane bound ASYN (Fig. 13).

A DEER study [141] measuring distances of up to 8.7 nm showed that the PD-linked ASYN mutations also remain capable of adopting both structures, and that the protein to lipid ratio determines whether the protein adopts the broken or extended helix conformation.

This ability of ASYN to adopt different structures can provide an explanation of the disparate results obtained using similar experimental techniques and often with only slight variations in experimental conditions reported in the literature [99, 133–136, 138, 139, 142].

A complementary approach studying protein–membrane interaction by SDSL EPR is to utilize spin-labeled lipids. In the case of ASYN, different restrictions of segmental motion in the chains of different lipids were observed upon ASYN binding. This observation indicates that ASYN associates at the interfacial region of the bilayer where it may favor a local concentration of certain phospholipids [143–146].

3.1.4 Lipid-Induced Aggregation

Using singly spin labeled mutants, *inter*molecular distances are also accessible by DEER. It was shown that ASYN may influence the membrane structure and even disrupt membranes. Under those conditions intra- and intermolecular distance measurements by DEER allowed for the conclusion that ASYN forms aggregates once in contact with SUVs [99, 147].

The simplest model for such an aggregate was proposed based on the distance constraints (Fig. 14). In these aggregates, two ASYN molecules are in close contact, but they could form part of a larger aggregate in which the proteins are arranged in an ordered fashion.

4 Concluding Remarks

SDSL EPR has developed as a powerful tool in order to study structure and dynamics of bio-macromolecules. Mobility and distance measurements being sensitive to dynamics on the picosecond to microsecond timescales, covering the

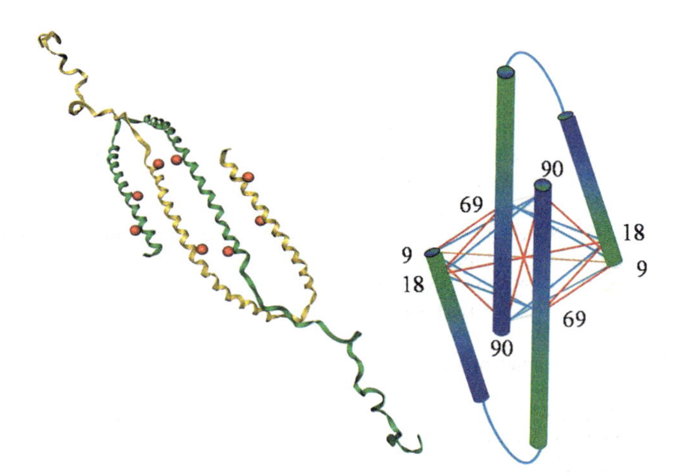

Fig. 14 *Left*: ASYN forms well-defined aggregates with lipids. In these aggregates, two ASYN molecules are in close contact, but they could form part of a larger aggregate in which the proteins are arranged in an ordered fashion. Spin labels are depicted as *red circles*. *Right*: Cartoon representation of EPR distance constraints. Taken from [141]

dynamics of proteins in solution, or giving access to distances between spin labels in the nanometer range, are probably amongst the most important experimental approaches.

Advantages of SDSL EPR include straightforward labeling procedures using one type of label for a broad accessible range in distance measurements, small size of label-molecules featuring minimal perturbations of the system under investigation, virtually no limitation of the size of the complex under study, and high sensitivity. Sophisticated pulsed EPR methods for distance determination compel by their unsurpassed accuracy as well as their ability to extract distance *distributions* and therefore to detect coexisting structures. Since unlabeled diamagnetic molecules are EPR silent, distance constraints can be obtained background free in (frozen) solution.

IDPs form a unique protein category characterized by the absence of a well-defined structure and by remarkable conformational flexibility. As a result of these properties, and because of its high potential, SDSL EPR is amongst the most suitable methods to study IDPs and will gain increasing attention in this rapidly advancing field. Recent results on α-synuclein show that SDSL EPR has now reached a level where broad application unraveling structure and dynamics of IDPs is feasible.

Acknowledgements I am indebted to C. Jao, J. Freed, G. Jeschke, and R. Langen for permission to reproduce figures. I wish to thank Dr. Martina Huber and Prof. Dr. Vinod Subramaniam for a longstanding cooperation, Marco Wassmer, Martin Spitzbarth, and Christian Hintze for designing figures, and Gunnar Jeschke and the EPR people in Konstanz for fruitful discussions.

References

1. Hubbell WL, Altenbach C (1994) Investigation of structure and dynamics in membrane-proteins using site-directed spin-labeling. Curr Opin Struct Biol 4:566–573
2. Hubbell WL, McHaourab HS, Altenbach C, Lietzow MA (1996) Watching proteins move using site-directed spin labeling. Structure 4:779–783
3. Hubbell WL, Gross A, Langen R, Lietzow MA (1998) Recent advances in site-directed spin labeling of proteins. Curr Opin Struct Biol 8:649–656
4. Feix J, Klug C (1998) Site-directed spin labeling of membrane proteins and peptide-membrane interactions. In: Berliner LJ (ed) Biological magnetic resonance, vol 14. Plenum Press, New York
5. Likhtenshtein GI, Yamauchi J, Nakatsuji S, Smirnov AI, Tamura R (2008) Nitroxides. Wiley-VCH, Weinheim
6. Berliner LJ, Reuben J (1989) Spin labeling – theory and application. In: Biological magnetic resonance, vol 8. Academic, New York
7. Potapov A, Yagi H, Huber T, Jergic S, Dixon NE, Otting G, Goldfarb D (2010) Nanometer-scale distance measurements in proteins using Gd3+ spin labeling. J Am Chem Soc 132:9040–9048
8. Song Y, Meade TJ, Astashkin AV, Klein EL, Enemark JH, Raitsimring A (2011) Pulsed dipolar spectroscopy distance measurements in biomacromolecules labeled with Gd(III)

markers. J Magn Reson 210(1):59–68. doi:S1090-7807(11)00071-1 [pii] 10.1016/j.jmr.2011.02.010
9. Axel FS (1976) Biophysics with nitroxyl radicals. Biophys Struct Mech 2:181–218
10. Klug CS, Feix JB (2008) Methods and applications of site-directed spin labeling EPR spectroscopy. In: Terry A (ed) Biophysical tools for biologists: vol one in vitro techniques, vol 84. Methods in Cell Biology. Academic, New York
11. Braun P, Nägele B, Wittmann V, Drescher M (2011) Mechanism of multivalent carbohydrate-protein interactions studied by EPR spectroscopy. Angew Chem Int Ed, doi: 10.1002/anie.201104492
12. Junk MJN, Spiess HW, Hinderberger D (2010) The distribution of fatty acids reveals the functional structure of human serum albumin. Angew Chem Int Ed 49:8755–8759
13. Hubbell WL, Cafiso DS, Altenbach C (2000) Identifying conformational changes with site-directed spin labeling. Nat Struct Biol 7:735–739
14. Langen R, Oh KJ, Cascio D, Hubbell WL (2000) Crystal structures of spin labeled T4 lysozyme mutants: implications for the interpretation of EPR spectra in terms of structure. Biochemistry 39:8396–8405
15. Ogawa S, McConnel HM (1967) Spin-label study of hemoglobin conformations in solution. Proc Natl Acad Sci USA 58:19–26
16. Jahnke W, Rudisser S, Zurini M (2001) Spin label enhanced NMR screening. J Am Chem Soc 123:3149–3150
17. Lawrence JJ, Berne L, Ouvrierbuffet JL, Piette LH (1980) Spin-label study of histone H1-DNA interaction – comparative properties of the central part of the molecule and the N-amino and C-amino tails. Eur J Biochem 107:263–269
18. McHaourab HS, Lietzow MA, Hideg K, Hubbell WL (1996) Motion of spin-labeled side chains in T4 lysozyme, correlation with protein structure and dynamics. Biochemistry 35:7692–7704
19. Alexander RS, Nair SK, Christianson DW (1991) Engineering the hydrophobic pocket of carbonic anhydrase-II. Biochemistry 30:11064–11072
20. Becker CFW, Lausecker K, Balog M, Kalai T, Hideg K, Steinhoff HJ, Engelhard M (2005) Incorporation of spin-labelled amino acids into proteins. Magn Reson Chem 43:34–39
21. Karim CB, Zhang Z, Thomas DD (2007) Synthesis of TOAC spin-labeled proteins and reconstitution in lipid membranes. Nat Protoc 2:42–49
22. Toniolo C, Valente E, Formaggio F, Crisma M, Pilloni G, Corvaja C, Toffoletti A, Martinez GV, Hanson MP, Millhauser GL, George C, Flippen-Anderson JL (1995) Synthesis and conformational studies of peptides containing TOAC, a spin-labelled Cα, α-disubstituted glycine. J Pept Sci 1:45–57
23. Nakaie CR, Goissis G, Schreier S, Paiva ACM (1981) pH-Dependence of electron-paramagnetic-res spectra of nitroxides containing ionizable groups. Braz J Med Biol Res 14:173–180
24. Klare JP, Steinhoff HJ (2009) Spin labeling EPR. Photosynth Res 102:377–390
25. Bettio A, Gutewort V, Poppl A, Dinger MC, Zschornig O, Arnold K, Toniolo C, Beck-Sickinger AG (2002) Electron paramagnetic resonance backbone dynamics studies on spin-labelled neuropeptide Y analogues. J Pept Sci 8:671–682
26. Stoll S, Schweiger A (2006) EasySpin, a comprehensive software package for spectral simulation and analysis in EPR. J Magn Reson 178:42–55
27. Blank A, Talmon Y, Shklyar M, Shtirberg L, Harneit W (2008) Direct measurement of diffusion in liquid phase by electron spin resonance. Chem Phys Lett 465:147–152
28. Drescher M, Kaplan N, Dormann E (2006) Conduction-electron drift velocity measurement via electron spin resonance. Phys Rev Lett 96:037601
29. Dalton L (1985) EPR and advanced EPR studies of biological systems. CRC Press, Boca Raton
30. Berliner LJ, Reuben J (1998) Spin labeling – the next millenium. In: Biological magnetic resonace, vol 14. Academic, New York
31. Budil DE, Earle KA, Freed JH (1993) Full determination of the rotational diffusion tensor by electron-paramagnetic resonance at 250 GHz. J Phys Chem 97:1294–1303

32. Brustolon M (2008) What can be studied with electron paramagnetic resonance? In: Electron paramagnetic resonance. Wiley, Hoboken, New Jersey
33. Strancar J, Kavalenka A, Urbancic I, Ljubetic A, Hemminga MA (2010) SDSL-ESR-based protein structure characterization. Eur Biophys J Biophys Lett 39:499–511
34. Zhang ZW, Fleissner MR, Tipikin DS, Liang ZC, Moscicki JK, Earle KA, Hubbell WL, Freed JH (2010) Multifrequency electron spin resonance study of the dynamics of spin labeled T4 lysozyme. J Phys Chem B 114:5503–5521
35. Liang ZC, Freed JH (1999) An assessment of the applicability of multifrequency ESR to study the complex dynamics of biomolecules. J Phys Chem B 103:6384–6396
36. Hanson GR, Gates KE, Noble CJ, Griffin M, Mitchell A, Benson S (2004) XSophe-Sophe-XeprView (R). A computer simulation software suite (v. 1.1.3) for the analysis of continuous wave EPR spectra. J Inorg Biochem 98:903–916
37. Bax A (1989) Two-dimensional NMR and protein-structure. Annu Rev Biochem 58:223–256
38. Belle V, Fournel A, Woudstra M, Ranaldi S, Prieri F, Thome V, Currault J, Verger R, Guigliarelli B, Carriere F (2007) Probing the opening of the pancreatic lipase lid using site-directed spin labeling and EPR spectroscopy. Biochemistry 46:2205–2214
39. Lund J, Dalton H (1985) Further characterization of the FAD and FE2S2 redox centers of component-C, the NADH – acceptor reductase of the soluble methane monooxygenase of methylococcus-capsulatus (BATH). Europ J Biochem 147:291–296
40. Isas JM, Langen R, Haigler HT, Hubbell WL (2002) Structure and dynamics of a helical hairpin and loop region in annexin 12: a site-directed spin labeling study. Biochemistry 41:1464–1473
41. Margittai M, Fasshauer D, Pabst S, Jahn R, Langen R (2001) Homo- and heterooligomeric SNARE complexes studied by site-directed spin labeling. J Biol Chem 276:13169–13177
42. Jao CC, Der-Sarkissian A, Chen J, Langen R (2004) Structure of membrane-bound alpha-synuclein studied by site-directed spin labeling. Proc Natl Acad Sci USA 101:8331–8336
43. Alexander N, Bortolus M, Al-Mestarihi A, McHaourab H, Meilerl J (2008) De novo high-resolution protein structure determination from sparse spin-labeling EPR data. Structure 16:181–195
44. Gross A, Columbus L, Hideg K, Altenbach C, Hubbell WL (1999) Structure of the KcsA potassium channel from Streptomyces lividans: a site-directed spin labeling study of the second transmembrane segment. Biochemistry 38:10324–10335
45. Domingo Köhler S, Weber A, Howard SP, Welte W, Drescher M (2010) The proline-rich domain of TonB possesses an extended polyproline II-like conformation of sufficient length to span the periplasm of Gram-negative bacteria. Prot Sci 19:625–630
46. Steinhoff HJ (1990) Residual motion of hemoglobin-bound spin labels and protein dynamics – viscosity dependence of the rotational correlation times. Eur Biophys J 18:57–62
47. Constantine KL (2001) Evaluation of site-directed spin labeling for characterizing protein-ligand complexes using simulated restraints. Biophys J 81:1275–1284
48. Jeschke G, Polyhach Y (2007) Distance measurements on spin-labelled biomacromolecules by pulsed electron paramagnetic resonance. Phys Chem Chem Phys 9:1895–1910
49. Steinhoff HJ, Suess B (2003) Molecular mechanisms of gene regulation studied by site-directed spin labeling. Methods 29:188–195
50. Rabenstein MD, Shin YK (1995) Determination of the distance between 2 spin labels attached to a macromolecule. Proc Natl Acad Sci USA 92:8239–8243
51. Jeschke G, Bender A, Paulsen H, Zimmermann H, Godt A (2004) Sensitivity enhancement in pulse EPR distance measurements. J Magn Reson 169:1–12
52. Godt A, Schulte M, Zimmermann H, Jeschke G (2006) How flexible are poly(para-phenyleneethynylene)s? Angew Chem Int Ed 45:7560–7564
53. Borbat PP, Davis JH, Butcher SE, Freed JH (2004) Measurement of large distances in biomolecules using double-quantum filtered refocused electron spin-echoes. J Am Chem Soc 126:7746–7747
54. Berliner LJ, Eaton SS, Eaton GR (2002) Distance measurements in biological systems by EPR. In: Biological magnetic resonance, vol 19. Academic, New York

55. Jeschke G, Koch A, Jonas U, Godt A (2002) Direct conversion of EPR dipolar time evolution data to distance distributions. J Magn Reson 155:72–82
56. Bowman MK, Maryasov AG, Kim N, DeRose VJ (2004) Visualization of distance distribution from pulsed double electron-electron resonance data. Appl Magn Reson 26:23–39
57. Jeschke G, Panek G, Godt A, Bender A, Paulsen H (2004) Data analysis procedures for pulse ELDOR measurements of broad distance distributions. Appl Magn Reson 26:223–244
58. Milov AD, Tsvetkov YD, Formaggio F, Oancea S, Toniolo C, Raap J (2004) Solvent effect on the distance distribution between spin labels in aggregated spin labeled trichogin GA IV dimer peptides as studied by pulsed electron-electron double resonance. Phys Chem Chem Phys 6:3596–3603
59. Chiang YW, Borbat PP, Freed JH (2005) The determination of pair distance distributions by pulsed ESR using Tikhonov regularization. J Magn Reson 172:279–295
60. Chiang YW, Borbat PP, Freed JH (2005) Maximum entropy: a complement to Tikhonov regularization for determination of pair distance distributions by pulsed ESR. J Magn Reson 177:184–196
61. Jeschke G, Chechik V, Ionita P, Godt A, Zimmermann H, Banham J, Timmel CR, Hilger D, Jung H (2006) DeerAnalysis2006 – a comprehensive software package for analyzing pulsed ELDOR data. Appl Magn Reson 30:473–498
62. Zhou Z, DeSensi SC, Stein RA, Brandon S, Dixit M, McArdle EJ, Warren EM, Kroh HK, Song LK, Cobb CE, Hustedt EJ, Beth AH (2005) Solution structure of the cytoplasmic domain of erythrocyte membrane band 3 determined by site-directed spin labeling. Biochemistry 44:15115–15128
63. Nakamura M, Ueki S, Hara H, Arata T (2005) Calcium structural transition of human cardiac troponin C in reconstituted muscle fibres as studied by site-directed spin labelling. J Mol Biol 348:127–137
64. Fajer PG (2005) Site directed spin labelling and pulsed dipolar electron paramagnetic resonance (double electron-electron resonance) of force activation in muscle. J Phys Condens Mat 17:S1459–S1469
65. Jeschke G, Abbott RJM, Lea SM, Timmel CR, Banham JE (2006) The characterization of weak protein-protein interactions: evidence from DEER for the trimerization of a von Willebrand factor A domain in solution. Angew Chem Int Ed 45:1058–1061
66. Jeschke G (2002) Distance measurements in the nanometer range by pulse EPR. ChemPhysChem 3:927–932
67. Jeschke G, Spiess HW (2006) Distance measurements in solid-state NMR and EPR spectroscopy. In: Dolinšek J, Vilfan M, Žumer S (eds) Novel NMR and EPR techniques, vol 684. Lecture Notes in Physics, Springer, Berlin/Heidelberg
68. Dockter C, Volkov A, Bauer C, Polyhach Y, Joly-Lopez Z, Jeschke G, Paulsen H (2009) Refolding of the integral membrane protein light-harvesting complex II monitored by pulse EPR. Proc Natl Acad Sci USA 106:18485–18490
69. Qu KB, Vaughn JL, Sienkiewicz A, Scholes CP, Fetrow JS (1997) Kinetics and motional dynamics of spin-labeled yeast iso-1-cytochrome c.1. Stopped-flow electron paramagnetic resonance as a probe for protein folding/unfolding of the C-terminal helix spin-labeled at cysteine 102. Biochemistry 36:2884–2897
70. Pannier M, Veit S, Godt A, Jeschke G, Spiess HW (2000) Dead-time free measurement of dipole-dipole interactions between electron spins. J Magn Reson 142:331–340
71. Martin RE, Pannier M, Diederich F, Gramlich V, Hubrich M, Spiess HW (1998) Determination of end-to-end distances in a series of TEMPO diradicals of up to 2.8 nm length with a new four-pulse double electron electron resonance experiment. Angew Chem Int Ed 37:2834–2837
72. Persson M, Harbridge JR, Hammarstrom P, Mitri R, Martensson LG, Carlsson U, Eaton GR, Eaton SS (2001) Comparison of electron paramagnetic resonance methods to determine distances between spin labels on human carbonic anhydrase II. Biophys J 80:2886–2897

73. Banham JE, Baker CM, Ceola S, Day IJ, Grant GH, Groenen EJJ, Rodgers CT, Jeschke G, Timmel CR (2008) Distance measurements in the borderline region of applicability of CW EPR and DEER: a model study on a homologous series of spin-labelled peptides. J Magn Reson 191:202–218
74. Jeschke G (2002) Determination of the nanostructure of polymer materials by electron paramagnetic resonance spectroscopy. Macromol Rapid Commun 23(4):227–246
75. Riplinger C, Kao JPY, Rosen GM, Kathirvelu V, Eaton GR, Eaton SS, Kutateladze A, Neese F (2009) Interaction of radical pairs through-bond and through-space: scope and limitations of the point-dipole approximation in electron paramagnetic resonance spectroscopy. J Am Chem Soc 131(29):10092–10106
76. Margittai M, Langen R (2004) Template-assisted filament growth by parallel stacking of tau. Proc Natl Acad Sci USA 101:10278–10283
77. Margittai M, Langen R (2006) Side chain-dependent stacking modulates tau filament structure. J Biol Chem 281:37820–37827
78. Molin YN, Salikhov KM, Zamaraev KI (1980) Spin exchange. Springer, Berlin
79. Altenbach C, Oh KJ, Trabanino RJ, Hideg K, Hubbell WL (2001) Estimation of inter-residue distances in spin labeled proteins at physiological temperatures: experimental strategies and practical limitations. Biochemistry 40:15471–15482
80. Scarpelli F, Drescher M, Rutters-Meijneke T, Holt A, Rijkers DTS, Killian JA, Huber M (2009) Aggregation of transmembrane peptides studied by spin-label EPR. J Phys Chem B 113:12257–12264
81. Steinhoff HJ, Radzwill N, Thevis W, Lenz V, Brandenburg D, Antson A, Dodson G, Wollmer A (1997) Determination of interspin distances between spin labels attached to insulin: comparison of electron paramagnetic resonance data with the X-ray structure. Biophys J 73 (6):3287–3298
82. Schweiger A, Jeschke G (2005) Principles of pulse electron paramagnetic resonance. Oxford University Press, Oxford, reprinted 2005 edn
83. Milov AD, Ponomarev AB, Tsvetkov YD (1984) Electron electron double-resonance in electron-spin echo – model biradical systems and the sensitized photolysis of decalin. Chem Phys Lett 110(1):67–72
84. Milov AD, Salikohov KM, Shirov MD (1981) Application of endor in electron-spin echo for paramagnetic center space distribution in solids. Fiz Tverd Tela 23(4):975–982
85. Jeschke G (2011) DeerAnalysis2011 user manual. http://wwweprethzch/software/DeerAnalysis_2011_manual.pdf. Accessed 7 Apr 2011
86. Richard W, Bowman A, Sozudogru E, El-Mkami H, Owen-Hughes T, Norman DG (2010) EPR distance measurements in deuterated proteins. J Magn Reson 207:164–167
87. Zou P, Mchaourab HS (2010) Increased sensitivity and extended range of distance measurements in spin-labeled membrane proteins: Q-band double electron-electron resonance and nanoscale bilayers. Biophys J 98(6):L18–L20. doi:DOI 10.1016/j.bpj.2009.12.4193
88. Höfer P, Heilig R, Schmalbein D (2003) The superQ-FT accessory for pulsed EPR, ENDOR and ELDOR at 34 GHz. Bruker SpinReport 152(153):37–43
89. Larsen RG, Singel DJ (1993) Double electron-electron resonance spin-echo modulation – spectroscopic measurement of electron-spin pair separations in orientationally disordered solids. J Chem Phys 98:5134–5146
90. Lovett JE, Bowen AM, Timmel CR, Jones MW, Dilworth JR, Caprotti D, Bell SG, Wong LL, Harmer J (2009) Structural information from orientationally selective DEER spectroscopy. Phys Chem Chem Phys 11:6840–6848
91. Torok M, Milton S, Kayed R, Wu P, McIntire T, Glabe CG, Langen R (2002) Structural and dynamic features of Alzheimer's A beta peptide in amyloid fibrils studied by site-directed spin labeling. J Biol Chem 277:40810–40815
92. Jeschke G, Sajid M, Schulte M, Godt A (2009) Three-spin correlations in double electron-electron resonance. Phys Chem Chem Phys 11:6580–6591

93. Hilger D, Jung H, Padan E, Wegener C, Vogel KP, Steinhoff HJ, Jeschke G (2005) Assessing oligomerization of membrane proteins by four-pulse DEER: pH-dependent dimerization of NhaA Na+/H + antiporter of E-coli. Biophys J 89:1328–1338
94. Upadhyay AK, Borbat PP, Wang J, Freed JH, Edmondson DE (2008) Determination of the oligomeric states of human and rat monoamine oxidases in the outer mitochondrial membrane and octyl beta-D-glucopyranoside micelles using pulsed dipolar electron spin resonance spectroscopy. Biochemistry 47:1554–1566
95. Bode BE, Margraf D, Plackmeyer J, Durner G, Prisner TF, Schiemann O (2007) Counting the monomers in nanometer-sized oligomers by pulsed electron – electron double resonance. J Am Chem Soc 129:6736–6745
96. Jeschke G, Schlick S (2006) Spatial distribution of stabilizer-derived nitroxide radicals during thermal degradation of poly(acrylonitrile-butadiene-styrene) copolymers: a unified picture from pulsed ELDOR and ESR imaging. Phys Chem Chem Phys 8:4095–4103
97. Domingo Köhler S, Spitzbarth M, Diederichs K, Exner TE, Drescher M (2011) A short note on the analysis of distance measurements by electron paramagnetic resonance. J Magn Reson 208:167–170
98. Pannier M, Schops M, Schadler V, Wiesner U, Jeschke G, Spiess HW (2001) Characterization of ionic clusters in different ionically functionalized diblock copolymers by CW EPR and four-pulse double electron-electron resonance. Macromolecules 34:5555–5560
99. Drescher M, Veldhuis G, van Rooijen BD, Milikisyants S, Subramaniam V, Huber M (2008) Antiparallel arrangement of the helices of vesicle-bound alpha-synuclein. J Am Chem Soc 130:7796–7797
100. Sajid M, Jeschke G, Wiebcke M, Godt A (2009) Conformationally unambiguous spin labeling for distance measurements. Chemistry 15:12960–12962
101. Borbat PP, McHaourab HS, Freed JH (2002) Protein structure determination using long-distance constraints from double-quantum coherence ESR: study of T4 lysozyme. J Am Chem Soc 124:5304–5314
102. Sale K, Sar C, Sharp KA, Hideg K, Fajer PG (2002) Structural determination of spin label immobilization and orientation: a Monte Carlo minimization approach. J Magn Reson 156:104–112
103. Fajer P, Likai S, Liu YS, Perozo E, Budil D, Sale K (2004) Molecular modeling tools for dipolar EPR. Biophys J 86:191A
104. Polyhach Y, Bordignon E, Jeschke G (2011) Rotamer libraries of spin labelled cysteines for protein studies. Phys Chem Chem Phys 13(6):2356–2366. doi:Doi 10.1039/C0cp01865a
105. Polyhach Y, Jeschke G (2010) Prediction of favourable sites for spin labelling of proteins. Spectroscopy 24:651–659
106. Hubbell WL, Altenbach C, Hubbell CM, Khorana HG (2003) Rhodopsin structure, dynamics, and activation: a perspective from crystallography, site-directed spin labeling, sulfhydryl reactivity, and disulfide cross-linking. Adv Protein Chem 63:243–290
107. Altenbach C, Froncisz W, Hyde JS, Hubbell WL (1989) Conformation of spin-labeled melittin at membrane surfaces investigated by pulse saturation recovery and continuous wave power saturation electron-paramagnetic resonance. Biophys J 56:1183–1191
108. Altenbach C, Flitsch SL, Khorana HG, Hubbell WL (1989) Structural studies on transmembrane proteins 2 spin labeling of bacteriorhodopsin mutants at unique cysteines. Biochemistry 28:7806–7812
109. Percival PW, Hyde JS (1975) Pulsed EPR spectrometer 2. Rev Sci Instrum 46:1522–1529
110. Yin JJ, Pasenkiewiczgierula M, Hyde JS (1987) Lateral diffusion of lipids in membranes by pulse saturation recovery electron-spin-resonance. Proc Natl Acad Sci USA 84:964–968
111. Lacapere JJ, Pebay-Peyroula E, Neumann JM, Etchebest C (2007) Determining membrane protein structures: still a challenge! Trends Biochem Sci 32:259–270
112. Torres J, Stevens TJ, Samso M (2003) Membrane proteins: the 'Wild West' of structural biology. Trends Biochem Sci 28:174

113. Torres J, Stevens TJ, Samso M (2003) Membrane proteins: the 'Wild West' of structural biology. Trends Biochem Sci 28:137–144
114. Tompa P (2002) Intrinsically unstructured proteins. Trends Biochem Sci 27:527–533
115. Timsit Y, Allemand F, Chiaruttini C, Springer M (2006) Coexistence of two protein folding states in the crystal structure of ribosomal protein L20. EMBO Rep 7:1013–1018
116. Eliezer D (2009) Biophysical characterization of intrinsically disordered proteins. Curr Opin Struct Biol 19:23–30
117. Dyson HJ, Wright PE (2002) Coupling of folding and binding for unstructured proteins. Curr Opin Struct Biol 12:54–60
118. Biswas R, Kühne H, Brudvig GW, Gopalan V (2001) Use of EPR spectroscopy to study macromulecular structure and function. Sci Prog 84:45–68
119. Morin B, Bourhis JM, Belle V, Woudstra M, Carriere F, Guigliarelli B, Fournel A, Longhi S (2006) Assessing induced folding of an intrinsically disordered protein by site-directed spin-labeling electron paramagnetic resonance spectroscopy. J Phys Chem B 110:20596–20608
120. Belle V, Rouger S, Costanzo S, Liquiere E, Strancar J, Guigliarelli B, Fournel A, Longhi S (2008) Mapping alpha-helical induced folding within the intrinsically disordered C-terminal domain of the measles virus nucleoprotein by site-directed spin-labeling EPR spectroscopy. Proteins 73:973–988
121. Murakami K, Hara H, Masuda Y, Ohigashi H, Irie K (2007) Distance measurement between Tyr10 and Met35 in amyloid beta by site-directed spin-labeling ESR spectroscopy: implications for the stronger neurotoxicity of A beta 42 than A beta 40. Chembiochem 8:2308–2314
122. Iurascu MI, Cozma C, Tomczyk N, Rontree J, Desor M, Drescher M, Przybylski M (2009) Structural characterization of β-amyloid oligomer-aggregates by ion mobility mass spectrometry and electron spin resonance spectroscopy. Anal Bioanal Chem 395:2509–2519
123. Sepkhanova I, Drescher M, Meeuwenoord NJ, Limpens R, Koning RI, Filippov DV, Huber M (2009) Monitoring Alzheimer amyloid peptide aggregation by EPR. Appl Magn Reson 36:209–222
124. Lundberg KM, Stenland CJ, Cohen FE, Prusiner SB, Millhauser GL (1997) Kinetics and mechanism of amyloid formation by the prion protein H1 peptide as determined by time-dependent ESR. Chem Biol 4:345–355
125. Igarashi R, Sakai T, Hara H, Tenno T, Tanaka T, Tochio H, Shirakawa M (2010) Distance determination in proteins inside Xenopus laevis oocytes by double electron-electron resonance experiments. J Am Chem Soc 132:8228–8229
126. Drescher M, Godschalk F, Veldhuis G, van Rooijen BD, Subramaniam V, Huber M (2008) Spin-label EPR on alpha-synuclein reveals differences in the membrane binding affinity of the two antiparallel helices. Chembiochem 9:2411–2416
127. Chen M, Margittai M, Chen J, Langen R (2007) Investigation of alpha-synuclein fibril structure by site-directed spin labeling. J Biol Chem 282:24970–24979
128. Der-Sarkissian A, Jao CC, Chen J, Langen R (2003) Structural organization of alpha-synuclein fibrils studied by site-directed spin labeling. J Biol Chem 278:37530–37535
129. Masuda M, Dohmae N, Nonaka T, Oikawa T, Hisanaga SI, Goedert M, Hasegawa M (2006) Cysteine misincorporation in bacterially expressed human alpha-synuclein. FEBS Lett 580:1775–1779
130. George JM, Jin H, Woods WS, Clayton DF (1995) Characterization of a novel protein regulated during the critical period for song learning in the zebra finch. Neuron 15:361–372
131. Davidson WS, Jonas A, Clayton DF, George JM (1998) Stabilization of alpha-synuclein secondary structure upon binding to synthetic membranes. J Biol Chem 273:9443–9449
132. Ulmer TS, Bax A, Cole NB, Nussbaum RL (2005) Structure and dynamics of micelle-bound human alpha-synuclein. J Biol Chem 280:9595–9603
133. Chandra S, Chen XC, Rizo J, Jahn R, Sudhof TC (2003) A broken alpha-helix in folded alpha-synuclein. J Biol Chem 278:15313–15318
134. Bussell R, Eliezer D (2003) A structural and functional role for 11-mer repeats in alpha-synuclein and other exchangeable lipid binding proteins. J Mol Biol 329:763–778

135. Bussell R, Ramlall TF, Eliezer D (2005) Helix periodicity, topology, and dynamics of membrane-associated alpha-synuclein. Prot Sci 14:862–872
136. Jao CC, Hegde BG, Chen J, Haworth IS, Langen R (2008) Structure of membrane-bound alpha-synuclein from site-directed spin labeling and computational refinement. Proc Natl Acad Sci USA 105:19666–19671
137. Borbat P, Ramlall TF, Freed JH, Eliezer D (2006) Inter-helix distances in lysophospholipid micelle-bound alpha-synuclein from pulsed ESR measurements. J Am Chem Soc 128:10004–10005
138. Bortolus M, Tombolato F, Tessari I, Bisaglia M, Mammi S, Bubacco L, Ferrarini A, Maniero AL (2008) Broken helix in vesicle and micelle-bound alpha-synuclein: insights from site-directed spin labeling-EPR experiments and MD simulations. J Am Chem Soc 130:6690–6691
139. Georgieva ER, Ramlall TF, Borbat PP, Freed JH, Eliezer D (2008) Membrane-bound alpha-synuclein forms an extended helix: long-distance pulsed ESR measurements using vesicles, bicelles, and rodlike micelles. J Am Chem Soc 130:12856–12857
140. Robotta M, Braun P, van Rooijen B, Subramaniam V, Huber M, Drescher M (2011) Direct evidence of coexisting horseshoe and extended helix conformations of membrane-bound alpha-synuclein. Chemphyschem 12:267–269
141. Georgieva ER, Ramlall TF, Borbat PP, Freed JH, Eliezer D (2010) The lipid-binding domain of wild type and mutant alpha-synuclein compactness and interconversation between the broken and extended helix forms. J Biol Chem 285:28261–28274
142. Trexler AJ, Rhoades E (2009) Alpha-synuclein binds large unilamellar vesicles as an extended helix. Biochemistry 48:2304–2306
143. Ramakrishnan M, Jensen PH, Marsh D (2003) Alpha-synuclein association with phosphatidylglycerol probed by lipid spin labels. Biochemistry 42:12919–12926
144. Bussell R, Ramlall TF, Eliezer D (2005) Helix periodicity, topology, and dynamics of membrane-associated alpha-synuclein. Protein Sci 14(4):862–872. doi:Doi 10.1110/Ps.041255905
145. Kamp F, Beyer K (2006) Binding of alpha-synuclein affects the lipid packing in bilayers of small vesicles. J Biol Chem 281(14):9251–9259. doi:DOI 10.1074/jbc.M512292200
146. Ramakrishnan M, Jensen PH, Marsh D (2006) Association of alpha-synuclein and mutants with lipid membranes: spin-label ESR and polarized IR. Biochemistry 45(10):3386–3395. doi:10.1021/bi052344d
147. Drescher M, van Rooijen BD, Veldhuis G, Subramaniam V, Huber M (2010) A stable lipid-induced aggregate of alpha-synuclein. J Am Chem Soc 132:4080–4081
148. Bordignon E (2011) Site-directed spin labeling of membrane proteins. Top Curr Chem. doi:10.1007/128_2011_243

Top Curr Chem (2012) 321: 121–158
DOI: 10.1007/128_2011_243

Published online: 7 September 2011

Site-Directed Spin Labeling of Membrane Proteins

Enrica Bordignon

Abstract EPR spectroscopy of site-directed spin labeled membrane proteins is at present a common and valuable biophysical tool to study structural details and conformational transitions under conditions relevant to function. EPR is considered a complementary approach to X-ray crystallography and NMR because it provides detailed information on (1) side chain dynamics with an exquisite sensitivity for flexible regions, (2) polarity and water accessibility profiles across the membrane bilayer, and (3) distances between two spin labeled side chains during protein functioning. Despite the drawback of requiring site-directed mutagenesis for each new piece of information to be collected, EPR can be applied to any complex membrane protein system, independently of its size. This chapter describes the state of the art in the application of site-directed spin labeling (SDSL) EPR to membrane proteins, with specific focus on the different types of information which can be obtained with continuous wave and pulsed techniques.

Keywords Conformational changes · Distances · Dynamics · EPR · Membrane proteins · Spin labeling · Structural biology · Water accessibility

Contents

E. Bordignon
Laboratory of Physical Chemistry, ETH Zurich, Wolfgang-Pauli-str. 10, 8093 Zurich, Switzerland
e-mail: enrica.bordignon@phys.chem.ethz.ch

1 Introduction

In 1989, 4 years after the first high resolution X-ray structure of a membrane protein [1], a paper on site-directed spin labeling (SDSL) EPR on a transmembrane protein appeared in Biochemistry [2], followed by a second report in Science [3] titled "Transmembrane protein structure: spin labeling of bacteriorhodopsin mutants." Hubbell suggested that a combination of EPR, site-directed mutagenesis and nitroxide labeling could serve as a new biophysical tool to unravel structural details of membrane proteins. This first EPR study focused on the different dynamics (mobilities) of 18 consecutive spin labeled cysteines in the protein and their distinct water accessibility. The EPR analysis could clearly distinguish between a short water exposed loop and the transmembrane α-helical segment of the protein. In the same year the first 3D structure of bacteriorhodopsin (BR) was solved by electron microscopy [4]. The X-ray structure of BR appeared only later [5]. In 1994 in a review article, Hubbell further confirmed the potentiality of the technique, stating that "the technology is now in place for an expanded range of applications of SDSL" [6]. Indeed, to date, a plethora of structural information has been obtained by SDSL EPR on a variety of membrane proteins, confirming the statement of the pioneer of the technique. The increased availability of structures with atomistic details from X-ray and NMR did not relegate SDSL EPR to proteins of unknown structure or to minor aspects of the spin labeled side chains accessibility and dynamics. On the contrary, it enormously favored its application to conformational transitions between different protein states, dimerization–oligomerization processes, characterization of protein dynamics during the conformational changes under physiological conditions, and recently also modeling of overall conformational transitions based on interspin distance constraints.

This chapter will introduce the most important aspects of SDSL EPR applied to membrane proteins corroborated by prominent examples from the literature and will address the newest developments of the technique.

2 Spin Labeling of Membrane Proteins

To date, the most commonly used spin labels for membrane proteins are pyrroline-type nitroxide radicals. Several labeling strategies and spin labels are available at present. However, the extreme specificity for the free thiol of cysteines, a stoichiometric reaction at most sites, and the relatively small size and flexibility of the modified side chain make the (1-oxyl-2,2,5,5-tetramethyl-Δ3-pyrroline-3-methyl) methanethiosulfonate spin label (MTSSL) the most popular choice for SDSL applications. 3-(2-Iodoacetamido)-2,2,5,5-tetramethyl-1-pyrrolidinyloxy radical labels (IAP) are also used in combination with cysteines. The advantage of IAP vs MTSSL is that the covalent bond with the cysteine is irreversibly formed and this may be more suitable for particular situations in which the MTSSL tends to be released from the labeled sites.

The labeling strategies for membrane proteins are similar to those applicable to water soluble proteins (a cys-less background is also required prior to the cys site-specific mutagenesis); however more care must be taken in identifying sites which are accessible for the spin label. In particular, sites buried in the middle of the transmembrane core of the protein may not be labeled at all, or once labeled may influence the protein stability. Sites located at the interface between the membrane and the bulk water, e.g., in the short loops connecting the transmembrane helices, or at cytoplasmic or periplasmic domains, are usually highly accessible and can be targeted for first tests on the protein under investigation (Fig. 1a). The freely downloadable software MMM [http://www.epr.ethz.ch/software/index] can be used to predict the accessibility towards spin labels (currently MTSSL or IAP) for each site of the protein prior to the site-specific mutagenesis if a structural template of the protein exists [7].

Experimentally, the spin labeling procedure is carried out on proteins solubilized in detergent, to maximize the reaction yield. In a few cases (protein- and detergent-dependent) the spin labels may bind aspecifically to the cys-less protein, or to the detergent micelles. It is then good practice to spin label prior to the SDSL study the cys-less protein and eventually also the empty micelles, to verify that noncovalent aspecific labeling is negligible. Otherwise, one needs to change the spin label to protein ratio during the labeling procedure or eventually the type of detergent to minimize this unwanted effect. It is worth mentioning that the labels trapped in the micelles are removed by the reconstitution procedure, unless the protein is measured in detergent-solubilized form.

2.1 *Spin Labeling Protocols for Membrane Proteins*

In the following, a general protocol for spin labeling membrane proteins with MTSSL (also valid for IAP) is described, with particular attention to steps which may cause problems or may need to be slightly modified according to the protein

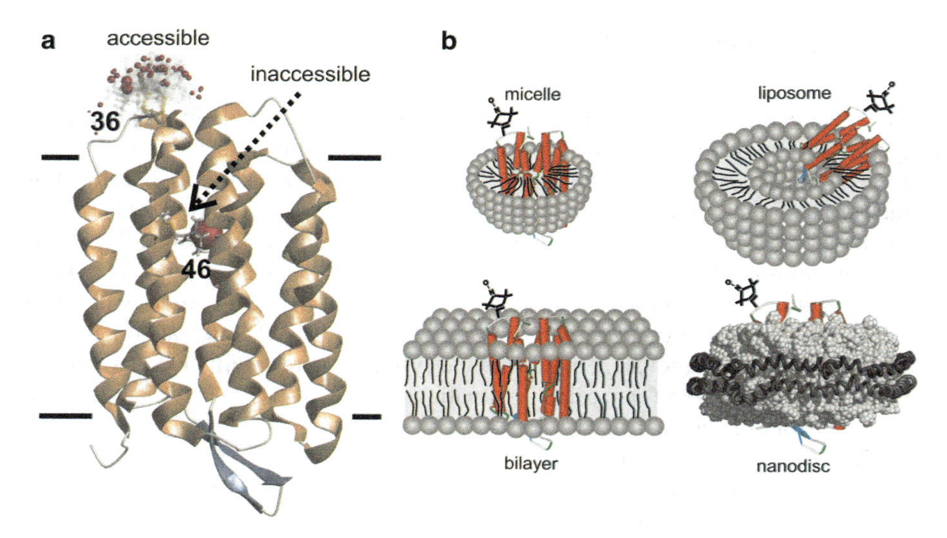

Fig. 1 Membrane-mimicking environments. (**a**) Bacteriorhodopsin spin labeled at position 36 and 46 with the software MMM. (**b**) Schematic drawings of bacteriorhodopsin inserted in micelle, liposome, membrane bilayer and nanodisc. The liposome is not drawn to scale

stability. Here it is assumed that the expression and purification of the cys-less membrane protein of interest has been optimized prior to the spin labeling. The membrane protein engineered to have one cysteine residue to be spin labeled is usually prepared in the detergent-solubilized form and its purity is >98%. Knowledge of the protein concentration is mandatory for the spin labeling protocol, and for the subsequent determination of the spin labeling efficiency. For labeling, the detergent-solubilized protein is diluted to a concentration of about 10–30 μM. In some cases the protein is incubated for 1 h at room temperature with 10 mM DTT (1,4-dithio-DL-threitol) to reduce eventually oxidized cysteines. DTT must be removed before starting the labeling procedure (1 μM final residual concentration) to avoid nitroxide reduction and to allow the formation of the disulfide bond between MTSSL and the cys residue. Immediately after DTT removal, tenfold molar excess of MTSSL is added from a 100 mM stock solution (usually in DMSO, stored at −80 °C). If the final concentration of MTSSL in the water solution exceeds 500 μM, spin label biradicals can form which are not anymore available for the labeling of the protein. Usually, the protein-MTSSL solution is shaken overnight at 4 °C, after which the labeling should be complete. Labeling can also be performed for 4–5 h at room temperature, depending on the protein stability. Note that very accessible sites can be labeled after few minutes of incubation. In general, the labeling efficiency increases with (1) increasing the label-to-protein molar excess (e.g., up to 40-fold); (2) increasing the temperature (from 4 to 37 °C, depending on the protein stability); (3) slightly denaturating the protein (valid only for very stable proteins, e.g., 40% DMSO was used to label position 46 in bacteriorhodopsin, Fig. 1a); and (4) preincubating the protein with reducing agents. After the spin labeling, one proceeds with the removal of the excess free label in the buffer. Different methods are possible: (1) dialysis – slow but efficient, requires at least 500 mL of detergent-containing buffer, which could be expensive; (2) centrifugal filter devices – fast and efficient, but some proteins precipitate on the membrane after centrifugation; (3) desalting columns – very fast and efficient, performed by centrifugation or gravity, requires a larger volume than centrifugal devices.

Once the free label is removed, the protein is concentrated to the desired final concentration. At this stage the labeling efficiency can be tested. Continuous wave EPR is highly sensitive, enabling measurements on as little as 100 pmol of protein (10 μL of 10 μM spin labeled protein). For optimal signal-to-noise ratio, concentrations in the order of 100 μM are preferable. The residual free label is clearly distinguishable in the EPR spectrum from the bound fraction due to its narrow three-line spectrum typical of isotropic fast motional regimes. In the case that the spectrum still contains a disturbing fraction of free label, further washing is necessary. The labeling efficiency can be calculated directly from the continuous wave EPR spectrum by second integral analysis and comparison with the spectrum of a standard nitroxide radical of known concentration (typically a 300 μM water solution of TEMPOL). The ratio between the spin concentration and the protein concentration gives directly the labeling efficiency for single cys mutants. Labeling efficiencies in the 70–100% range are considered good; problematic sites allows only for 0–50% efficiency. Efficiencies >100% pose some questions on the specificity of the labeling, which must be further tested with the cys-less background protein. For interspin distance determination (two spin labeled cysteines per protein) an average efficiency of at least 60–70% per cysteine is advisable.

2.2 Reconstitution Protocols

After successful spin labeling, the detergent-solubilized protein can be either used directly or reconstituted into liposomes or other bilayer-mimicking environments (Fig. 1b). Reconstitution protocols are strictly protein-specific and several methods can be found in the literature based on detergent extraction by biobeads, dilution methods, etc. An important issue in membrane protein reconstitution in general and most specifically for EPR analysis is the homogeneity of the distribution of the proteins inside the membranes. In particular, lipid to protein ratios >500 are advisable to avoid protein crowding which may lead, for example, to unwanted interprotein interspin distances in the 2–8 nm range. Extrusion of the liposomes before or after reconstitution can be performed to increase the homogeneity of the sample. The type of lipids plays an important role in either determining the efficiency of the insertion or contributing to the directionality of the insertion process. In fact, membrane proteins can insert into liposomes both with no preferential orientation (e.g., 50% of the proteins exposes the cytoplasmic regions to the lumen and the other 50% the periplasmic region) or with remarkable preference for one specific orientation with respect to the lumen (e.g., proteins having long cytoplasmic domains tend to insert in liposomes leaving those domains exposed to the bulk water).

An alternative approach consists in reconstitution of the membrane protein in bilayer-mimicking environments, i.e., self-assembling lipid bilayer nanodiscs (Fig. 1b). Nanodiscs consist of a small portion of membrane bilayer that has been solubilized by the addition of two amphipathic proteins, the membrane scaffold proteins (MSP) derived from the apolipoprotein A-1 [8–10]. Details of the preparation can also be found at http://sligarlab.life.uiuc.edu/nanodisc/protocols.html. These proteins wrap around the hydrophobic core of the lipids, effectively creating a soluble portion of membrane.

The advantages of this method are: (1) different nanodiscs diameters can be obtained with different MSP sequences; (2) the protein is in a bilayer-mimicking environment; (3) both the cytoplasmic and the periplasmic exposed regions of the protein are easily accessible with ligands, nucleotides, interacting proteins, etc., without the problems associated with the directional incorporation of the proteins in liposomes and the compartmentalization between lumen and bulk water. The disadvantages are: (1) the reconstitution in nanodiscs is protein-specific and requires optimization of the protocol and of the MSP to be used; (2) the maximal concentrations achievable are less than 100 μM.

3 Dynamics of Spin Labeled Side Chains

Room temperature EPR spectra of spin labels attached at protein sites are extremely sensitive to even minor changes in the reorientational motion of the spin label with respect to the external magnetic field. The reorientation of the spin label in the protein is generally anisotropic and it is a complex function of the spin label molecular structure and the primary, secondary, tertiary, and eventually quaternary structure of the protein under investigation. Complex distributions of motional states can often be observed in spectra exhibiting more than one spectral component. Simulations of the CW spectra are performed using molecular dynamics simulations [11–15] or a slow motion theory based on the stochastic Liouville equation (SLE). The theoretical model developed by Freed based on the SLE (SRLS, Slowly Relaxing Local Structure) is the most advanced to date [16, 17]. Two principle dynamic modes are included in SRLS, the global tumbling of the protein and the internal spin label motion consisting of side chain isomerizations and protein backbone fluctuations. In the case of medium-size membrane proteins embedded in micelles or in membranes, the overall tumbling of the object generally exceeds the range of rotational correlation times measurable by EPR (>100 ns). Thus, one can consider the protein as "immobile" on the EPR timescale, and the spin label motion only due to internal rotation around the dihedral angles of the label and to the protein backbone fluctuations. Even with the reduced number of rotational motions to be considered in the case of membrane proteins, a comprehensive theoretical description of the spectra has not yet been achieved. The SRLS model was shown to be able to separate the effects of faster internal modes of motion from slower overall motions only if spectra at multiple frequencies are available. Even with the multifrequency SRSL approach, up to three distinct spectral components are often required to account for all observed spectral features, showing the intricate and composite interaction between the label and the protein environment.

The complications in the spectral fitting partially stem from the fact that MTSSL is a rather long and flexible label characterized by five rotatable bonds. Clearly a more "rigid" label, with the N–O group closely related to the backbone atom positions would alleviate this problem, but it would also create steric constraints at the spin labeled site, which may disturb the protein fold. In fact, it is exactly its intrinsic flexibility that makes MTSSL a nonperturbing probe at the majority of the sites in proteins.

For a semiquantitative description of spin label mobility, simple empirical parameters extracted from the spectra at room temperature are widely used. Experimentally, to detect room temperature EPR spectra, 10–20 μL of sample are inserted in glass capillaries (0.9 mm outer diameter) and the spectra are detected with a sweep width of 15 mT, 0.15 mT maximal modulation amplitude (corresponding to the natural linewidth of a freely tumbling MTSSL in water) and 0.5–1 mW microwave power (the power saturation curve must be measured at least once for the cavity in use). The mobility scale is based on (1) the inverse of the spectral second moment and (2) the inverse of the central EPR linewidth. Both parameters are increasing functions of the spin label mobility and can be used to give insights into the protein secondary structure, based on the available correlation between mobility and protein structure obtained from a large number of studies, mainly on α-helical proteins [18–20]. Although the semiquantitative analysis is usually enough to draw conclusions on secondary structure and conformational changes, interpretative problems may arise if spectra show complicated features with multiple components, which in the literature are often identified as "immobile" and "mobile" components.

The high sensitivity for relatively fast motions (rotational correlation times in the 0.1–3 ns range) extends the structural analysis beyond the usually "silent" flexible regions, which are increasingly identified as relevant for signal transduction. In general, the extent of labeling is correlated to the accessibility of the site towards the spin label; thus very mobile positions show spin labeling efficiency close to 100%. Nevertheless, very immobile positions can be labeled due to the flexible nature of MTSSL, and the plasticity of the side chains surrounding a buried spin labeled site. Care must be taken when spectra close to the rigid limit are detected at room temperature; in fact the label can be trapped in one specific orientation relative to the backbone. This makes general statements about side chain mobility or distances with respect to a second spin labeled site ambiguous. Moreover, the protein structure might have been affected by the introduced steric hindrance. Such interpretative problems can be solved by analyzing the overall pattern of side chain mobilities in an extended protein segment, i.e., by performing a nitroxide site scan on consequent residues.

It is worth noting that macromolecular crystallography, in contrast to SDSL EPR, requires solution conditions that may also alter the conformational sampling of a macromolecule, especially in dynamic regions. Interestingly, a recent SDSL EPR study examined a conformational equilibrium in the N-terminal energy coupling motif of BtuB, the *E. coli* outer membrane transporter for vitamin B_{12} [21]. EPR indicated that this segment is in equilibrium between folded and unfolded forms and that the substrate binding shifts this equilibrium toward the unfolded form. EPR spectra obtained from the same spin labeled mutant in protein crystals indicated that this unfolding transition is blocked in the crystal. SDSL may as well influence protein structures or conformational transitions if the label is attached to sites relevant for function. To avoid this possibility, spin labeling should not target side chains directly involved in active sites, or buried in tight protein pockets. However, the protein's activity after spin labeling should always be measured.

In the following paragraphs a selection of relevant SDSL EPR studies performed on several membrane proteins is reviewed. The data presented, far from being comprehensive, were chosen to emphasize in detail all information which can be derived from mobility, accessibility, and distance measurements.

3.1 EPR-Derived Molecular Architecture of the Potassium Channel KcsA

Details of the overall architecture and of the gating mechanism of the potassium channel KcsA were elucidated in the last few years with SDSL by the group of Perozo. Here, one long site scan (63 single cysteines spin labeled with MTSSL) published in 1998 on this membrane protein is reviewed to highlight the potentiality of the analysis of the nitroxide mobility to infer structural details of long protein segments [22]. Some caveats arising from the use of homo-oligomeric proteins are addressed. Potassium channels are integral membrane proteins which can be found virtually in all living organisms. They catalyze an efficient flow of K^+ across the membrane bilayer, with a fundamental role in generation and modulation of the electrical excitability in cells [23].

The data available at the time of the EPR research on the potassium channel KcsA were: (1) mechanistic information derived from electrical measurements; (2) tetrameric organization of the monomer (160 residue polypeptide); (3) α-helical nature of the polypeptide chain; (4) stabilization of the open conformation by low pH; (5) newly available X-ray structure from the MacKinnon's group [24], with the PDB file not yet released. The systematic EPR scan of the membrane embedded region of the channel via mobility, accessibility and distance analysis was used to generate a three-dimensional architecture of the two transmembrane (TM) helices of the channel and to elucidate their topology with respect to the membrane–water interface. Interestingly, 66 cys mutants were prepared in the protein stretch 22–52 and 86–120 where the two TM segments were thought to be located (Fig. 2a) and only three of them were found not to be expressed, confirming that cysteine mutagenesis is tolerated without relevant structural rearrangements in most cases; thus they are good probes to report on local protein properties. All the cys mutants were tested for structural and functionality studies prior to the EPR study. It is important to note that functionality tests are mandatory for newly engineered mutants, and it is recommended to perform them before and after labeling to avoid interpretative problems. A nonfunctional or nonperfectly folded spin labeled mutant cannot be used for EPR analysis.

Each single cysteine mutant was spin labeled and analyzed by RT CW EPR spectroscopy. As an example of a nitroxide scan, spectra detected in the first (22–32) and second (86–109) segments are shown in Fig. 2b. Interestingly, being a homotetramer, each single cysteine mutant provides four spins per channel. If the distance between the four spin labels in the tetramer is on the order of 2 nm or higher, no dipolar interaction appears; thus the features of the room temperature spectra purely reflect the spin label dynamics.

In contrast, for distances <2 nm, the evident spectral broadening due to spin exchange and dipolar coupling interferes with the mobility analysis (e.g., position 108, Fig. 4b, denoted with an asterisk). To obtain mobility information, an approach called "spin dilution" is used, which consists in under-labeling the protein (e.g., in this study a 1:10 ratio spin label:channel was used) to obtain a spectrum which reflects mostly the dynamics. Alternatively, to minimize asymmetry, it is possible to label the protein with a mixture of MTSSL and its diamagnetic analog (several molecules are available, e.g., the compound A167900 from Toronto Research, Canada).

In Fig. 4d, the effect of spin dilution is illustrated. Spectral broadening is clearly evident in the spectral wings of the fully labeled sample. The shortest distance between two adjacent monomers based on the X-ray structure (PDB 18L8) can now be simulated with the software MMM (http://www.epr.ethz.ch/software/index) based on a rotamer library approach [7] showing a mean distance of 1.4 nm, in agreement with the dipolar broadening observed.

The structural analysis on KcsA was performed based on the mobility of each spin labeled side chain in the protein segments under investigation. It is worth recognizing in Fig. 4b that most of the CW RT spectra show multiple spectral components, characterized by different "mobility" (a few examples are highlighted by arrows). This is a very general property of the R1 side chain in proteins. The components reflect the anisotropy of the spin label reorientational motion, but their appearance could also have other causes. They could arise from a slow equilibrium between two different protein conformations or the presence of asymmetric sites in the protein. The molecular interpretation of different spectral components is cumbersome. Multifrequency EPR [17], temperature analysis of the CW spectra [27], pulse saturation recovery techniques [28], or high pressure EPR [29] can help unravel the possible origins of the spectral components. In the case of KcsA, the spin labels motional information was quantitatively extracted from the inverse central linewidth (ΔH_0^{-1}, mobility parameter) and was corroborated by the measure of the accessibility of the spin labeled side chains towards lipids (O_2

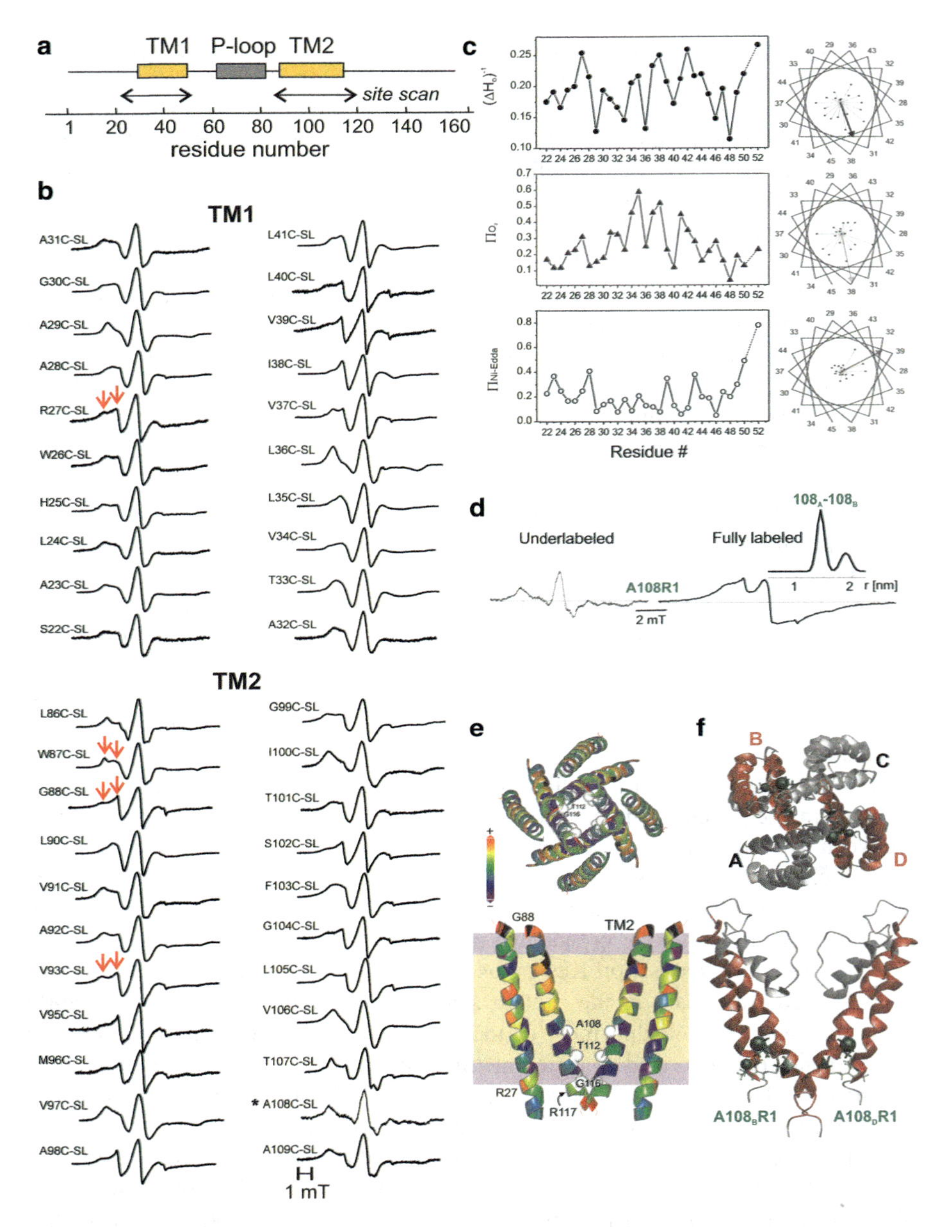

Fig. 2 A nitroxide scan on KcsA. (**a**) Linear representation of the putative transmembrane topology of KcsA and the nitroxide scan (linear scale with *arrows*). (**b**) Room temperature CW EPR spectra for two regions in TM1 and TM2. Multiple nitroxide components are *highlighted* by *red arrows* in selected spectra. (**c**) Mobility and accessibility plots. Periodical pattern are visible. *On the right*, helical wheel representation showing the trends of the EPR parameters extracted from the spectra in a polar coordinate representation. (**d**) Example of dipolar broadening on position 108, and effect of underlabeling on the spectral shape. *On the right* the shortest distance

accessibility) and water (NiEDDA accessibility) by power saturation measurements (see Sect. 4 for a detailed description of the power saturation technique). Close examination of the spectral set shows a periodic behavior in the mobility and accessibility parameters (Fig. 2c). An ideal helix with 3.6 residues per turn yields an angle of 360°/3.6 = 100° between successive residues. If a nitroxide scan reveals the α-helical periodicity, the secondary structure of the segment under investigation is unveiled. In this case both the mobility and the O_2 accessibility clearly show this pattern. The helical wheel representations are useful to correlate the periodical pattern of α-helical segments with the orientation of the helices with respect to the interfaces (Fig. 2c). In this representation the "wheel" is a view along the long axis of the helix, with numbered positions and residues indicated by vectors pointing out of the helix. The norm of each vector is chosen to be proportional, for example, to the mobility or the accessibility parameter at the specific site. The EPR data obtained by mobility and accessibility were complemented by the dipolar broadening observed in between few pairs of residues (Fig. 2d). Using the symmetry constraints of the tetramer and biochemical information on the analogous Shaker channel, the EPR data provided the three-dimensional architecture of the two TM helices in KcsA (Fig. 2e). The agreement between the EPR-derived model and the X-ray structure of KcsA (Fig. 2f) shows the potential of a nitroxide scan to obtain the architecture of a membrane protein.

3.2 Light-Induced Conformational Changes in Bacteriorhodopsin Detected via Transient Mobility Changes

The dynamics of spin labeled side chains are strictly related to the secondary and tertiary structure of the protein under investigation. If functional conformational transitions are induced in proteins with substrates, ligands, pH, nucleotides, ions, lipids, etc., the concomitant changes in the spin label dynamics describe the protein rearrangements. SDSL EPR can follow the proteins "in action," and being a technique applicable to proteins in physiological environments, it gives unique potentiality to unveil the rearrangements of membrane proteins during their function, e.g., substrate-ions transport, light activation, oligomerization, ligand binding, etc. In the case of the seven-transmembrane helix protein bacteriorhodopsin, light excitation induces transient isomerization of the retinal molecule bound in the core of the protein through a Schiff base. It is the light-induced change in the conformation of the retinal molecule which results in a conformational change of the surrounding protein and the proton pumping action during a so-called

Fig. 2 (continued) simulated with MMM on the X-ray structure (PDB 18L8). (**e**) EPR-derived molecular architecture of the channel, colored according to the water accessibility scale (*red*, high water accessibility; *blue*, low). *Upper panel*, intracellular view; *bottom panel*, lateral view of two monomers. (**f**) Crystal structure of the channel (PDB 18L8). Panels **a–e** adapted from [25]

"photocycle." It is possible to follow repetitively by EPR the changes in spin label mobility during the photocycle by transient detection. The first application of transient EPR on spin labeled proteins was published in 1994 by Hubbell and Steinhoff [30]. In this case the EPR spectrum is detected in the dark-adapted state, and the spectral amplitude at each *B* field position is monitored after light excitation for the time necessary for the dark state to be recovered at the end of the cycle. By plotting the maximal transient change in spectral intensity for each *B* field of the spectrum, one can obtain the "difference" light *minus* dark spectrum (2 *minus* 3 in the simulated example of Fig. 3a). An increase or decrease in mobility is associated with characteristic features in the difference spectra; thus the changes in mobility can be directly related to the protein conformational transitions (in Fig. 3a a simulated mobility increase is shown). In bacteriorhodopsin (Fig. 3b) EPR transient responses to light were observed for some spin labeled side chains during the photocycle. The biggest response was found at position 101 (helix *C*), located five residues away from the Asp which changes its protonation state during the photocycle. The transient EPR change could be unambiguously assigned to a decrease in mobility (Fig. 3c, highlighted by arrows, the opposite changes with respect to those simulated in Fig. 3a). Further EPR characterization performed in the group of Steinhoff revealed that helix *F* is moving outwards during the light-induced activation [31]. The same light-induced opening mechanism was later found by EPR also in the related protein sensory rhodopsin II (SRII) [32–34]. SRII is an archaeal seven-transmembrane helix retynilidene protein found in combination with a membrane embedded transducer protein in a 2:2 complex. Remarkably, the complex topology and stochiometry was revealed by EPR [35] before the release of the X-ray structure [36]. The two TM helices of the transducer in the core of the complex are followed towards the cytoplasm by two HAMP-domains and a methyl-accepting signaling domain. Upon light excitation, the retinal in SRII isomerizes, helix *F* opens in analogy to BR, and transfers the signal to the transducer which in turn signals this information to the flagellar components of the cell via a rotation of the TM2 helix of the transducer (Fig. 3c). Both the SRII light-induced conformational change and the related transducer signaling were detected by transient EPR at room temperature, and corroborated by interspin distance changes in different protein states trapped by freezing the complex during illumination. Later in the chapter an analogous light-induced opening of the G-protein coupled receptor (GPCR) rhodopsin will be presented as elucidated by pulse EPR techniques.

4 Accessibility, Polarity and Proticity

Side chains in membrane proteins are inherently facing different types of environments: the neighboring side chains of the protein, the bulk water, or the lipid alkyl chains. The protein topography with respect to the water and membrane interfaces is an important aspect in membrane protein studies and is not directly

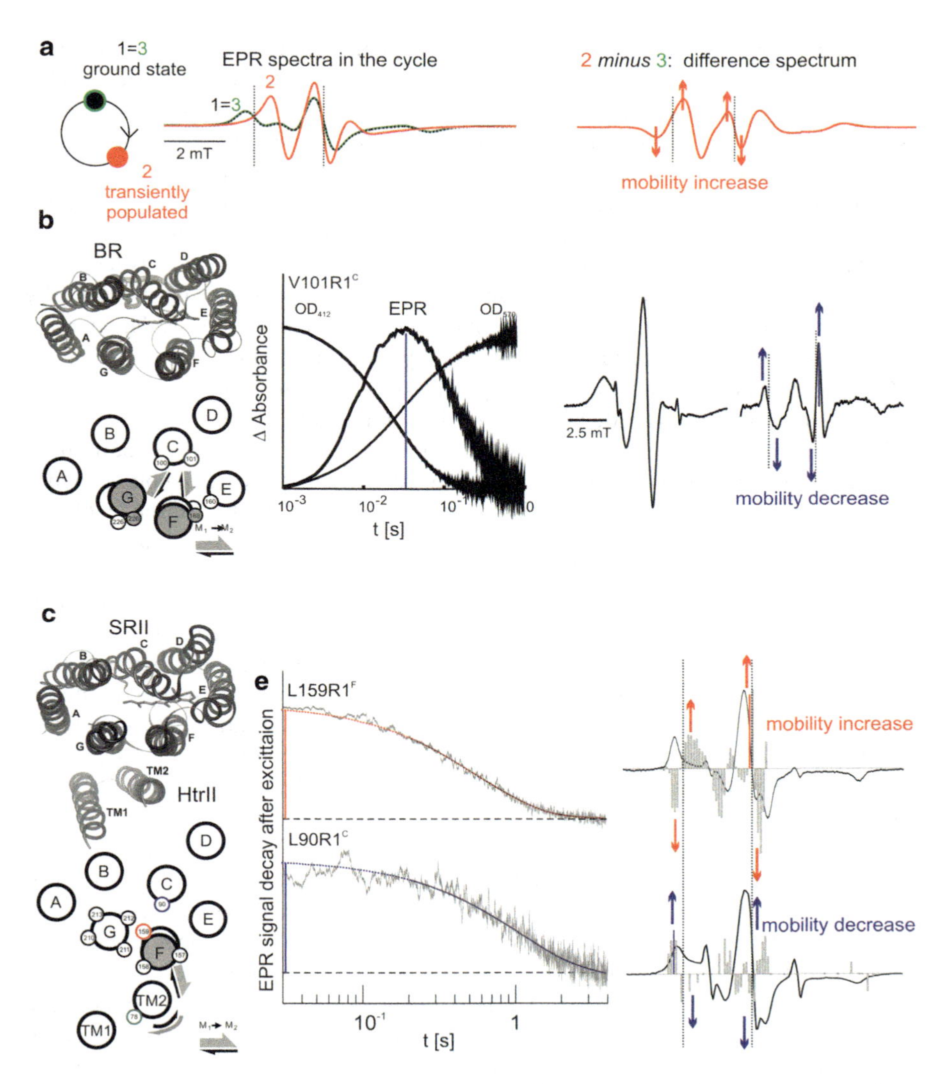

Fig. 3 Light-induced helix movements in BR and SRII. (**a**) Schematic representation of the conformational changes detected by EPR during the photocycle. The difference spectrum reflects the increase in mobility in state 2 with respect to the ground state. *Arrows* highlight the characteristic changes. (**b**) *Left*, periplasmic view of BR in ribbon (*up*) and schematic (*down*) representations. The EPR-detected movements are highlighted by *arrows*. *Central panel*, EPR detection of transient change in mobility at one fixed B field position, superimposed to the optical traces detected at different wavelengths. *Right panel*, spectrum of the dark-adapted state and difference spectrum characteristic of a mobility decrease. Adapted from [30]. (**c**) Analogous scheme for SRII in complex with the transducer HtrII. *Central panel*, normalized transient traces for two positions in SRII. *Right panel*, dark-adapted spectra with the difference spectra superimposed (*stick* representation). Adapted from [33]

addressable by X-ray crystallography. The most common EPR technique to obtain information on side chain accessibility at room temperature is power saturation CW EPR, which consists in detecting spectra at increasing microwave power to induce saturation [37, 38]. The amplitude of the central line is plotted vs the square root of the incident microwave power, and the saturation curve is analyzed to obtain the empirical $P_{1/2}$ value (the power at which the intensity of the line is half of what it should be in the absence of saturation) which is inversely proportional to the product of the effective relaxation times T_1 and T_2. The saturation behavior is characteristic of the local environment of the spin label and can be modified by addition of paramagnetic species soluble in water (chromium oxalate CrOX, charged or NiEDDA, neutral) or preferentially partitioning in membrane bilayers (O_2). Experimentally, to obtain a CW power saturation curve and extract the $P_{1/2}$ (and subsequently the Π values [20, 38]), the sample is loaded into gas-permeable TPX (polymethylpentene) capillaries. Loop-gap resonators are preferentially used which allow for sample volumes less than 10 μL and provide a homogeneous microwave field. The first saturation curve is detected in the presence of N_2. Subsequently, for oxygen accessibility experiments, nitrogen is replaced by air (21% oxygen) or 100% O_2. For water accessibility, a negligible volume of a concentrated NiEDDA (nickel-ethylenediamiediacetic acid) stock solution is added to the sample to obtain the desired final concentration (usually 20 mM, to be decreased if the $P_{1/2}$ is difficult to extract). Collisions with the paramagnetic reagents increase the nitroxide relaxation rates via Heisenberg exchange (predominant mechanism) which is proportional to the reagent concentration. The difference in $P_{1/2}$ (or Π) in the presence and absence of paramagnetic species reflects the relative collision frequency, or the relative "accessibility" of the spin labels towards the exchange reagent. This method allows fast measurements on commercial CW spectrometers on a small amount of sample. However, indirect effects of paramagnetic quenchers are measured instead of the direct effects of water and in the case of multiple spectral components only averaged accessibility can be obtained.

A second technique consists in the direct measure of the changes induced by paramagnetic agents in the spin-lattice relaxation T_1 by saturation recovery [39]. In this case an intense saturation pulse of microwave is delivered at room temperature at a frequency corresponding to the central EPR line and the return of the spectral intensity is monitored with a weak CW observing microwave field. With this relatively complex setup it is possible to measure T_1 selectively on different spectral components.

A third approach is to detect the accessibility of the nitroxide towards deuterated water molecules by comparing the amplitude of the ESEEM oscillations at cryogenic temperatures [40]. The advantage of this direct measurement of accessibility is accompanied by the disadvantage of using cryogenic temperatures, and by the difficulty in distinguishing protein-buried from membrane-embedded sites and in disentangling the effects arising from deuterium atoms from water, exchangeable protons, or deuterated glycerol (the latter is used as cryoprotectant).

A developing approach measures by NMR the local hydration dynamics around nitroxides at room temperature via Overhauser dynamic nuclear polarization (DNP).

Key steps to pave the way for the application to spin labeled proteins were performed by the group of Han [41].

X-band nitroxide EPR spectra in the rigid limit ($T < 200$ K) also convey information on the nitroxide microenvironment. In fact, the A_{zz} principal value of the ^{14}N hyperfine tensor (half of the splitting between the positive low field peak and the negative high field peak of the X-band CW spectrum) is proportional to the polarity of the nitroxide microenvironment (for MTSSL the A_{zz} value changes from 3.3 to 3.7 mT going from an apolar to a polar environment). This information can be complemented by detection of the propensity of the nitroxide to form H-bond (proticity) via the g_{xx} principal value of the g tensor. Sufficiently precise measurements of g_{xx} require the continuous wave detection of the low temperature nitroxide spectrum at W band (3.4 T/95 GHz).

4.1 Alternating Access in the ABC Transporter MsbA

MsbA is a putative lipid A flippase from *E. coli* [42]. It belongs to the ubiquitous protein superfamily of ABC transporters, active transporters which couple the vectorial translocation of a variety of substrates across the membranes with ATP hydrolysis [43]. ABC transporters share a common architecture consisting of two conserved cytoplasmic modules which harness ATP energy for biological work and two highly divergent transmembrane domains (TMDs) which provide the translocation channel for each substrate. MsbA has sequence similarity to the subclass of exporters linked to the development of multidrug resistance and cancer. A systematic SDSL EPR analysis on MsbA was published in 2009 by the group of Mchaourab [26], which shed light on the overall conformational changes during the nucleotide cycle. Here, the mobility and accessibility study performed on the transporter is reviewed, with particular emphasis on the water accessibility information obtained by continuous wave power saturation (examples of power saturation curves for a water exposed residue are presented in Fig. 4e).

MsbA is a homodimer of a polypeptide consisting of fused nucleotide binding domains (NBDs) and TMDs, a common feature of bacterial ABC exporters. The substrate is picked up from the inner leaflet of the membrane, and exported to the outside of the membrane by a conformational change of the transporter. Crystal structures of MsbA exist in the apo inward-facing conformation (PDB 3B5W, Cα atoms only) and in the AMPPNP-bound outward-facing conformation (PDB 3B60, 3.7 Å resolution). AMPPNP (5′-adenylyl imidodiphosphate) is a nonhydrolyzable analog of ATP, which traps the transporter in the closed state. The apo state of the transporter is defined by a V-shaped chamber open to the cytoplasm and the inner leaflet of the membrane, with the two NBDs separated by about 5 nm (Fig. 4a). Addition of AMPPNP induces a tight packing of the NBDs which sandwiches the two nucleotides and induces an opening of the TMDs towards the outer leaflet of the bilayer.

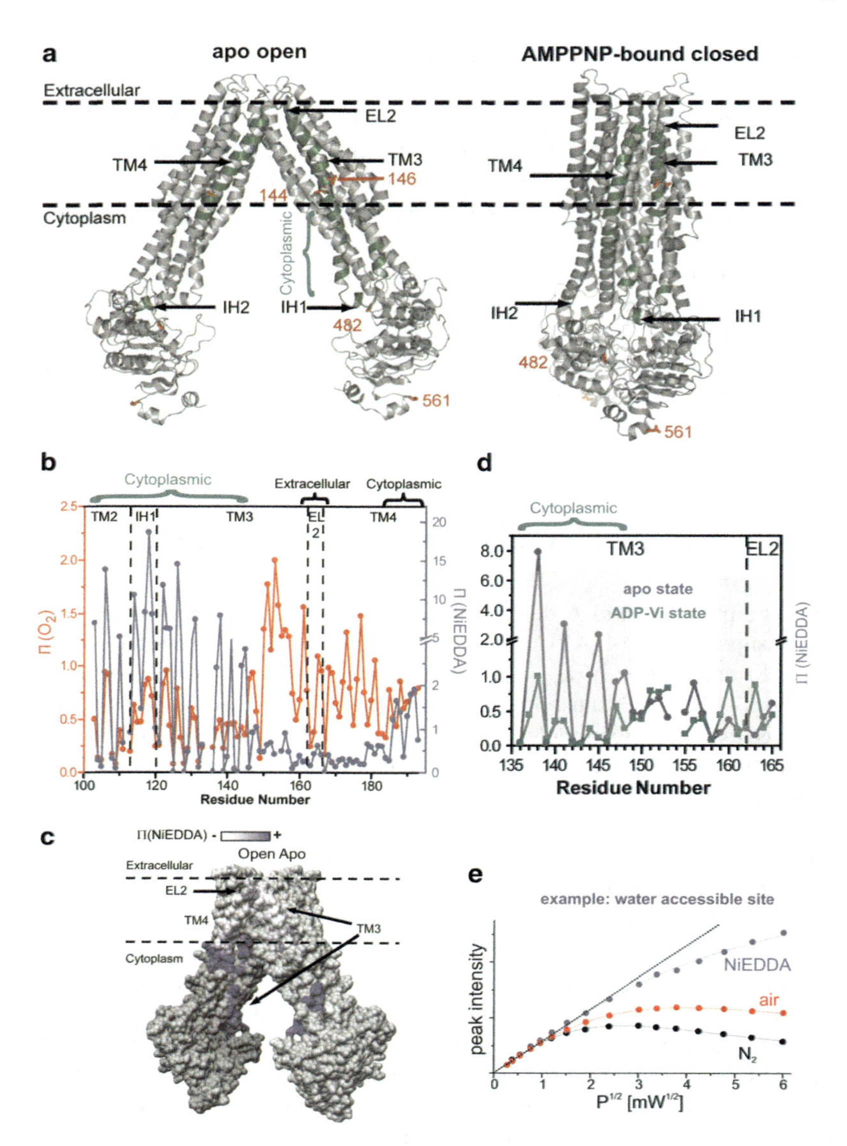

Fig. 4 Accessibility data on MsbA reveals the alternating-access mechanism. (**a**) Structural model of the open (provided by Chang) and closed states (PDB 3B60). The cytoplasmic region in the nitroxide scan (*green*) is highlighted. (**b**) Accessibility profile of the apo-state. The cytoplasmic region shows high water accessibility, in line with the structure. (**c**) Plot of the water accessibility in water the structure. (**d**) Changes in water accessibility in the cytoplasmic region upon transition to the ADP-Vi intermediate. (**e**) Example of three saturation curves for a water-exposed side chain to extract the accessibility parameter. Adapted from [26]

To “obtain a complementary perspective on the nature and amplitude of MsbA conformational motion,” collision frequencies of 90 consecutive spin labeled side chains (103–193, highlighted in green in the MsbA ribbon representation of Fig. 4a) with the apolar molecular oxygen and the water soluble NiEDDA were determined in MsbA reconstituted in liposomes and compared to the available X-ray structures. Large collision frequencies with NiEDDA are found in the apo state at solvent-exposed sites from positions 103 to 140, in line with their cytoplasmic location. The sequence-specific patterns reveal the α-helical character of the transporter. The accessibility becomes progressively smaller approaching the bilayer, in line with the X-ray structure. In contrast, the O_2 accessibility increases towards the middle of the bilayer, following the oxygen gradient in the membrane. The NiEDDA accessibility trend mapped to the structure of the apo open conformation validates the crystal model, and helps to define the membrane–water boundaries which are absent in the crystal (Fig. 4c, dashed lines).

Addition of ADP and vanadate traps the transporter in the post ATP-hydrolysis intermediate state; in fact the vanadate acts as γ-phosphate analog and along with ADP forms a complex mimicking the transition state of ATP hydrolysis. In this state opposite changes in the NiEDDA accessibility are detected at both sides of the TM3 helix spanning the bilayer (cytoplasmic fragment of TM3 presented in Fig. 4d). The result of this study highlights an alternation in the hydration profile between the two sides of the bilayer, corroborating the alternating-access mechanism for substrate export.

It is worth noting that the first systematic EPR analysis on Msba was published in 2005 [44] and the data were compared to the X-ray structures of MsbA in the open and closed states published from 2001 to 2005, which were retracted in 2007 due to data misinterpretation caused by an error in in-house data reduction software. The comparison of the EPR data with the corrected crystal structures reviewed here [26] resolved the earlier incongruencies brought about by wrong helix assignment.

4.2 *Water Penetration Profile in the Bacteriorhodopsin Channel*

A detailed polarity analysis can be performed on membrane proteins at cryogenic temperatures extracting the A_{zz} and g_{xx} parameters from low temperature W-band CW EPR spectra. In the temperature regime below 200 K the reorientational correlation time of an otherwise unrestricted spin label side chain exceeds 100 ns, i.e., the nitroxide may be considered as immobilized on the EPR timescale. The study of bacteriorhodopsin (already introduced in Sect. 3.2) spin labeled with MTSSL in the proton channel is a remarkable application of high field polarity–-proticity analysis on a membrane protein [45, 46]. Experimentally, for low temperature W-band spectra, 1 μL of sample is inserted in quartz capillaries (0.5 mm inner diameter) and the spectra are detected with a sweep width of 40 mT, 0.25 mT modulation amplitude (due to the broad spectral linewidth), and 5 μW microwave power (to avoid saturation at 160 K, the power saturation curve must be measured to avoid lineshape artifacts prior to the high field experiments). The variation of A_{zz} (3.3–3.7 mT from an apolar aprotic to a polar protic solvent) and g_{xx} (2.0091–2.0081 from an apolar aprotic to a polar protic solvent) is observable at W band. The g_{xx} and A_{zz} variation along the bacteriorhodopsin channel is seen as a shift of the position of the low field maximum and of the hyperfine splitting in the high field region, respectively (Fig. 5a). Spectral fitting can be performed with the function "pepper" from Easyspin [48]. A distribution of H-bonds around the NO group, and planar deviations of the NO group, make the g_{xx} region of the W-band spectra rather broad (g_{xx} strain). Additionally, multiple components can be clearly identified for several protein positions in the g_{xx} region, arising from nitroxide subpopulations having 0, 1, and 2 H-bonds [49]. Plots of g_{xx} and A_{zz} vs nitroxide position along the bacteriorhodopsin proton channel (Fig. 5a, right panel) reveal a characteristic variation in the polarity of the nitroxide microenvironment along the channel. Residue S162R1 (R1 denotes the spin labeled side chain) is located in the E–F loop at the cytoplasmic surface, whereas residue K129 is positioned in the D–E loop on the extracellular surface. The high polarity detected is in agreement with the structure. The polarity of positions 100, 167, and 171 is significantly less and reaches its minimum at position 46 in the vicinity of the retinal. The plot directly reflects the hydrophobic barrier which the proton has to overcome on its way through the proton channel. The 2D plot g_{xx} vs A_{zz} is also used to discriminate sites with different polarity and proticity [50, 51].

4.3 Pulsed EPR Determination of Water Accessibility in LCHIIb

An alternative approach to measure at cryogenic temperatures the proximity of (deuterated) water molecules to spin labeled side chains consists in extracting the modulation depth of the deuterium ESEEM (Electron Spin Echo Envelope Modulation) oscillations for spin labeled membrane proteins in (partially) deuterated solvents. This technique is highly sensitive and requires only 10–20 nmol of singly spin labeled protein per sample and relies on the reproducible and precise quantification of the nitroxide hyperfine couplings to deuterium nuclei in deuterated water molecules. The technique provides estimates of the distance and number of nuclear spins in the proximity of an electron spin in the 3–6 Å range [52]. This new method was first applied in 2009 to the detergent-solubilized light harvesting complex (LHCIIb) of photosystem II of green plants [47]. The light-harvesting (or antenna) complex of plants is an array of protein and chlorophyll molecules embedded in the thylakoid membrane which transfer energy to one chlorophyll molecule at the reaction center of a photosystem. LHCIIb is the main antenna of photosystem II and consists of a transmembrane protein and several cofactors, such as chlorophyll *a* and *b*, carotenoids, and lipids that are noncovalently bound to it. Crystal structures of the complex, except for the first few residues of the N-terminal domain of the protein, are available [53, 54]. Hence, spin labeled sites accessible towards the water (e.g., S52 at the interface with the stromal side of the thylakoid membrane) or

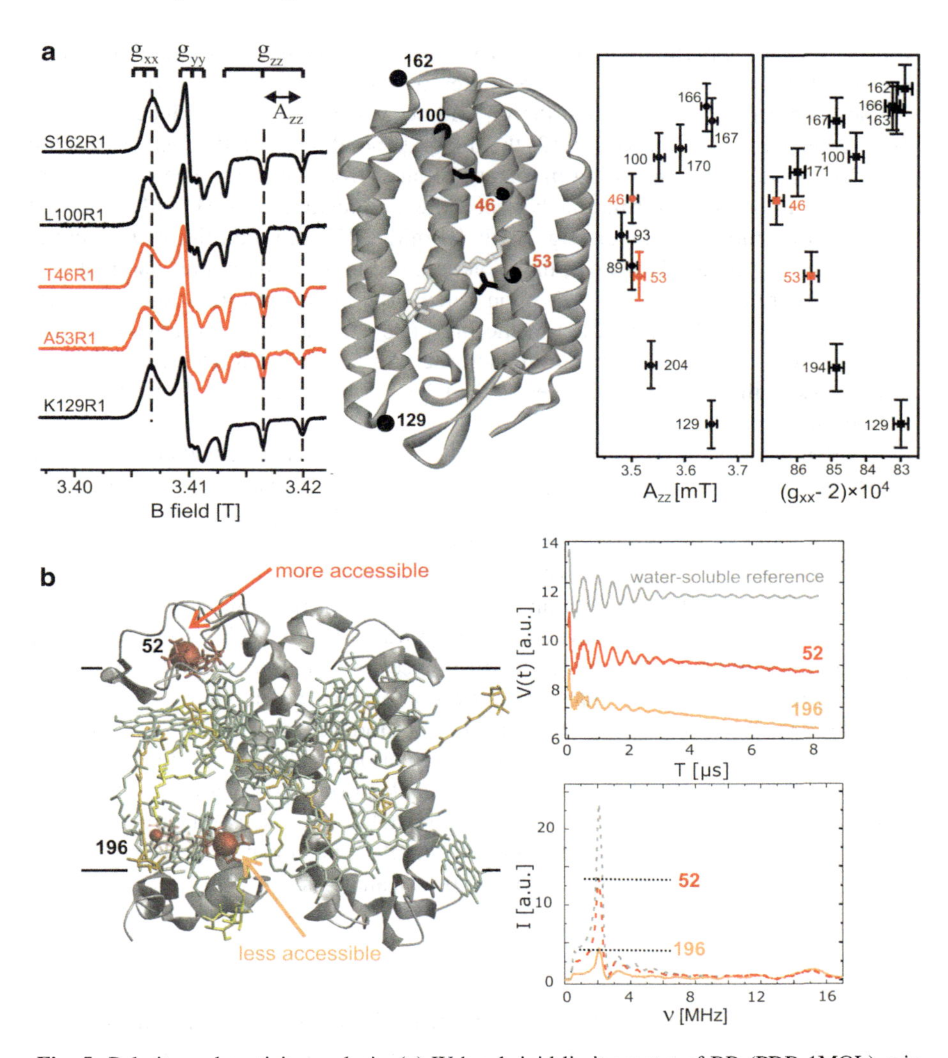

Fig. 5 Polarity and proticity analysis. (**a**) W-band rigid limit spectra of BR (PDB 1MOL) spin labeled in the proton channel. The changes in g_{xx} and A_{zz} are highlighted by *vertical lines*. *Right panel*, plots of g_{xx} and A_{zz} vs nitroxide position showing the hydrophobic barrier in the channel. Adapted from [45]. (**b**) Structure of LHCIIb (PDB 2BHW) with two IAP spin labels attached with MMM at positions 52 and 196. *Right panel*, ESEEM traces detected on a water soluble standard (*gray*) and on the two spin labeled positions. *Bottom*, ESEEM modulation depths correlated with the water accessibility. Adapted from [47]

less accessible (e.g., V196 located towards the TMDs interior) can be predicted (Fig. 5b, PDB 2BHW). LHCIIb used in this study was spin labeled with IAP (iodoacetamido proxyl spin label). It is worth pointing out that the preparation of the antenna complex requires the mixing of the purified apo-protein solubilized in detergent with a solution of pigments and lipids in slightly reducing conditions (necessary for the pigments' stability). IAP was chosen as best labeling probe based on the following criteria: (1) first attempts to use iodoacetamido TEMPO (6-membered ring) revealed that the piperidine- was more easily reduced than the analogous pyrrolidine-nitroxide; (2) the disulfide bond between the cys residue and MTSSL was unstable, resulting in release of label from the sites. The reversibility of the disulfide bond formation between the cys residue and MTSSL makes it possible to remove the label (e.g., with 10 mM DTT, dithiothreithol) to recover the single cys mutant. However, the weakness of the disulfide bonds also makes it possible to release the label (1) upon relevant protein conformational changes involving the spin label site, (2) in the presence of a neighboring unlabeled free cysteine, and (3) upon temperature increase. Although the MTSSL is the most widely used, the choice of the best spin label probe is necessarily protein- and conditions-dependent.

Experimentally, the three-pulse ESEEM experiment is performed at 50 K with a $\pi/2$-τ-$\pi/2$-T-$\pi/2$-τ-echo pulse sequence with standard phase cycling. The pulse length of the $\pi/2$ pulse is usually 16 ns and the constant interpulse delay $\tau = 344$ ns, corresponding to a proton blind spot. Suppression of the proton modulations at the blind spot improves precision of the fitting of the deuterium modulation. The second interpulse delay, T, with an initial value of 80 ns, is incremented in steps of 8 ns. The three-pulse ESEEM time domain traces for two selected positions are presented in Fig. 5b (right panel). The differences in water accessibility are clearly seen in the normalized primary data by comparing the depth of the ESEEM modulations, defined as the peak-to-peak amplitude between the first maximum and the first minimum of the deuterium modulation (higher for 52 than for 196). Different methods to extract quantitatively the deuterium modulation depth taking into account its dependence on the interpulse delay are discussed in the original work [47]. The ESEEM water accessibility method is shown to be robust and to complement the more "traditional" accessibility analysis.

5 Interspin Distance Determination: Modeling Structures and Conformational Changes of Membrane Proteins

The interaction between two spin labels attached to either a single polypeptide or adjacent polypeptides (e.g., protein complexes, oligomers, etc.) is composed of static dipolar interaction, which can be modulated by the residual motion of the spin label side chains, and exchange interaction [55]. By measuring the spin–spin interaction at cryogenic temperature in the so-called "rigid limit" case (<200 K), the residual motion of the label is minimized and the exchange and dipolar interactions can be precisely extracted. Interspin distances in the 1–5 nm range are routinely measured by EPR on membrane proteins with an accuracy varying with the mean distance and distance distribution to be measured and with the chosen experimental conditions. The short distances (1–2 nm) are extracted from continuous wave EPR spectra in the rigid limit via lineshape analysis [56, 57]. The long distances (1.7–6 nm) can be measured by pulse EPR techniques, namely Double Electron Electron Resonance (DEER, also known as PELDOR, pulsed electron-electron double resonance) [58, 59] or Double-Quantum EPR [60]. The interspin distance determination is a powerful strategy to deduce proximity of selected secondary structural elements in membrane proteins or protein complexes of unknown structure, and to follow conformational changes correlated with the physiological activity of membrane proteins.

Since the linewidth of the spectra is a steep function of the interspin distance, empirical or semiempirical parameters such as spectral amplitude ratios or spectral second moment values were used to extract distances semiquantitatively and to answer structural questions in the past. In the following, the software available to extract distances and the caveats hidden in the analysis are presented.

The dipolar broadening produced by interspin distances in the 1.5–2 nm range is extracted from low temperature CW spectra by convolution or deconvolution methods given that the spectra of the two noninteracting spin labels are available. In the case of homodimeric proteins, the noninteracting spectrum can be approximated by spin dilution methods (e.g., the protein is spin labeled with 20% MTSSL and 80% nondiamagnetic analog, leading to approximately 4% of double labeled homodimers in the final sample). In the case of higher order homo-oligomers, molecular engineering can help to solve the problems of overlapping distances in CW lineshape analysis (discussed in the next section).

The most popular pieces of software available to extract short distances from CW EPR spectra are DIPFIT (from Steinhoff's group) [57], ShortDistances (from Altenbach, https://sites.google.com/site/altenbach/labview-programs/short-distances) or user-based fitting routines with Easyspin [48].

In DIPFIT one first needs to simulate the rigid limit spectrum of the noninteracting species (to extract the linewidth parameters – pseudo Voigtian lineshape – which will be kept constant in the fitting of the dipolar spectrum). The software performs a convolution of the simulated noninteracting spectrum with a Pake function (in the strong coupling regime) characterized by a distance r and Gaussian distance distribution with a standard deviation (σ_r) to fit the dipolar-broadened spectrum. The distance distribution obtained can be considered reliable if the mean distance is in the 1.4–2.0 nm range. For $r > 2$ nm, the accuracy of the convolution methods progressively decreases, as the dipolar splitting gets closer to the natural EPR linewidth, and for $r < 1.4$ nm the presence of exchange coupling influences the accuracy of the distances extracted solely based on the dipolar terms [61].

A problem arises if the spin labeling efficiency is not 100%, a case generally encountered. DIPFIT allows the fraction of the noninteracting spectrum to be fitted. However, the three main parameters to be varied, namely the distance, its distribution, and the fraction of noninteracting species, are strictly correlated. The knowledge of the spin labeling efficiency is mandatory for an accurate determination for the distance (and it should be >60% for reliable distance extraction).

The software ShortDistances is also solely based on dipolar terms, requires the noninteracting spectrum as template, and performs the convolution based on a Pake function (strong coupling regime) which allows for different distance distribution functions to be used. Due to the large amount of parameters to be fitted, a simple Gaussian distribution is to be preferred, because more complicated functions not supported by modeling or structural information may lead to results difficult to interpret. The possibility to extract distances from room temperature spectra, which is in principle allowed in ShortDistances cannot be generalized to arbitrary pairs of interacting spins, and each case must be evaluated thoroughly before convolution with a Pake function to simulate interspin distances which are modulated by the dynamics of the side chains [62]. Decreasing the temperature to 160 K is the suggested general procedure to solve at least the problem of the residual spin label mobility.

In general, convolution of the noninteracting spectrum provides only an approximation to the lineshape of the dipolar-broadened spectrum even in the rigid limit. First, this corresponds to treating the dipole–dipole interaction as a small perturbation, neglecting the pseudosecular term. This approximation is valid only if the dipole–dipole interaction is much smaller than the typical frequency difference between the two coupled spins. Second, orientation correlation between the spin–spin vector and the molecular frame of the nitroxide is neglected. Both these approximations can be relaxed by simulating the dipole–dipole coupled spectrum with Easyspin [48]. However, such an approach is computationally much more expensive than Pake convolution. Furthermore, the CW EPR lineshape at a single frequency may not be sufficient to extract both the interspin distance and the relative orientation of the labels. In any case these complications introduce errors in distances extracted by Pake convolution methods, which have not yet been systematically studied by theoretical approaches.

Many of the problems encountered by lineshape analysis are not present when using DEER [58] for the longer distance range determination. In fact there is no need to have the spectrum of each noninteracting spin system to obtain a reliable interspin distance determination (one needs only one or two singly spin labeled proteins to confirm the background decay function to be used throughout the study on the same protein); the amount of noninteracting species in the sample does not contribute to the form factor $F(t)$, and thus not to the accuracy of the distance extracted (only the modulation depth is affected); a large range of distances can be determined with high precision including the 4–5 nm width of the membrane bilayer "barrier." The first DEER application to membrane proteins appeared in 2004 on the membrane protein Na^+/proline transporter PtuP showing that it was possible to measure the distance between two labels positioned across the membrane bilayer (about 5 nm) [63]. Jeschke suggested DEER as a powerful method to investigate structure and conformational changes of integral membrane proteins reconstituted in liposomes. In fact, since then, a tremendous number of publications using DEER as main driving technique for structural investigation on membrane proteins followed. The most common software used to extract distance distribution from DEER traces is DeerAnalysis [64] (http://www.epr.ethz.ch/software/index). Other software is available from Freed's (http://www.acert.cornell.edu/index_files/acert_ftp_links. php) and Fajer's (http://www.sb.fsu.edu/~fajer/Programs/DEFit/defit.html) labs.

One of the major problems facing distance determination by pulsed EPR on spin labeled membrane proteins is the short relaxation time T_m (generally around 1 μs). Solvent deuteration is routinely used (10–20 vol% deuterated glycerol as cryoprotectant for membrane-embedded or detergent-solubilized samples, or deuterated buffer) to slow down the nitroxide relaxation, thus extending the range of distance measurement and sensitivity. However, the problems connected to the too fast relaxation time called for a systematic analysis of all factors modifying the T_m [65]. Recently, deuteration of water-soluble histone proteins was shown to extend T_m to a considerable degree (up to 30 μs) [66].

Other sensitivity problems arise from the low concentrations usually available for membrane proteins, and the molecular crowding in the membrane bilayer leading to very fast background decay functions which impair the DEER analysis, especially for long distances. This problem can be minimized by increasing the lipid to protein ratio (at least 500–1,000 lipid per protein). All these factors call for the development of even more sensitive techniques, e.g., Q-band DEER [67, 68] at high microwave power, and new strategies, e.g., spin labels containing metal ions which can also be used in combination with nitroxide labels [69–71]. The DEER experiment is rather expensive in terms of measurement time, typically one trace with good signal to noise on membrane proteins requires 12–24 h, and thus even a small sensitivity increase (e.g., a factor 2 in S/N) can speed up a protein study four times.

The presence of multiple spins in the same protein or protein complex creates interpretative problems in DEER, due to the appearance of artifact peaks in the distance distribution. Methods to identify artifact distances in multispin systems are under development [72]. Strategies to overcome problems due to the superposition of distances in the same polypeptide can make use of orthogonal labeling (e.g., using different spin labels binding to cys and unnatural amino acids [73]). In the easier case of protein complexes with spin labels in different subunits, one can use different nitroxide isotopes, as demonstrated on model compounds [74].

Having an accurate distance determination between the NO groups of the spin labeled side chains still does not directly convey the structural information at the level of the backbone of the protein which would be required for modeling structures and complexes at high resolution. To correlate the spin–spin distance constraints to the backbone–backbone distances requires modeling. At the present time, modeling approaches combining sparse accurate distance constraints to macromolecular structures are under development [7, 75].

5.1 *CW-Detected Distances Unravel the Structure of KcsA Intracellular Gate in the Open State*

Ion channels gate by transducing a variety of physical stimuli into a series of structural rearrangements controlling the access of permeant ions to a centrally-located water pore. SDSL EPR makes it possible to address the structural details of the gating mechanism by accurate determination of distance constraints during ion translocation. The chosen example of CW lineshape analysis deals with the treatment of interspin distance constraints in a "problematic" homotetrameric potassium channel KcsA (already introduced in Sect. 3.1) [76]. Spin labeling one single cys mutant of KcsA will lead to interspin distances between neighboring and diagonally related nitroxide side chains (Fig. 6a), which can be difficult to disentangle in the CW spectra. In the case of KcsA, the clear appearance of very short distances in the room temperature CW spectra for some positions in the intracellular C-terminal region of helix TM2 (e.g., position 108, Fig. 2d) was used to model the molecular architecture of the channel. The decrease of broadening under conditions stabilizing the open state also suggested the general opening mechanism of the gate. However, a quantitative description of the gate helix movement based on a nitroxide scan and interspin distance constraints faced the problem that both neighboring and diagonally placed side chains affected simultaneously the EPR lineshape, making the assignment of the distances to the correlated pairs almost impossible. To avoid this problem, molecular biology strategies were employed. Tandem dimer constructs were engineered, linking a cys-free monomer with a cys-containing one, in order to create a "pseudo-dimeric" channel, without dipolar interaction between neighboring subunits (Fig. 6a). In this way the distances extracted quantitatively from CW EPR spectra at low temperature (to avoid residual motions which modulate the dipolar interaction) could be precisely assigned to the diagonally located side chains. The structure of the open gate was inferred from ten distance constraints in the intracellular region (100–119 segment, Fig. 6b). Comparison of low temperature spectra can be done either normalizing the spectra to the maximum of their amplitudes, or spin-normalizing them (Fig. 6c). The most relevant effect can be seen in the spin-normalized comparison, with short distances inducing a relevant amplitude decrease.

Figure 6b shows the spin-normalized rigid limit spectra detected in the closed, nonconducting channel conformation (thick line, pH 7.0) and under conditions favoring channel opening (thin line, pH 4.0). Site-specific amplitude changes are clearly visible, correlated to interspin distance increase or decrease going from the closed to the open state. The right panel of Fig. 6b shows the first integral (absorption) of the spectra, with the fit obtained with convolution methods (Gaussian distribution of distances), starting from the spectra of the noninteracting spins (obtained by spin dilution) according to [56]. The pattern of distances was analyzed in terms of possible conformational changes starting from the available structure of KcsA in the closed state to obtain the conformation of the TM2 helix in the open channel. The pattern of distance changes is illustrated by helical wheel models

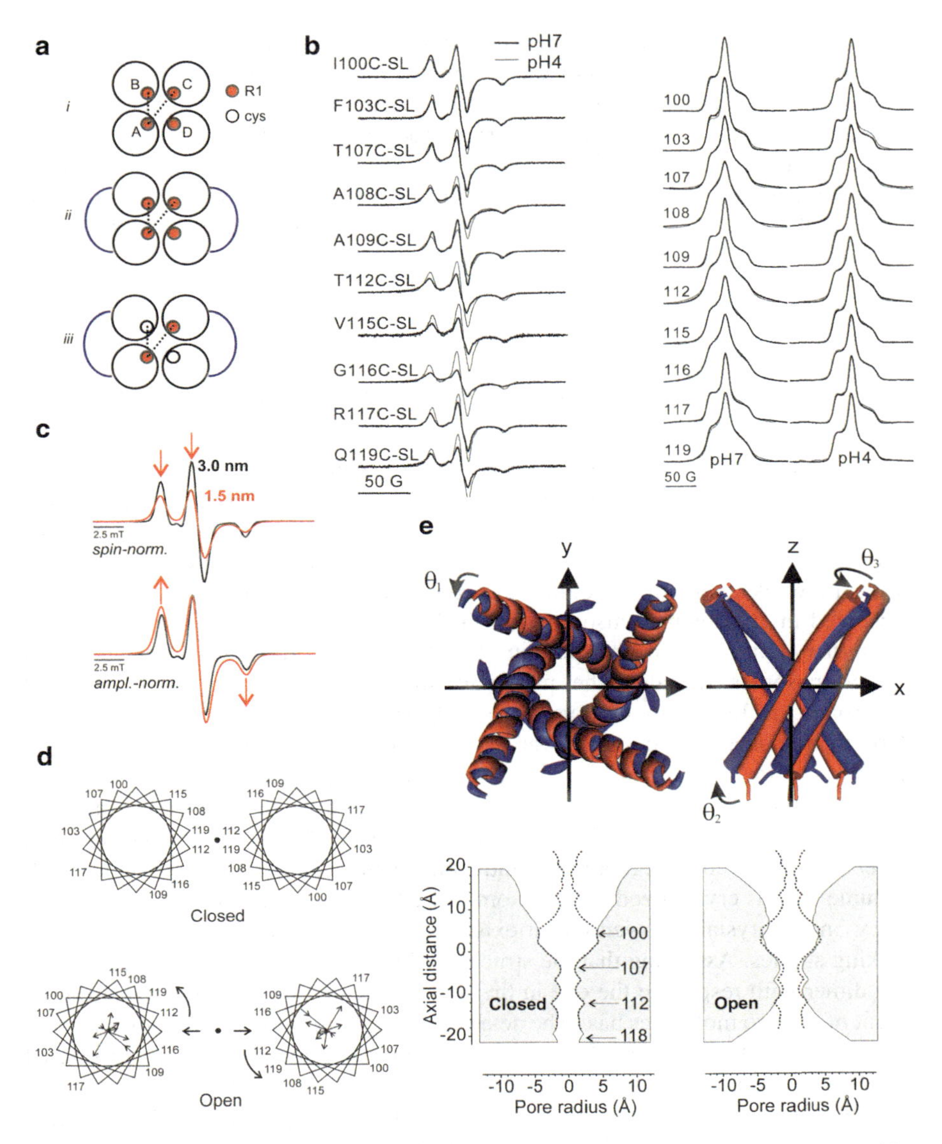

Fig. 6 Structural details obtained by lineshape analysis. (**a**) In KcsA, labeling at any residue position renders a tetramer with potentially four spin labels. (i) Tandem dimer construct (ii) with cys residues in both protomers (control used to evaluate the effects of the intersubunit linker) and (iii) with only one of the protomers containing a cys (used in the analysis). (**b**) Rigid-limit X-band EPR spectra obtained at pH 7 (*thick line*, closed state) and at pH 4 (*thin line*, open state). *Right panel*, absorption spectra obtained from integration and relative fits obtained with convolution superimposed. (**c**) Simulated spin- and amplitude-normalized spectra for the two interspin distances in the figure (100% spin labeling efficiency). (**d**) Helical wheel representation of residues 100–119. Both closed (*top*) and open (*bottom*) states are represented as pairs of helical wheel

(Fig. 6d), and gives the first insights into a possible rotation and translation movement of TM2 away from the symmetry axis. Perozo developed a simple computational approach (restraint-driven Cartesian transformation, ReDCaT) based on the exhaustive sampling of rigid-body movement in Cartesian space to determine the type, direction, and magnitude of the conformational changes in TM2 using limited distance information. The EPR-driven model of the conformational change of the TM2 helices from the closed to the open channel is presented in Fig. 6e, and supports a scissoring-type motion.

5.2 High Resolution Structure of an Na^+/H^+ Antiporter Dimer

Many processes in living cells involve the transient or permanent formation of complexes between membrane proteins. However, X-ray crystallography often reveals only the structure of a monomer which under physiological conditions forms dimers or oligomers. This is due to the fact that packing interactions compete with the weak interactions holding the complex together. Vice versa, complexes observed in crystals may also just reflect the packing conditions and not the real quaternary structure of the protein. SDSL EPR can be easily performed in proteoliposomes, revealing the quaternary structure of a protein in physiological environments. In the case of dimers, the analysis of the distances between the singly labeled monomers is straightforward. SDSL EPR in combination with modeling tools was shown to unveil the structure of the dimeric complex of the Na^+/H^+ antiporter NhaA using the available X-ray structure of the monomer and nine distance constraints obtained by DEER [77]. NhaA, the Na^+/H^+ antiporter from *E. coli*, is involved in regulation of intracellular pH, cellular Na^+ content and cell volume. It was crystallized as a monomer (PDB 1ZCD), but cryoelectron microscopy on 2D crystals showed that it exists as a dimer, as also supported by cross-linking studies. Assuming that the structure of the two monomers is unchanged in the dimer with respect to the one in the X-ray structure, only the relative arrangement of the two molecules has to be determined to obtain a high resolution structure of the dimer. This corresponds to solving a problem of docking two proteins based on distance constraints. Just by detecting interspin distances between singly spin labeled sites in NhaA, the dimeric nature of the antiporter was immediately confirmed (Fig. 7a). A combined DEER-X-ray approach was applied to solve the docking of the proteins in the dimer. A similar approach is used to obtain the chemotaxis receptor-kinase assembly by Freed [78, 79]. To obtain the NhaA dimer,

---

Fig. 6 (continued) diagrams. The *arrows* inside the wheels in the open state represent distance changes from close to open (outward is increase and inward, decrease). (**e**) Modeling the conformational rearrangements in TM2. *Up*, graphical representation showing the type and extent of individual helical movements (*red* = closed and *blue* = open), *bottom*, calibrated cross-sectional representation of the conformational changes. Adapted from [76]

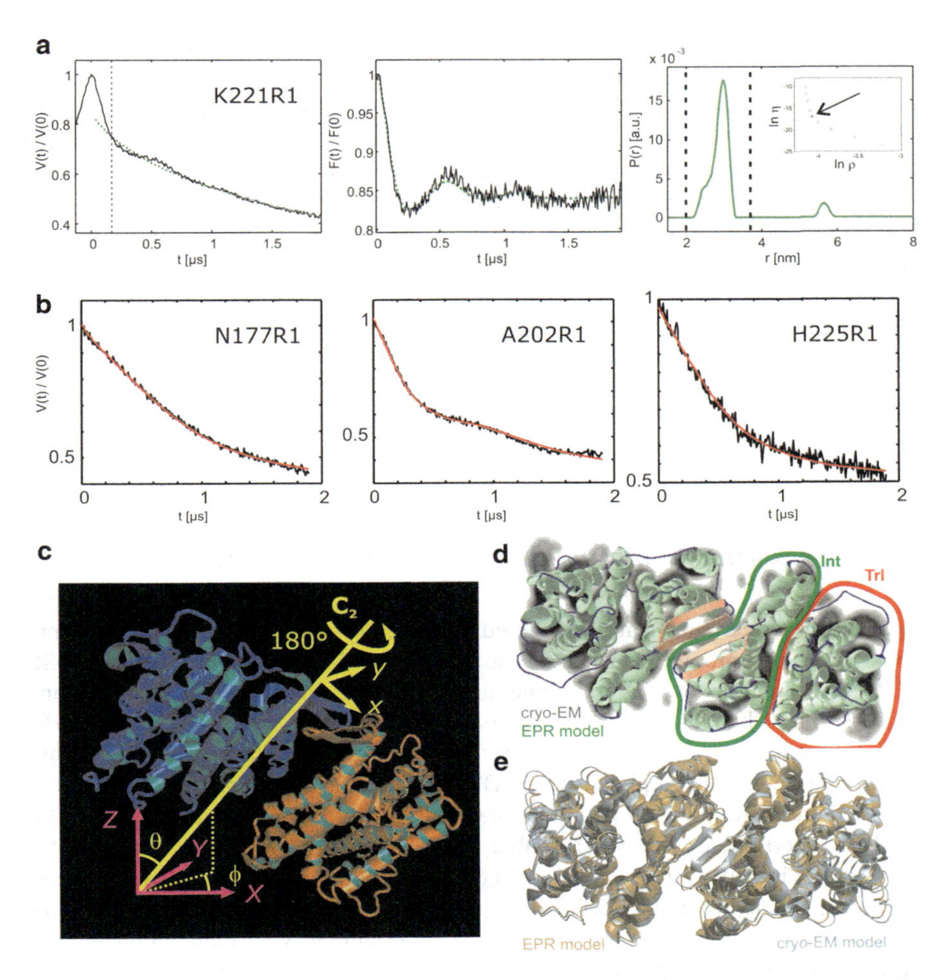

Fig. 7 Modeling a high-resolution dimeric structure. (**a**) Estimate of the mean distance on the example of spin label K-221R1. The primary DEER trace ($V(t)/V(0)$), the form factor and the distance distribution obtained by Tikhonov regularization (the L curve is shown in the *inset*) with the software DeerAnalysis are presented. (**b**) Fits of primary experimental DEER data (*black lines*) by simulated data (*red lines*) corresponding to the final structure of the NhaA dimer and a distribution of spin label conformations modeled by a rotamer library. (**c**) C_2 symmetry axis of the dimer created with the EPR constraints. (**d**) Comparison between the EPR structure and the electron density projection to the membrane plane obtained by cryo-EM on 2D crystals. (**e**) Comparison between the EPR structure and the dimer modeled on the refined cryo-EM data (PDB 3FI1). Adapted from [77]

the primary data $V(t)$ are fitted directly (Fig. 7b), and the information on the distance distribution are used in the docking approach to match a modeled conformational distribution of the spin labels.

In the first stage of the modeling the two component molecules are treated as rigid bodies. The relative arrangement of a dimer requires in principle six parameters (three Euler angles for the rotation and three components of a translation vector). In this case, the C_2 symmetry of the resulting homodimer reduces the necessary parameters to four (Fig. 7c). The nine distance constraints experimentally obtained can then be considered enough to produce a reliable docking of the two proteins. Details of the grid search algorithm and the final energy minimization which yield the dimeric arrangement of NhaA are thoroughly described in the original work [77]. The dimer obtained by DEER distance constraints in proteoliposomes is presented in Fig. 7d, superimposed to the low resolution cryo-EM density showing the rather good agreement, in particular in the interfaces. More recently, a higher resolution dimeric model based on the monomeric structure was produced (PDB 3FI1) using a better resolved 2D cryo-EM crystal. The new model is shown to be in good agreement with the EPR-model (Fig. 7e), confirming the high potentiality of EPR-driven modeling for membrane protein complexes in quasi-native environments.

5.3 Following Membrane Insertion of Soluble Proteins

Water soluble proteins can be triggered by a variety of factors to insert spontaneously in the membrane bilayer (e.g., toxins, apoptotic proteins, etc.). With SDSL EPR one can observe in real time the membrane insertion via mobility and accessibility changes. Moreover, interspin distances can be measured to model the conformation of the active membrane-bound state of the protein. The case presented here is the application of SDSL EPR to the proapoptotic Bcl-2 protein Bax, triggered to insert into the membrane bilayer by the proapoptotic protein Bid [80]. Bax and Bid are two water soluble ~20-kDa proteins. NMR structures exist for both in the inactive water soluble form (Fig. 8a, one NMR model for each protein) showing the hydrophobic helical hairpin protected in the core of the proteins colored in orange in Fig. 8a. Despite the high similarity in their globular fold, their function is complementary: Bid needs to be cleaved by caspases to be activated and the resulting C-terminal 15 kDa fragment (p15 or t-Bid) in turns activates Bax. The interaction is thought to involve the BH3 domain of Bid (colored in yellow in Fig. 8a) which can dislodge the BH3 domain of Bax (colored in yellow) [81]. The conformational change in Bax leads to insertion of the hydrophobic hairpin (orange in Fig. 8a) in the membrane. Once inserted, Bax forms oligomers which permeabilize the outer mitochondrial membrane. Through the membrane pores, the formed cytochrome c is released, triggering cell death (apoptosis). The graph in Fig. 8b shows how it is possible to follow the conformational change of Bax via mobility changes of the spin labeled side chains over time. The very mobile spectrum characteristic of the inactive form in water slowly changes towards a more immobile spectrum when Bax becomes spontaneously membrane-embedded (black trace) or changes immediately (within the 2 min needed to insert the tube in the spectrometer) after addition of detergents (red trace). Addition of increasing amounts of cleaved active Bid (p15/7-Bid) is shown to accelerate Bax

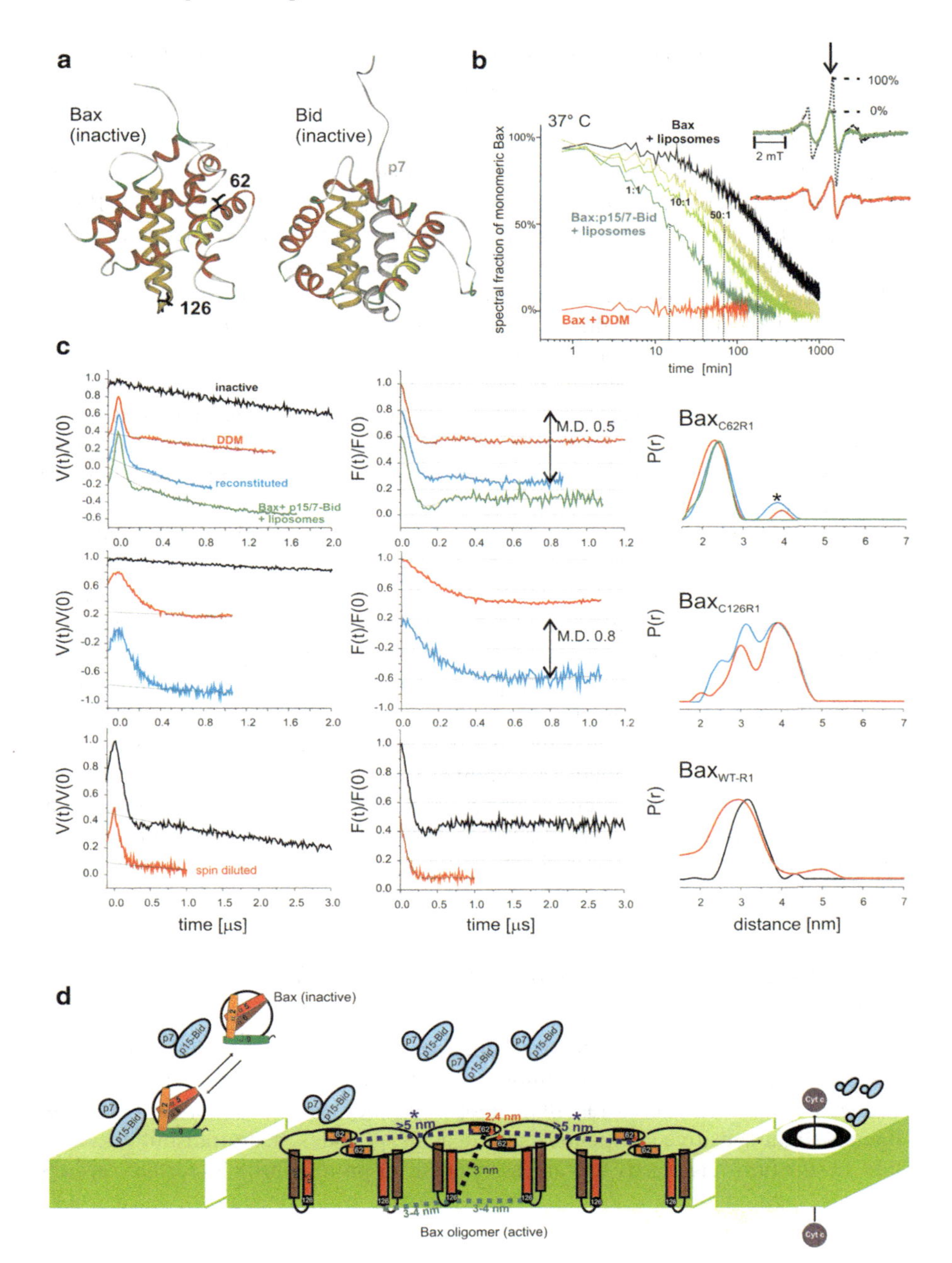

Fig. 8 EPR detection of membrane insertion of spin labeled proteins. (**a**) NMR models of inactive Bax and Bid (PDB 1F16, 2BID, respectively). The buried helical hairpin is shown in *orange*, and the BH3 domain in *yellow*. Positions spin labeled in Bax are *highlighted*. (**b**) Kinetics of Bax conformational changes at 37 °C. Spectra of double labeled Bax were recorded each 43 s at 37 °C

conformational changes. Overall, the fraction of Bax molecules inserted is independent of the stochiometric ratio between Bid and Bax, indicating the catalytic role of Bid in Bax insertion. Singly labeled Bax in the inactive monomeric form does not show dipolar oscillations in the DEER traces, as expected (Fig. 8c, black trace in the left panel). Dipolar oscillations appear as soon as detergent is added (red), the protein is reconstituted in liposomes (cyan), or it is triggered by Bid to insert into liposomes (green). Large Bax oligomers are formed, as judged by size exclusion chromatography and cryo-EM performed on the spin labeled samples. The narrow 2.4 nm distance between the two BH3 domains of Bax (position 62) reveals the formation of the dimer, which is the building-block of the pore-forming oligomers. One expects to detect both the intradimer distances and the interdimer distances in an oligomer. However, due to the fast relaxation of the nitroxides in membrane-embedded proteins, only relatively short DEER traces could be detected (max 1.8 μs) yielding for position 62 a clear intradimer distance (2.4 nm) and only a glimpse of the interdimer distances (asterisk in Fig. 8c, right panel), which are suggested to be >5 nm.

To prolong the nitroxide relaxation time, 10% deuterated glycerol was used in the sample, but this was not sufficient at the small protein concentration available to detect traces longer than ~2 μs, as would have been necessary for the determination of distances around 5 nm. Interestingly, deuterated water was found to decrease Bax stability and thus it could not be used. For position 126, the large modulation depth of the $F(t)$ (~0.8) clearly reveals the presence of the multispin system due to the oligomer formation [82]. The distances obtained cannot however be assigned to an intra- or interdimer interaction, which are both suggested to be in the 3–4 nm range. Analysis of the doubly labeled Bax required spin dilution (Bax spin labeled with MTSSL and with diamagnetic analog in a 1:3 molar ratio) to obtain the intramonomer distance in the oligomer. The interspin distances were used to model an active "quasi" oligomeric state of Bax in the membrane (Fig. 8d).

5.4 Distance Mapping Reveals Light Activation of Rhodopsin

GPCRs convey 80% of signal transduction across cell membranes [83]. The receptors consist of seven transmembrane helices (analogous to bacteriorhodopsin) connected by intra- and extracellular loops and they can be activated by diverse ligands (light-sensitive compounds, amino acids, peptides, neurotransmitters, etc.). GPCR in turn can activate an associated G-protein which further affects

Fig. 8 (continued) with different triggering agents. The intensity of the central EPR line vs incubation time (logarithmic scale) is plotted. The *inset* shows the spectra at time zero (*black dotted*) and at the end of the incubation time. (**c**) DEER traces and distance distribution obtained with DeerAnalysis in different conditions, as written in the figure. The *asterisk* highlights the "glimpse" of long distances reflecting the interdimer interaction. (**d**) Model of three adjacent Bax dimers, based on the interspin distances detected. Adapted from [80]

intracellular signaling proteins or target functional proteins. GPCRs are highly druggable, in the sense that one third of all drugs are targeted to them; thus the knowledge at the molecular level of their activation mechanism is highly valuable. The family of rhodopsins in GPCRs is the largest family, and rhodopsin was the first GPCR crystallized in 2000 [84]. To date, a few other GPCRs have been crystallized, both in the inactive and active forms, paving the way to the understanding of the ligand-induced mechanism of activation. The first evidence for rhodopsin activation came from SDSL EPR studies in Hubbell's lab. Transient EPR recordings of mobility changes showed real-time structural movements of the helix TM6 upon light excitation, analogously to what was observed in bacteriorhodopsin [85]. In 2008 DEER was used for a high-resolution distance mapping of the conformational changes induced by light monitored to unveil the activation mechanism [86]. In Fig. 9a the ribbon representation of rhodopsin is shown looking down at the cytoplasmic surface which recognizes the G protein transducin. The 9 spin labeled R1 side chains used in the study are shown with the 16 pairs of interspin distances detected via DEER. Figure 9b shows the normalized form factors $F(t)$ in the inactive (black) and active (red) states, with the dipolar spectra and the distance distributions obtained.

The capability of DEER to measure long distances made it possible to label selectively surface-exposed sites in rhodopsin, yielding distances in the 2–5 nm range. In this way the perturbation of the protein fold was minimized. Moreover the relatively mobile sites showed only minor dynamic changes in the active compared to the inactive state, suggesting that changes in the spin label conformation that could affect the measured distance change did not occur. Thus, changes in interspin distances between pairs of these nitroxides reliably report relative movements of the helices to which they are attached. This advantage in terms of reliability of the distance changes surely compensates for the broader distance distributions which may appear between the solvent-exposed sites. To find the relative positions of the nitroxides in the inactive and active states relative to the structure of inactive rhodopsin (PDB 1GZM), all sets of interspin distances were taken into account with a global geometry optimization. The calculated relative positions of the spin labeled side chains are shown in Fig. 9c as blue (inactive) and red (active) clouds. The activation of rhodopsin was found to be mainly accompanied by a radial outward movement of helix TM6 by 5 Å. The EPR results showed that the outward displacement of TM6 is a hallmark of rhodopsin activation, a finding which can probably be generalized to all GPCRs.

6 Conclusions

SDSL EPR is sensitive to flexible regions of proteins and to dynamical changes, and can be used to measure water accessibility profiles and accurate distances between spin labeled side chains. The wealth of information that can thus be obtained makes SDSL EPR a direct tool to access conformational changes of proteins. The bridge

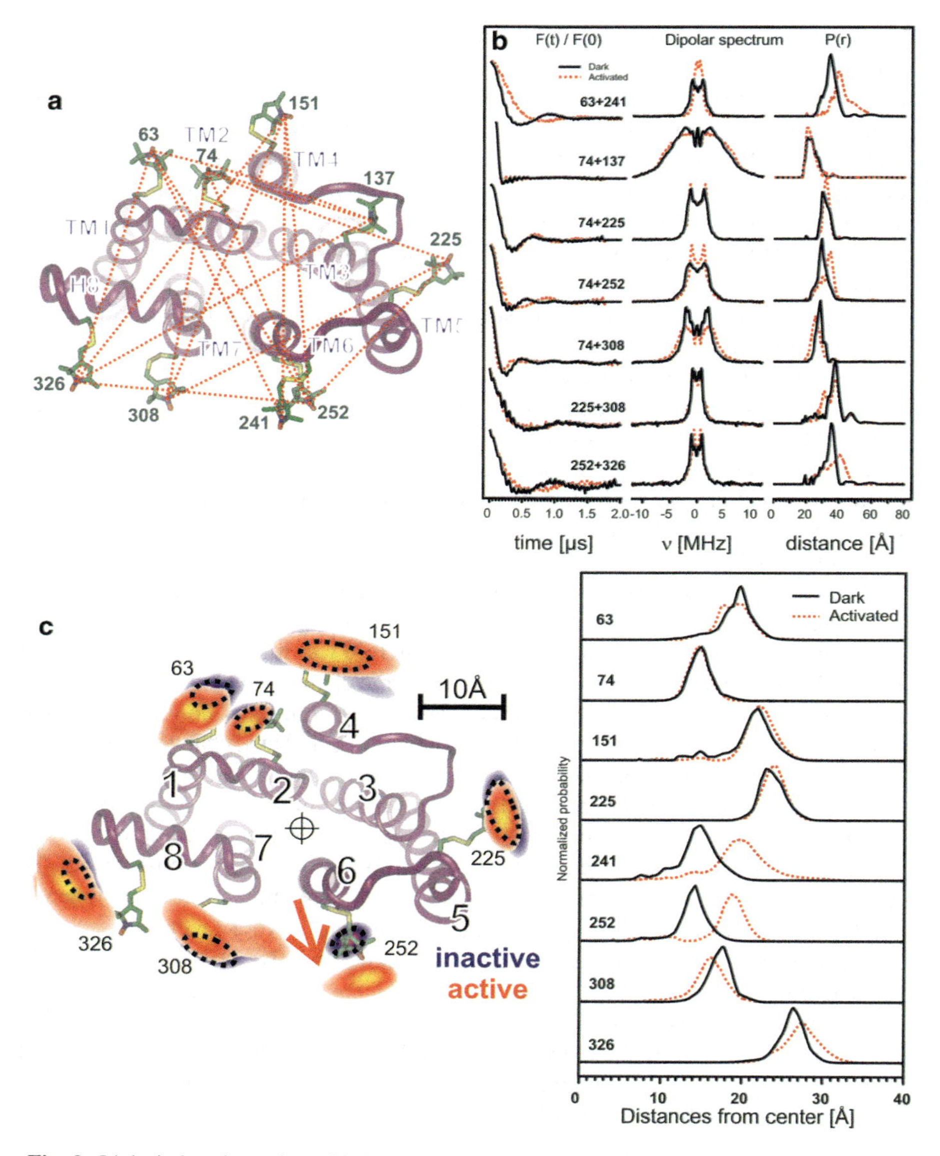

Fig. 9 Light-induced opening of helix TM6 in rhodopsin. (**a**) View of the cytoplasmic face of inactive rhodopsin (PDB entry 1GZM) showing modeled R1 side chains for all sites investigated. (**b**) Representative selection of DEER results in the inactive (*black*) and active (*red*) states. *Left*, normalized dipolar evolution after removal of the exponential background. *Middle*, DC centered Fourier transform of the data. *Right*, distribution calculated by Tikhonov regularization. (**c**) Projection contours of the spin locations calculated from the measured distance distributions for representative sites (*blue*, inactive; *red*, active). The outward movement of TM6 is highlighted by an *arrow*. Right, after defining a central reference point halfway between the dark location of 74 and 252, the radial distribution profiles along a line connecting the center and the most probable location is calculated for each site. Adapted from [86]

between EPR and a "de novo" structural determination of proteins is the modeling of the relative positions of the backbone atoms of a protein based on the spectroscopic information, which is still the limiting factor with respect to the more developed NMR technique.

References

1. Deisenhofer J, Epp O, Miki K, Huber R, Michel H (1985) Structure of the protein subunits in the photosynthetic reaction centre of Rhodopseudomonas viridis at 3Å resolution. Nature 318 (6047):618–624
2. Altenbach C, Flitsch SL, Khorana HG, Hubbell WL (1989) Structural studies on transmembrane proteins. 2. Spin labeling of bacteriorhodopsin mutants at unique cysteines. Biochemistry 28(19):7806–7812
3. Altenbach C, Marti T, Khorana HG, Hubbell WL (1990) Transmembrane protein-structure – spin labeling of bacteriorhodopsin mutants. Science 248(4959):1088–1092
4. Henderson R, Baldwin JM, Ceska TA, Zemlin F, Beckmann E, Downing KH (1990) Model for the structure of bacteriorhodopsin based on high-resolution electron cryo-microscopy. J Mol Biol 213(4):899–929
5. PebayPeyroula E, Rummel G, Rosenbusch JP, Landau EM (1997) X-ray structure of bacteriorhodopsin at 2.5 angstroms from microcrystals grown in lipidic cubic phases. Science 277 (5332):1676–1681
6. Hubbell WL, Altenbach C (1994) Investigation of structure and dynamics in membrane-proteins using site-directed spin-labeling. Curr Opin Struct Biol 4(4):566–573
7. Polyhach Y, Bordignon E, Jeschke G (2011) Rotamer libraries of spin labelled cysteines for protein studies. Phys Chem Chem Phys 13(6):2356–2366
8. Denisov IG, Grinkova YV, Lazarides AA, Sligar SG (2004) Directed self-assembly of monodisperse phospholipid bilayer nanodiscs with controlled size. J Am Chem Soc 126 (11):3477–3487
9. Skar-Gislinge N, Simonsen JB, Mortensen K, Feidenhans'l R, Sligar SG, Lindberg Møller B, Bjørnholm T, Arleth L (2010) Elliptical structure of phospholipid bilayer nanodiscs encapsulated by scaffold proteins: casting the roles of the lipids and the protein. J Am Chem Soc 132(39):13713–13722
10. Grinkova YV, Denisov IG, Sligar SG (2010) Engineering extended membrane scaffold proteins for self-assembly of soluble nanoscale lipid bilayers. Protein Eng Des Select 23 (11):843–848
11. Steinhoff HJ, Hubbell WL (1996) Calculation of electron paramagnetic resonance spectra from Brownian dynamics trajectories: application to nitroxide side chains in proteins. Biophys J 71(4):2201–2212
12. Beier C, Steinhoff H-J (2006) A structure-based simulation approach for electron paramagnetic resonance spectra using molecular and stochastic dynamics simulations. Biophys J 91 (7):2647–2664
13. Oganesyan VS (2011) A general approach for prediction of motional EPR spectra from molecular dynamics (MD) simulations: application to spin labelled protein. Phys Chem Chem Phys 13:4724–4737
14. Sezer D, Freed JH, Roux B (2009) Multifrequency electron spin resonance spectra of a spin-labeled protein calculated from molecular dynamics simulations. J Am Chem Soc 131 (7):2597–2605
15. Oganesyan VS (2007) A novel approach to the simulation of nitroxide spin label EPR spectra from a single truncated dynamical trajectory. J Magn Reson 188(2):196–205

16. Liang ZC, Freed JH (1999) An assessment of the applicability of multifrequency ESR to study the complex dynamics of biomolecules. J Phys Chem B 103(30):6384–6396
17. Zhang Z, Fleissner MR, Tipikin DS, Liang Z, Moscicki JK, Earle KA, Hubbell WL, Freed JH (2010) Multifrequency electron spin resonance study of the dynamics of spin labeled T4 lysozyme. J Phys Chem B 114(16):5503–5521
18. Isas JM, Langen R, Haigler HT, Hubbell WL (2002) Structure and dynamics of a helical hairpin and loop region in annexin 12: a site-directed spin labeling study. Biochemistry 41 (5):1464–1473
19. Mchaourab HS, Lietzow MA, Hideg K, Hubbell WL (1996) Motion of spin-labeled side chains in T4 lysozyme, correlation with protein structure and dynamics. Biochemistry 35 (24):7692–7704
20. Bordignon E, Steinhoff HJ (2007) Membrane protein structure and dynamics studied by site-directed spin-labeling ESR. In: Hemminga MA, Berliner LJ (eds) ESR spectroscopy in membrane biophysics, vol 27. Biological magnetic resonance. Springer, New York, pp 129–164
21. Freed DM, Horanyi PS, Wiener MC, Cafiso DS (2010) Conformational exchange in a membrane transport protein is altered in protein crystals. Biophys J 99(5):1604–1610
22. Perozo E, Cortes DM, Cuello LG (1998) Three-dimensional architecture and gating mechanism of a K+ channel studied by EPR spectroscopy. Nat Struct Biol 5(6):459–469
23. Hille B (2001) Ion channels of excitable membranes, 3rd edn. Sinauer Associates, Sunderland
24. Doyle DA, Morais Cabral J, Pfuetzner RA, Kuo A, Gulbis JM, Cohen SL, Chait BT, MacKinnon R (1998) The structure of the potassium channel: molecular basis of K+ conduction and selectivity. Science 280(5360):69–77
25. Perozo E, Cortes DM, Cuello LG (1998) Molecular architecture of the Streptomyces K^+ channel first transmembrane segment. A site directed spin-labeling study. Biophys J 74(2):A44
26. Zou P, Mchaourab HS (2009) Alternating access of the putative substrate-binding chamber in the ABC transporter MsbA. J Mol Biol 393(3):574–585
27. Doebber M, Bordignon E, Klare JP, Holterhues J, Martell S, Mennes N, Li L, Engelhard M, Steinhoff HJ (2008) Salt-driven equilibrium between two conformations in the HAMP domain from *N. pharaonis*: the language of signal transduction? J Biol Chem 283(42):28691–28701
28. Bridges MD, Hideg K, Hubbell WL (2010) Resolving conformational and rotameric exchange in spin-labeled proteins using saturation recovery EPR. Appl Magn Reson 37(1–4):363
29. McCoy J, Hubbell WL (2011) High-pressure EPR reveals conformational equilibria and volumetric properties of spin-labeled proteins. Proc Natl Acad Sci USA 108(4):1331–1336
30. Steinhoff HJ, Mollaaghababa R, Altenbach C, Hideg K, Krebs M, Khorana HG, Hubbell WL (1994) Time-resolved detection of structural-changes during the photocycle of spin-labeled bacteriorhodopsin. Science 266(5182):105–107
31. Radzwill N, Gerwert K, Steinhoff HJ (2001) Time-resolved detection of transient movement of helices F and G in doubly spin-labeled bacteriorhodopsin. Biophys J 80(6):2856–2866
32. Wegener AA, Chizhov I, Engelhard M, Steinhoff HJ (2000) Time-resolved detection of transient movement of helix F in spin-labelled pharaonis sensory rhodopsin II. J Mol Biol 301(4):881–891
33. Bordignon E, Klare JP, Holterhues J, Martell S, Krasnaberski A, Engelhard M, Steinhoff HJ (2007) Analysis of light-induced conformational changes of *Natronomonas pharaonis* sensory rhodopsin II by time resolved electron paramagnetic resonance spectroscopy. Photochem Photobiol 83(2):263–272
34. Klare JP, Bordignon E, Engelhard M, Steinhoff HJ (2004) Sensory rhodopsin II and bacteriorhodopsin: light activated helix F movement. Photochem Photobiol Sci 3(6):543–547
35. Wegener AA, Klare JP, Engelhard M, Steinhoff HJ (2001) Structural insights into the early steps of receptor-transducer signal transfer in archaeal phototaxis. EMBO J 20(19):5312–5319
36. Gordeliy VI, Labahn J, Moukhametzianov R, Efremov R, Granzin J, Schlesinger R, Büldt G, Savopol T, Scheidig AJ, Klare JP, Engelhard M (2002) Molecular basis of transmembrane signalling by sensory rhodopsin II-transducer complex. Nature 419(6906):484–487

37. Altenbach C, Greenhalgh DA, Khorana HG, Hubbell WL (1994) A collision gradient-method to determine the immersion depth of nitroxides in lipid bilayers – application to spin-labeled mutants of bacteriorhodopsin. Proc Natl Acad Sci USA 91(5):1667–1671
38. Altenbach C, Froncisz W, Hemker R, Mchaourab H, Hubbell WL (2005) Accessibility of nitroxide side chains: absolute Heisenberg exchange rates from power saturation EPR. Biophys J 89(3):2103–2112
39. Pyka J, Ilnicki J, Altenbach C, Hubbell WL, Froncisz W (2005) Accessibility and dynamics of nitroxide side chains in T4 lysozyme measured by saturation recovery EPR. Biophys J 89 (3):2059–2068
40. Volkov A, Dockter C, Polyhach Y, Paulsen H, Jeschke G (2010) Site-specific information on membrane protein folding by electron spin echo envelope modulation spectroscopy. J Phys Chem Lett 1(3):663–667
41. Pavlova A, McCarney ER, Peterson DW, Dahlquist FW, Lew J, Han S (2009) Site-specific dynamic nuclear polarization of hydration water as a generally applicable approach to monitor protein aggregation. Phys Chem Chem Phys 11(31):6833–6839
42. Reyes CL, Ward A, Yu J, Chang G (2006) The structures of MsbA: insight into ABC transporter-mediated multidrug efflux. FEBS Lett 580(4):1042–1048
43. Locher KP (2009) Structure and mechanism of ATP-binding cassette transporters. Philos Trans R Soc Lond B Biol Sci 364(1514):239–245
44. Dong J, Yang G, McHaourab HS (2005) Structural basis of energy transduction in the transport cycle of MsbA. Science 308(5724):1023–1028
45. Möbius K, Savitsky A, Wegener C, Plato M, Fuchs M, Schnegg A, Dubinskii AA, Grishin YA, Grigor'ev IA, Kühn M, Duché D, Zimmermann H, Steinhoff HJ (2005) Combining high-field EPR with site-directed spin labeling reveals unique information on proteins in action. Magn Reson Chem 43:S4–S19
46. Wegener C, Savitsky A, Pfeiffer M, Möbius K, Steinhoff HJ (2001) High-field EPR-detected shifts of magnetic tensor components of spin label side chains reveal protein conformational changes: the proton entrance channel of bacteriorhodopsin. Appl Magn Reson 21 (3–4):441–452
47. Volkov A, Dockter C, Bund T, Paulsen H, Jeschke G (2009) Pulsed EPR determination of water accessibility to spin-labeled amino acid residues in LHCIIb. Biophys J 96(3):1124–1141
48. Stoll S, Schweiger A (2006) EasySpin, a comprehensive software package for spectral simulation and analysis in EPR. J Magn Reson 178(1):42–55
49. Bordignon E, Brutlach H, Urban L, Hideg K, Savitsky A, Schnegg A, Gast P, Engelhard M, Groenen EJJ, Mobius K, Steinhoff HJ (2010) Heterogeneity in the nitroxide micro-environment: polarity and proticity effects in spin-labeled proteins studied by multi-frequency EPR. Appl Magn Reson 37(1–4):391–403
50. Plato M, Steinhoff HJ, Wegener C, Törring JT, Savitsky A, Möbius K (2002) Molecular orbital study of polarity and hydrogen bonding effects on the g and hyperfine tensors of site directed NO spin labelled bacteriorhodopsin. Mol Phys 100(23):3711–3721
51. Brutlach H, Bordignon E, Urban L, Klare JP, Reyer H-J, Engelhard M, Steinhoff HJ (2006) High-field EPR and site directed spin labeling reveal a periodical polarity profile: the sequence 88–94 of the phototransducer NpHtrII in complex with sensory rhodopsin, NpSRII. Appl Magn Reson 30:359–372
52. Jeschke G, Schweiger A (eds) (2001) Principles of pulse electron paramagnetic resonance. Oxford University Press, Oxford
53. Standfuss J, Terwisscha van Scheltinga AC, Lamborghini M, Kuhlbrandt W (2005) Mechanisms of photoprotection and nonphotochemical quenching in pea light-harvesting complex at 2.5 A resolution. EMBO J 24(5):919–928
54. Liu Z, Yan H, Wang K, Kuang T, Zhang J, Gui L, An X, Chang W (2004) Crystal structure of spinach major light-harvesting complex at 2.72 Å resolution. Nature 428(6980):287–292
55. Berliner LJ, Eaton SS, Eaton GR (eds) (2000) Distance measurements in biological systems by EPR. Biological magnetic resonance. Kluwer Academic/Plenum, New York

56. Rabenstein MD, Shin YK (1995) Determination of the distance between 2 spin labels attached to a macromolecule. Proc Natl Acad Sci USA 92(18):8239–8243
57. Steinhoff HJ, Radzwill N, Thevis W, Lenz V, Brandenburg D, Antson A, Dodson G, Wollmer A (1997) Determination of interspin distances between spin labels attached to insulin: comparison of electron paramagnetic resonance data with the X-ray structure. Biophys J 73 (6):3287–3298
58. Pannier M, Veit S, Godt A, Jeschke G, Spiess HW (2000) Dead-time free measurement of dipole-dipole interactions between electron spins. J Magn Reson 142(2):331–340
59. Milov AD, Ponomarev AB, Tsvetkov YD (1984) Electron-electron double resonance in electron spin echo: model biradical systems and the sensitized photolysis of decalin. Chem Phys Lett 110(1):67–72
60. Borbat PP, McHaourab HS, Freed JH (2002) Protein structure determination using long-distance constraints from double-quantum coherence ESR: study of T4 lysozyme. J Am Chem Soc 124(19):5304–5314
61. Banham JE, Baker CM, Ceola S, Day IJ, Grant GH, Groenen EJJ, Rodgers CT, Jeschke G, Timmel CR (2008) Distance measurements in the borderline region of applicability of CW EPR and DEER: a model study on a homologous series of spin-labelled peptides. J Magn Reson 191(2):202–218
62. Altenbach C, Oh KJ, Trabanino RJ, Hideg K, Hubbell WL (2001) Estimation of inter-residue distances in spin labeled proteins at physiological temperatures: experimental strategies and practical limitations. Biochemistry 40(51):15471–15482
63. Jeschke G, Wegener C, Nietschke M, Jung H, Steinhoff HJ (2004) Interresidual distance determination by four-pulse double electron-electron resonance in an integral membrane protein: the Na^{+}/proline transporter PutP of *Escherichia coli*. Biophys J 86(4):2551–2557
64. Jeschke G, Chechik V, Ionita P, Godt A, Zimmermann H, Banham J, Timmel CR, Hilger D, Jung H (2006) DeerAnalysis2006 – a comprehensive software package for analyzing pulsed ELDOR data. Appl Magn Reson 30(3–4):473–498
65. Dastvan R, Bode BE, Karuppiah MP, Marko A, Lyubenova S, Schwalbe H, Prisner TF (2010) Optimization of transversal relaxation of nitroxides for pulsed electron-electron double resonance spectroscopy in phospholipid membranes. J Phys Chem B 114(42):13507–13516
66. Ward R, Bowman A, Sozudogru E, El-Mkami H, Owen-Hughes T, Norman DG (2010) EPR distance measurements in deuterated proteins. J Magn Reson 207(1):164–167
67. Zou P, Mchaourab HS (2010) Increased sensitivity and extended range of distance measurements in spin-labeled membrane proteins: Q-band double electron-electron resonance and nanoscale bilayers. Biophys J 98(6):L18–L20
68. Höfer P, Heilig R, Schmalbein D (2000) The superQ-FT accessory for pulsed EPR, ENDOR and ELDOR at 34 GHz. Bruker Spin Rep 152(153):37–43
69. Lueders P, Jeschke G, Yulikov M (2011) Double electron–electron resonance measured between Gd^{3+} ions and nitroxide radicals. J Phys Chem Lett 2(6):604–609
70. Potapov A, Song Y, Meade TJ, Goldfarb D, Astashkin AV, Raitsimring A (2010) Distance measurements in model bis-Gd(III) complexes with flexible "bridge". Emulation of biological molecules having flexible structure with Gd(III) labels attached. J Magn Reson 205(1):38–49
71. Raitsimring AM, Gunanathan C, Potapov A, Efremenko I, Martin JM, Milstein D, Goldfarb D (2007) Gd3+ complexes as potential spin labels for high field pulsed EPR distance measurements. J Am Chem Soc 129(46):14138–14139
72. Jeschke G, Sajid M, Schulte M, Godt A (2009) Three-spin correlations in double electron-electron resonance. Phys Chem Chem Phys 11(31):6580–6591
73. Fleissner MR, Brustad EM, Kalai T, Altenbach C, Cascio D, Peters FB, Hideg K, Peuker S, Schultz PG, Hubbell WL (2009) Site-directed spin labeling of a genetically encoded unnatural amino acid. Proc Natl Acad Sci USA 106(51):21637–21642
74. Jeschke G, Zimmermann H, Godt A (2006) Isotope selection in distance measurements between nitroxides. J Magn Reson 180(1):137–146

75. Hirst SJ, Alexander N, McHaourab HS, Meiler J (2011) RosettaEPR: an integrated tool for protein structure determination from sparse EPR data. J Struct Biol 173(3):506–514
76. Liu YS, Sompornpisut P, Perozo E (2001) Structure of the KcsA channel intracellular gate in the open state. Nat Struct Biol 8(10):883–887
77. Hilger D, Polyhach Y, Padan E, Jung H, Jeschke G (2007) High-resolution structure of a Na^+/H^+ antiporter dimer obtained by pulsed election paramagnetic resonance distance measurements. Biophys J 93(10):3675–3683
78. Bhatnagar J, Borbat PP, Pollard AM, Bilwes AM, Freed JH, Crane BR (2010) Structure of the ternary complex formed by a chemotaxis receptor signaling domain, the CheA histidine kinase, and the coupling protein CheW as determined by pulsed dipolar ESR spectroscopy. Biochemistry 49(18):3824–3841
79. Bhatnagar J, Freed JH, Crane BR (2007) Rigid body refinement of protein complexes with long-range distance restraints from pulsed dipolar ESR. In: Methods in enzymology, vol 423. Academic, London, pp 117–133
80. Bleicken S, Classen M, Padmavathi PVL, Ishikawa T, Zeth K, Steinhoff HJ, Bordignon E (2010) Molecular details of BAX activation, oligomerization, and membrane insertion. J Biol Chem 285(9):6636–6647
81. Gavathiotis E, Suzuki M, Davis ML, Pitter K, Bird GH, Katz SG, Tu HC, Kim H, Cheng EHY, Tjandra N, Walensky LD (2008) BAX activation is initiated at a novel interaction site. Nature 455(7216):1076–1081
82. Bode BE, Margraf D, Plackmeyer J, Durner G, Prisner TF, Schiemann O (2007) Counting the monomers in nanometer-sized oligomers by pulsed electron-electron double resonance. J Am Chem Soc 129(21):6736–6745
83. Millar RP, Newton CL (2010) The year in G protein-coupled receptor research. Mol Endocrinol 24(1):261–274
84. Palczewski K, Kumasaka T, Hori T, Behnke CA, Motoshima H, Fox BA, Trong IL, Teller DC, Okada T, Stenkamp RE, Yamamoto M, Miyano M (2000) Crystal structure of rhodopsin: a G protein-coupled receptor. Science 289(5480):739–745
85. Altenbach C, Yang K, Farrens DL, Farahbakhsh ZT, Khorana HG, Hubbell WL (1996) Structural features and light-dependent changes in the cytoplasmic interhelical E-F loop region of rhodopsin: a site-directed spin-labeling study. Biochemistry 35(38):12470–12478
86. Altenbach C, Kusnetzow AK, Ernst OP, Hofmann KP, Hubbell WL (2008) High-resolution distance mapping in rhodopsin reveals the pattern of helix movement due to activation. Proc Natl Acad Sci USA 105(21):7439–7444

Top Curr Chem (2012) 321: 159–198
DOI: 10.1007/128_2011_300

Published online: 13 December 2011

Structure and Dynamics of Nucleic Acids

Ivan Krstić, Burkhard Endeward, Dominik Margraf, Andriy Marko, and Thomas F. Prisner

Abstract In this chapter we describe the application of CW and pulsed EPR methods for the investigation of structural and dynamical properties of RNA and DNA molecules and their interaction with small molecules and proteins. Special emphasis will be given to recent applications of dipolar spectroscopy on nucleic acids.

Keywords CW EPR spectroscopy · Dynamics · ENDOR · ESEEM · Hyperfine spectroscopy · In-cell EPR · Nucleic acids · PELDOR · RNA-ligand interactions · RNA-protein interactions · Structure

Contents

I. Krstić, B. Endeward, D. Margraf, A. Marko, and T.F. Prisner (✉)
Institute of Physical and Theoretical Chemistry Center of Biomolecular Magnetic Resonance, Goethe University Frankfurt, Frankfurt am Main, Germany
e-mail: prisner@chemie.uni-frankfurt.de

1 Introduction

Nucleic acid (NA) molecules play essential roles in such biological functions as translation and transcription, most importantly by storing the genetic code (DNA) and transferring it for protein synthesis (RNA). Additionally, they are involved in catalytic and regulatory functions in cells, as has been recently discovered in ribozymes, riboswitches, small regulatory RNAs, and the ribosome. In all these cases, the interactions of NAs with small molecules or proteins are of central importance for their function. Different from most proteins, NAs can adopt several tertiary structures, depending on their surrounding and external conditions. This large conformational flexibility is on the one hand an important and crucial factor for a detailed understanding of their specific function and on the other a limiting factor for classical structural methods, such as for example X-ray crystallography. Methods capable of investigating structural and dynamic aspects of macromolecules are therefore especially significant for the functional characterization of NAs. NMR is one of the most promising methods for working with NAs: it can determine the tertiary structure and the fast local dynamics of NAs with atomic resolution, but is at its limit for larger NA molecules or NA–protein complexes and slow conformational dynamics [1, 2]. Fluorescence Resonance Energy Transfer (FRET) can detect such long range distance changes by attaching two fluorophore-labels to the NA molecule [3, 4]. Its high sensitivity allows the performance of such measurements at room temperature down to a single molecule level; however, due to the linker flexibility and other unknown parameters, the derived distance information is mostly qualitative in nature. EPR spectroscopy is a method which can fill this gap between the local high resolution structures, derived by NMR spectroscopy, and the dynamical information obtained by FRET spectroscopy. Within the last few years the bandwidth of EPR methods used to investigate dynamics and structural constraints on NAs has strongly increased and has established EPR as a new and promising method in the field of nucleic acid research. In this review we want to summarize some of the more recent results obtained by EPR spectroscopy on NAs. We first review the most common nitroxide spin labeling strategies used for NAs, briefly introduce different EPR experiments used for the investigation of structure and conformational flexibility of NAs, and discuss selected applications which illustrate and highlight the potential of the EPR method on NAs. Finally, we conclude by pointing out some new trends and potential perspectives for the future.

2 Spin Labeling

Site-specific spin labeling of nucleic acids differs substantially from the procedure used for proteins; therefore the most common methods to attach nitroxide radicals to nucleic acids will be described and compared in this chapter. Several reviews

[5–7] and monographs [8, 9] can provide the reader with more in-depth information.

First side-directed spin labeling (SDSL) studies date back to the early 1970s. There, tRNAs were subjected to nitroxide labeling via either naturally occurring rare nucleobases, e.g., 2-thio-uridine, enzymatically introduced nucleobases such as 2-thiocytidine and 4-thio-uridine, or the 3′-ends [10–14]. It is worthwhile to note that labeling of the 3′-end, **1** (Fig. 1), causes strong structural perturbations due to the transformation of the sugar moiety to a morpholine derivative. Moreover, the general drawback of the above-mentioned investigations is their requirement of tRNA and, depending on the label, an introduction of a positive charge in the case of 2-thiocytidine. The more recent approaches can be subdivided into strategies

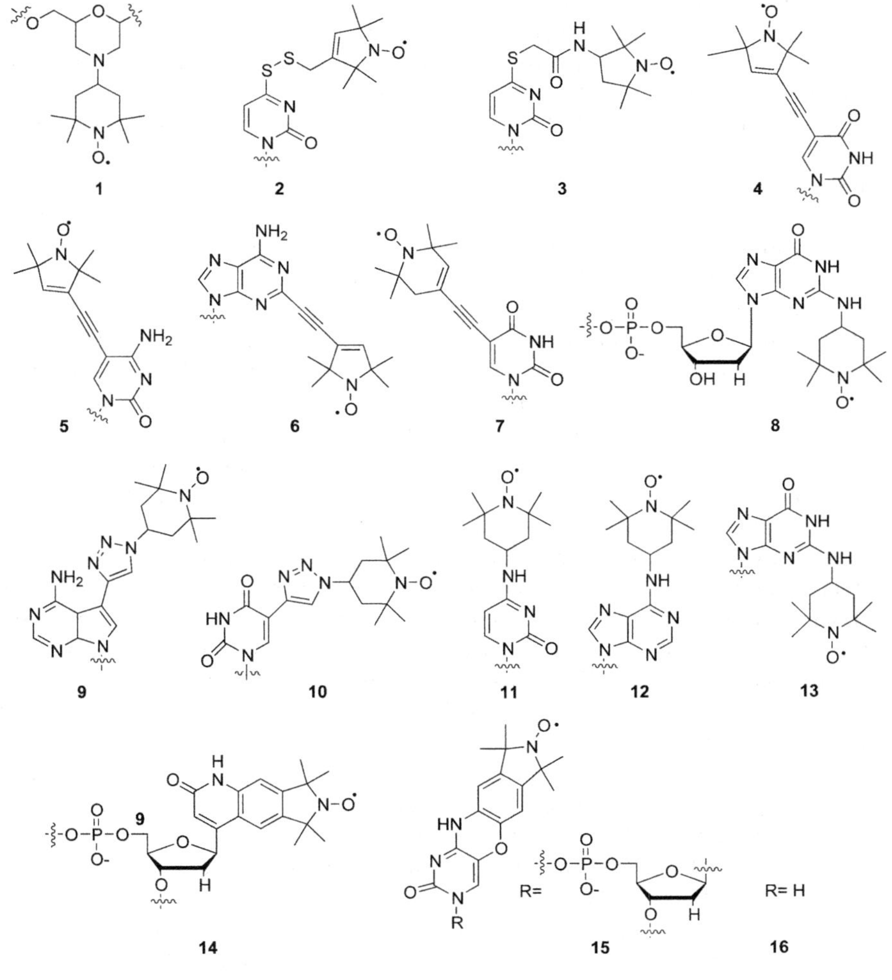

Fig. 1 Overview of labeled and modified nucleobases

where either the nucleobase or the sugar phosphate backbone is specifically modified by the nitroxide spin label.

2.1 Nucleobases

Approximately 25 years later, Hubbell and coworkers [15], as well as Varani and Ramos [16], developed a labeling strategy that involves reaction of 4-thiouridine substituted RNA with the nitroxides 2,2,5,5-tetramethylpyrroline-1-oxyl-3-methyl-methanethiosulfonate (MTSL) or 2,2,5,5-tetramethyl-pyrroline-1-oxyl-3-iodo-acetamide leading to **2** and **3** (Fig. 1), respectively. Both labeling strategies report almost quantitative yields as monitored via UV-spectroscopy, but compound **3** (Fig. 1) is chemically more stable than **2** due to thioether formation instead of a disulfide bridge generation. However, a crucial step is the choice of the labeling site to minimize disturbances of the intermolecular interface, which might arise from the loss of the imino proton in 4-thiouridine upon spin labeling. None of the authors found severe affection of the RNA structure at the chosen positions resulting from nitroxide modification. Yet the approach is limited to 4-thiouridine. Hopkins and Robinson circumvent the abstraction of the imino proton, thus lowering structural perturbations by usage of 5-iodouridine, 2,2,5,5-tetramethyl-pyrroline-1-oxyl-3-acetylene (TPA) and Pd-catalyzed Sonogashira chemistry during solid-state synthesis [17]. This protocol has been extended by Engels and coworkers to 5-iododesoxyuridine, 5-iodocytidine, and 2-iodoadenosine [18–21]; **4–6** (Fig. 1) depict the respective structures. This "on column coupling" shortens the time-consuming NAs synthesis of already spin-labeled phosphoramidites previously used in DNA dynamic studies [22] and involves simple purification. Another advantage is the coupling yield of 95% if a 2′-acid-labile orthoester strategy in combination with a new class of 5′-silyl ethers (Dharmacon®) instead of a standard *tert*-butyldimethylsilyl protection scheme is applied. The latter conditions led to quenching of the nitroxide radical and further byproducts. Also, the corresponding tetrahydropyridine analog of TPA, spin label **7** (Fig. 1), was recently introduced [23, 24]. It can be readily prepared in half the number of steps that are needed for its spectroscopic ancestor **4**. In addition, the length of the acetylene chain connecting the nitroxide to the base has been investigated [25] proving small amplitudes of internal motion for **4** as compared to a diacetylene tether. This is of great importance for PELDOR measurements discussed in Sect. 4.2.2. Alternatively, ethynyl side chain bearing derivatives of 7-deaza-2'deoxyadenosine or 2′-deoxyuridine may undergo a Cu^{+} catalyzed [3 + 2]-cycloaddition with 2,2,6,6-tetramethyl-piperidine-1-oxyl-4-azide (so-called "click chemistry") to yield ~60% of **9** and **10** (Fig. 1) [26]. Note that usage of Cu^{2+} instead of Cu^{+} leads to a reduction of the nitroxide to the hydroxylamine species.

Enzymatic strategies based on Klenow filling were also applied by Bobst and coworkers to incorporate different types of spin labels attached to the 5- or 4-position of uridine, or the 5-position of cytidine of RNA and DNA [9, 27, 28]. None of the labels affected the structure of the biomolecule when 1–2% nucleotides were spin labeled. Nitroxide labeled deoxyguanine, **8** (Fig. 1), can be created post-synthetically with 2,2,6,6-tetramethylpiperidine-1-oxyl-4-amine and a 2-fluoro-hypoxanthine containing oligodeoxynucleotide [29]. Melting point analysis indicates that nitroxide labeling of the guanine base does not destabilize the duplex. A more general strategy employed by Höbartner and coworkers allows direct attachment of the paramagnetic probe to the exocyclic amino groups of nucleobases cytosine and adenine in the major groove, and of guanine in the minor groove of Watson–Crick base-paired RNA duplexes. The convertible nucleosides O^4-(4-chlorophenyl)uridine, O^6-(4-chlorophenyl)inosine, and 2-fluoroinosine were labeled with 2,2,6,6-tetramethylpiperidine-1-oxyl-4-amine leading to compounds **11–13** (Fig. 1) [30]. Thermal destabilization was found to be in the order of 5–6°C per spin label for A, C, and G. It is worth noting that cleavage from the polystyrene support, deprotection of the phosphate backbone and nucleobases, as well as nucleobase substitution take place at the same time in this procedure.

Nitroxides may also be attached to the site of choice via more than one covalent bond, causing extremely rigid spin labels. As mentioned before, this improves the information content of PELDOR experiments. One route discussed in the literature is the synthesis of a quinolonyl based, heavily modified C-derived phosphoramidite **14** (Fig. 1). Because of the formation of a base pair with 2-aminopurine [31], its applications are limited. Nevertheless, to overcome the size limitations of synthesized oligonucleotides, **14** can be chemically incorporated into DNA and subjected to Klenow polymerase filling [32]. Another extremely rigid C-analog, **15** (Fig. 1), has been developed by Sigurdsson and coworkers [33], inspired by a phenoxazine based deoxycytidine derivative [34]. Thermal denaturation and circular dichroism (CD) experiments revealed a negligible effect on DNA duplex stability and conformation [35]. Moreover, simple treatment of **15** with sodium sulfide at 45°C for 14 h causes complete conversion to the corresponding amine [36]. This highly fluorescent species can therefore also be used for optical spectroscopy.

The same group has reported that removal of the sugar moiety in **15** leads to **16** (Fig. 1) showing unique properties [37] regarding specific interactions with abasic sites in DNA. At −30°C, 1 equiv. of **16** is reported to readily bind to the abasic site in a noncovalent manner until saturation occurs. The high specificity of binding was demonstrated via the small change of binding constants upon introduction of a tenfold excess of unmodified DNA. Minor nonspecific binding was found to be in the order of 5% at low temperatures. This novel labeling strategy is named noncovalent-SDSL and has promising applications on natural occurring DNA molecules.

2.2 *Sugar Phosphate Backbone*

Instead of altering nucleobases, several methods to address the sugar phosphate backbone of nucleic acids have been published. One approach, reported by Nagahara et al., is to incorporate a fully protected 2′-deoxyribonucleoside H-phosphonate into the oligonucleotide by means of an automated solid-phase

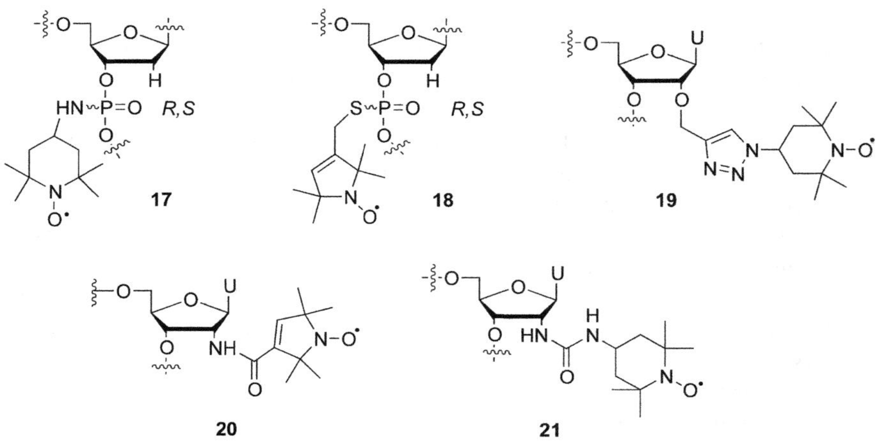

Fig. 2 Overview of modified sugar phosphate backbones

phosphoramidite protocol [38]. Note that usage of a 2′-deoxyribonucleoside is mandatory to avoid phosphodiester bond cleavage and that the negative phosphate charge present in unmodified internucleotide linkers is lacking. The following "in syringe" deprotection, reaction with 2,2,6,6-tetramethylpiperidine-1-oxyl-4-amine and reverse phase HPLC purification yields approximately 70% of both *R* and *S* diastereomers of crude **17** (Fig. 2). As CD measurements suggested differences in solution structures of the diastereomers, they need to be separated via further HPLC steps. As a universal group is labeled, the method is sequence independent and starting materials are commercially available. A similar yet more cost effective technique that substitutes phosphorothioates at a specific backbone position during chemical synthesis of RNA or DNA has been established by Hubbell, Qin, and coworkers [39–41]. MTSL is converted to a highly reactive species, referred to as R5, using sodium iodide and is brought to reaction with the corresponding phosphorothioate leading to racemic **18** (Fig. 2). The labeling efficiency is close to 100% and attachment of the nitroxide probe does not severely perturb the conformation of DNA or RNA duplexes. Additionally, separation of the diastereomers is not required prior to EPR distance measurements and arbitrary sequences can be subjected to spin labeling [42, 43]. One drawback is that the 2′-OH group neighboring the labeled phosphorus atom has to be removed to circumvent cleavage of the phosphodiester bond and that the negative charge is lost.

A different tactic takes advantage of a functionalized sugar system. One example from Flaender et al. is a synthesized 2′-*O*-propargyl-uridine phosphoramidite monomer that was chemically inserted within a DNA fragment [44]. The introduced acetylene unit reacts in a Cu^+ catalyzed [3 + 2]-cycloaddition (so-called "click chemistry") with 2,2,6,6-tetramethyl-piperidine-1-oxyl-4-azide to produce the triazole bridged molecule **19** (Fig. 2). This bioconjugation protocol is highly efficient and mild, but the label is flexible due to the tether length. Alternatively, commercially available 2′-amino substituted uridine ribonucleotides were subjected by DeRose and coworkers [45] to 2,2,5,5-tetramethyl-3-pyrroline-1-oxyl-3-succinimidyl-carboxylate to give **20** (Fig. 2). The yields varied between 20% and 50% depending on the length of the RNA. Also, thermal destabilization after labeling was observed. It was proposed to be caused by the rigid amide linker stabilizing a less favorable 2′-endo sugar pucker. Simply switching to 2,2,6,6-tetramethyl-piperidine-1-oxyl-4-isocyanate as spin bearing unit by Sigurdsson and coworkers results in the more flexible ureido linked compound **21** (Fig. 2) [46, 47]. This strategy circumvents disrupting RNA structure, labeling yields are ~90%, and the approach is generally applicable to DNA.

To summarize, a manifold of techniques capable of attaching a large variety of stable nitroxide radicals to nucleic acids exist. Great caution should be taken choosing a proper spin probe, the chemical or enzymatic method, and the label position of the system under study in order to avoid structural and functional perturbations. Finally, a general protocol or strategy to yield spin labeled RNA or DNA is not available – it always has to be adapted to the system.

3 Theoretical Background

As described in the introduction, EPR spectroscopy is a very powerful tool to study dynamics and structure of NAs and their interactions with metal ions, small molecules, or proteins. CW EPR spectroscopy at room temperature provides valuable information concerning local dynamics and global tumbling of the NA molecule, as well as the binding to other molecules (Fig. 3). The binding usually alters both the local and global modes of motion. By increasing viscosity of the environment the overall tumbling of the macromolecule gets reduced and CW EPR spectra are dominated by the base or nitroxide local motion. Hyperfine spectroscopy can be

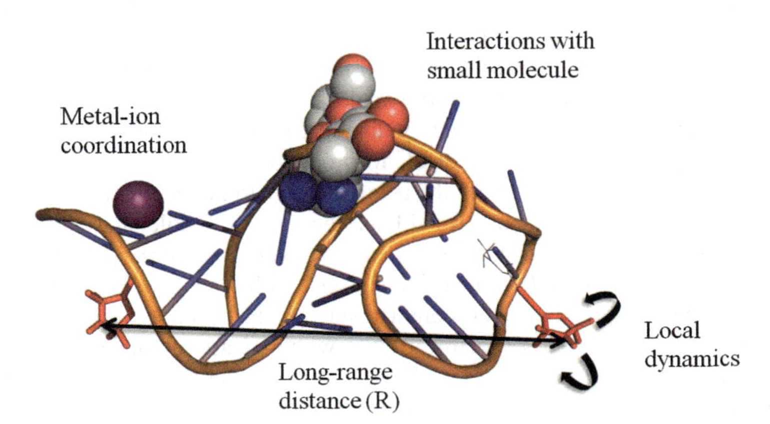

Fig. 3 EPR spectroscopy provides detailed information regarding metal ion coordination sphere, interaction with small molecules, local and global dynamics, as well as overall conformation of nucleic acid molecules

used to probe the local surrounding up to about 0.8 nm of the unpaired electron spin in great detail. This is especially interesting for metal ions like Mn^{2+} replacing physiologically crucial Mg^{2+} ions that play an important role in catalytic function and tertiary structure stabilization of ribozymes. Finally, dipolar spectroscopy can be applied to determine long range distances in NAs. This is a well established method to study complex structures of NAs and their changes upon binding to other molecules, and for the determination of the conformational flexibility of NAs under various conditions. In the following, a brief introduction to the basic concepts of these subclasses will be given. More detailed information regarding these methods can be found in standard textbooks of modern EPR [48].

3.1 CW EPR Spectroscopy

The shape of a nitroxide radical CW EPR spectrum reveals high sensitivity to its rotational motion in liquid solutions. Thus, it can be used as a valuable tool for monitoring the local and global dynamics of NAs on time scales defined by the EPR frequency.

Nitroxide EPR spectra are determined by Zeeman interactions of the unpaired electron and nuclear spins with the external magnetic field **B** and the hyperfine interaction of the electron spin with the nitrogen nuclear spin (I=1 for ^{14}N and I=1/2 for ^{15}N) [49]. In solid phase the magnetic energy of the electron-nucleus spin system can be described by the static Hamiltonian

$$\hat{H} = \mu_{\mathrm{B}}\mathbf{B}g\hat{\mathbf{S}} - \mu_{\mathrm{N}}g_{\mathrm{N}}\mathbf{B}\hat{\mathbf{I}} + \hat{\mathbf{I}}A\hat{\mathbf{S}} \tag{1}$$

where μ_{B} and μ_{N} are Bohr and nuclear magneton, respectively and $\boldsymbol{S}$ and $\boldsymbol{I}$ are the electron and nuclear spin vector operators, respectively. The g-tensor accounts for the anisotropy of the electron spin Zeeman interaction, whereas the A-tensor describes the field independent electron nuclear hyperfine interaction. The magnetic tensors g and A are both approximately diagonal in the coordinate frame reflecting the symmetry of the nitroxide radical. In order to obtain the Hamiltonian in the laboratory coordinate system with the z-axis parallel to the magnetic field $\boldsymbol{B}$, g- and A-tensors have both to be transformed from the nitroxide molecular axis system to the laboratory frame. Calculation of the energy levels and corresponding transition probabilities for all possible radical orientations with respect to the magnetic field with their appropriate statistical weight results in the so-called powder spectrum represented in Fig. 4. This spectrum has a width of approximately 70 Gauss at typical X-band frequencies (9.5 GHz) and represents the experimentally observed spectra of nitroxide radicals in a disordered frozen solution sample.

However, the method of obtaining the ensemble spectra by a superposition of spectra from molecules with all possible random orientations with respect to the external magnetic field, as used above for frozen samples, is not applicable for

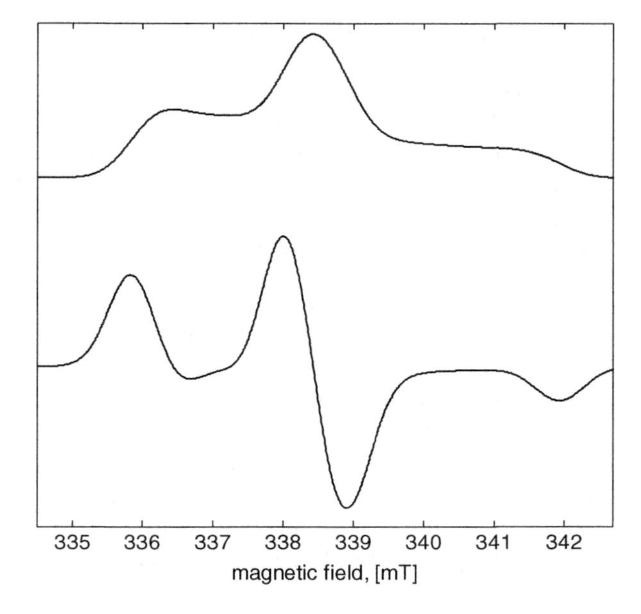

Fig. 4 Absorption (*top*) and derivative (*bottom*) powder EPR spectra of a nitroxide radical measured at X-band frequencies (9.5 GHz)

mobile radicals in liquid solution. In this case the effective values of the g- and A-tensors determining the Hamiltonian H in the laboratory frame are all time dependent and do not result in stationary solutions for the eigenvalues and eigenvectors of (1). In this case the density matrix has to be found for the system based on the Liouville/von Neumann equation with a time dependent Hamiltonian $H(t)$ to describe the spectrum [50, 51]:

$$\frac{\partial \hat{\rho}}{\partial t} = -\frac{i}{\hbar}\left[\hat{H}(t), \hat{\rho}\right] \tag{2}$$

The general solution of the density matrix for arbitrary rotational correlation time and ordering potential is non-trivial. In the case of very fast motion, meaning that the rotational correlation time of the radical is much shorter than the combined inverse anisotropies of the g- and A-tensor, the g- and A-tensors in (1) can be substituted by their averaged values $g_{\mathrm{iso}} = (g_{xx} + g_{yy} + g_{zz})/3$ and $A_{\mathrm{iso}} = (A_{xx} + A_{yy} + A_{zz})/3$, respectively. In the high field approximation the Hamiltonian H becomes diagonal, which allows solving the Liouville/von Neumann equation exactly. The time evolution of the transversal magnetization $M(t)$ is given by the trace of the product of the density matrix with the spin rising operator $\hat{S}_{+} = \hat{S}_x + \mathrm{i}\hat{S}_y$. Neglecting the nuclear Zeeman interaction, we obtain

$$M(t) \propto Tr\{\hat{S}_{+}\hat{\rho}\} = \exp\left(\mathrm{i}\frac{\mu_{\mathrm{B}} g_{\mathrm{iso}} B}{\hbar} t\right)\left(1 + 2\cos\left(\frac{A_{\mathrm{iso}}}{\hbar} t\right)\right)\exp\left(-\frac{t}{T_2}\right) \tag{3}$$

This expression describes the free induction decay (FID) signal of the spin system after an intense and short mw-pulse, which usually cannot be realized experimentally for a nitroxide. Nevertheless, for low microwave excitation power the CW EPR spectra are given as the Fourier transformation of the FID signal (3). For fast rotational motion this spectrum consists of three narrow hyperfine lines of equal intensities at the spectral positions $(\mu_B g_{iso} B - A_{iso})/\hbar$, $\mu_B g_{iso} B/\hbar$, and $(\mu_B g_{iso} B + A_{iso})/\hbar$. The width of the lines in this fast tumbling case is defined by $(\pi T_2)^{-1}$, where T_2 is the transversal relaxation time of the electron spin. For nitroxide radicals and magnetic field values of about 0.3 T this is valid for rotational correlation times faster than 10 ps.

If the radical motion slows down a gradual transition from a highly resolved three line spectrum of a liquid to the broad powder spectrum of a solid is observable (Fig. 5). For rotational correlation times τ_c in the range of 0.01–1 ns, the Lorenzian form of individual spectral lines can be described by the width parameter $T_2^{-1} = \overline{\Delta\omega^2}\tau_c$, being proportional to the ensemble average of the squared resonance frequency fluctuations $\Delta\omega$. Any dynamical process which causes the resonance frequency to change will therefore contribute to the line width. Approximating that anisotropy can be described by axial tensors [52], the modulation of the

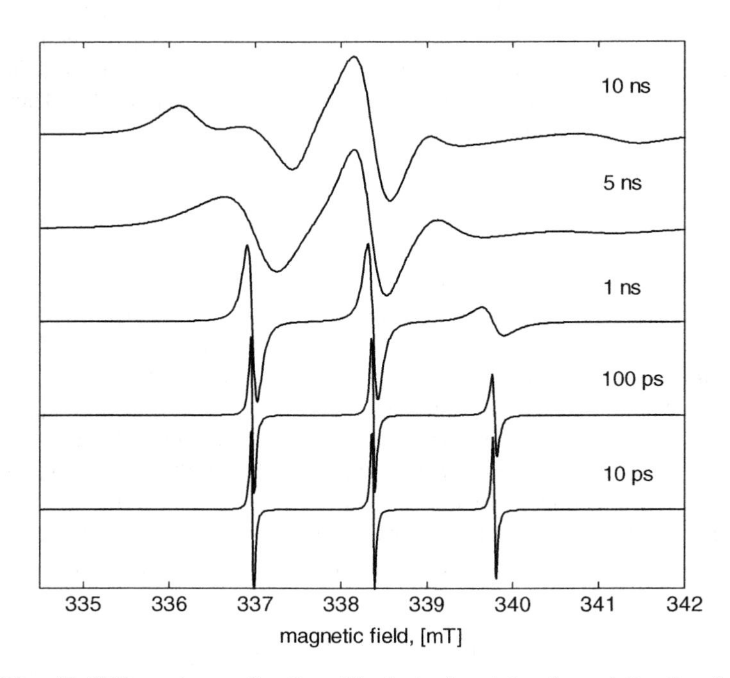

Fig. 5 Nitroxide EPR spectra as a function of the isotropic rotational correlation time for X-band frequencies. Simulation is performed by EasySpin [53]

anisotropic Zeeman, and hyperfine interactions by the rotational diffusion can be expressed by

$$\Delta\omega = \Delta g \frac{\mu_B B}{\hbar} + \frac{\Delta A}{\hbar} M_I \tag{4}$$

where M_I is the nuclear spin quantum number and Δg and ΔA represent the variation of Zeeman and hyperfine splitting, respectively. Inserting (4) in the expression for T_2^{-1} we obtain

$$\frac{1}{T_2} = a + bM_I + cM_I^2 \tag{5}$$

which predicts the line widths dependence on the nuclear spin quantum number M_I. The coefficients a, b, and c are defined as follows:

$$a = \overline{\Delta g^2} (\mu_B B/\hbar)^2 \tau_c \tag{6}$$

$$b = \overline{\Delta g \Delta A} \left(\mu_B B/\hbar^2\right) \tau_c \tag{7}$$

$$c = \overline{\Delta A^2} \tau_c / \hbar^2 \tag{8}$$

The narrowest line corresponds to $M_I = -1$ and the broadest to $M_I = +1$. This effect can be easily observed in Fig. 5.

To treat slow motion spectra in the rotational correlation time regime of nanoseconds, the time dependence of the Hamiltonian has to be taken explicitly into account. This is most conveniently done by the Stochastic Liouville Equation approach, which contains explicit superoperator expressions for the rotational diffusion of the molecule [54].

3.2 *Hyperfine Spectroscopy*

Radicals induced by illumination, spin labels covalently attached, or metal ions coordinated to a NA macromolecule may act as local sensors of the nuclei in their close surrounding. EPR is only sensitive to nuclei with a nuclear spin (such as for example ^{1}H, ^{13}C, ^{15}N, ^{31}P, and ^{55}Mn) within a distance up to 0.8 nm with respect to the unpaired electron spin. This is complementary to NMR, where such close-by nuclear spins are invisible because of paramagnetic broadening. Depending on distance and chemical nature, different interactions contribute to hyperfine spectroscopy. If the nuclei are very close to the unpaired electron, isotropic hyperfine interaction (A_{iso}) can be observed:

$$\hat{H}_{hfi} = A_{iso} \hat{\mathbf{I}}_z \hat{\mathbf{S}}_z \tag{9}$$

This interaction, which does not depend on the orientation of the molecule with respect to the external magnetic field, gives valuable information on the local structure, such as for example the bond strength and geometry of hydrogen bonds to the radical or on the covalency of metal ligand bonds.

The second interaction is the magnetic dipole–dipole interaction between the unpaired electron and nuclear spins. It is short-ranged due to its R^{-3} dependency and the "weak" magnetic moment of the interacting nuclei. This term is strongly anisotropic, meaning that it depends on the angle between the vector connecting the unpaired electron and the nuclear spin and the external magnetic field. Whereas this interaction is only a small perturbation for the electron spin at magnetic fields above 0.1 T with respect to the electron Zeeman interaction, it has a severe influence on the nuclear spins. The nuclear spin eigenstates are mixed by the additional hyperfine field arising from the strong magnetic moment of the unpaired electron leading to "forbidden" transitions further complicating CW EPR spectra. The according spin Hamiltonian can be expressed as

$$\hat{H}_{hfi} = \hat{\mathbf{I}}\, A\, \hat{\mathbf{S}} \approx A_{zz}\hat{\mathbf{I}}_z\hat{\mathbf{S}}_z + A_{xz}\hat{\mathbf{I}}_x\hat{\mathbf{S}}_z + A_{yz}\hat{\mathbf{I}}_y\hat{\mathbf{S}}_z \tag{10}$$

A third contribution of importance in hyperfine spectroscopy is the nuclear quadrupole moment for nuclei with $I > 1/2$, e.g., ^{2}H, ^{14}N, ^{17}O, and ^{33}S. They lead to an additional line broadening, arising from the shift of the nuclear spin eigenstates by the quadrupolar coupling:

$$\hat{H}_{qi} = \hat{\mathbf{I}}\, Q\, \hat{\mathbf{I}} \tag{11}$$

All these effects usually cause an inhomogeneous broadening of the EPR lines in frozen disordered samples or large and slowly tumbling macromolecules leading to broad and unresolved EPR spectra. In such cases, advanced pulsed and double resonance EPR methods have to be utilized to disentangle these interactions to nearby nuclei spins.

3.2.1 Electron Spin Echo Envelope Modulation

Pulsed EPR methods refocus the inhomogeneous line broadening, usually by a Hahn-echo sequence. Due to the mixed nuclear eigenstates the microwave pulses simultaneously drive allowed and forbidden transitions, leading to interferences between the excited transitions. This results in an oscillatory modulation of the electron spin echo signal intensity with a frequency defined by the nuclear spin energy splittings. These techniques are referred to as ESEEM (Electron Spin Echo Envelope Modulation) used on Hahn-echoes or stimulated echoes [55–57] or as the most popular 2D-variant HYSCORE (Hyperfine Sublevel Correlation Spectroscopy) [58]. The mixing of the nuclear spin states is most effective if the nuclear Zeeman splitting matches the hyperfine coupling; therefore this effect shows a strong dependence on the external magnetic field. Typically the effect vanishes for field strengths above 1 T, where the nuclear Zeeman-splitting exceeds the hyperfine splitting. Furthermore, it can only be observed in solids or frozen samples, because the anisotropy of the hyperfine interaction is averaged to zero in liquid solutions. The method allows one to obtain distances to close by nuclei. Quantitative simulation of the ESEEM effect can be used to determine the number of coupled nuclear spins [59]. The simplest experiment is the 2-pulse Hahn-echo ESEEM (Fig. 6). It has a low resolution as the signal decays with the short transverse relaxation time T_2 but it is very simple and thus very efficient for a small number of coordinated

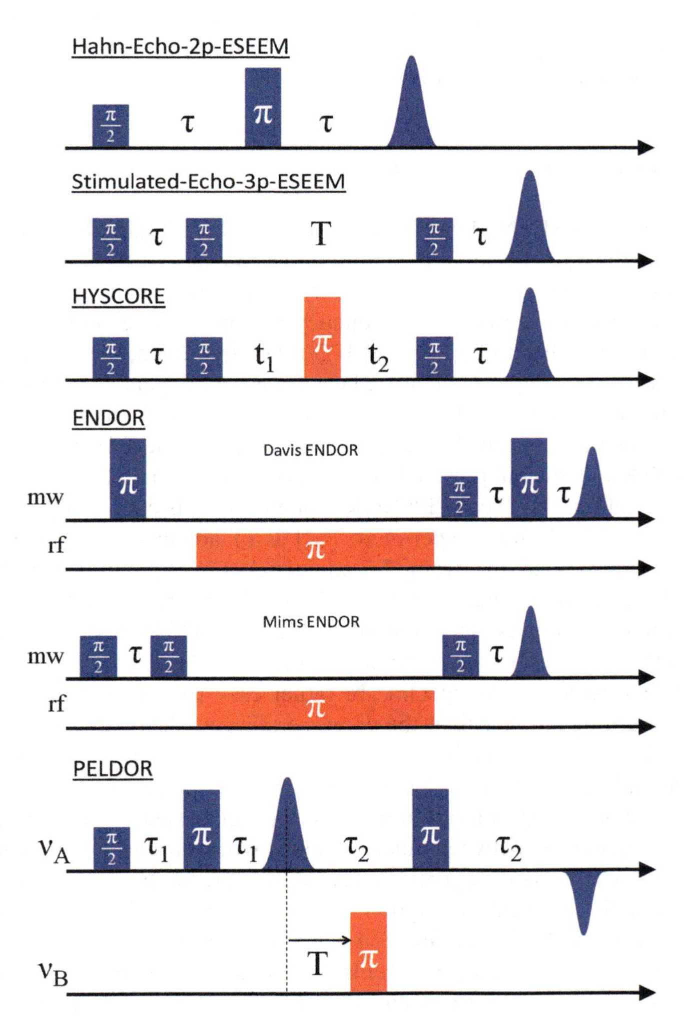

Fig. 6 Pulse-Sequences, different pulse sequences for hyperfine and dipolar spectroscopy (ESEEM, ENDOR, and PELDOR)

nuclear spins. The number of coupled nuclei spins is also extractable in favorable cases. The 3-pulse ESEEM, based on a stimulated echo sequence, has a higher spectral resolution. It suffers from so-called "blind spots"; certain hyperfine frequencies are suppressed due to the selected pulse delay between the first and second pulse. Repeating the 3-pulse ESEEM experiment with different pulse delays usually circumvents this drawback. In addition, 4-pulse ESEEM and, in the 2D version, HYSCORE further improve the spectral resolution of the ESEEM method. The off-diagonal signals in HYSCORE correlate hyperfine frequencies arising from one nuclear spin in the two different electron spin manifolds ($m_S = \pm 1/2$). Especially in these 2D experiments a high sensitivity is obtained via a magnetic field setting where the nuclear Zeemann splitting has approximately the same value as the hyperfine coupling (cancelation condition) [60]. Otherwise matched pulses can be applied to increase the modulation depth [61].

3.2.2 Electron Nuclear Double Resonance

In comparison to ESEEM, ENDOR (Electron Nuclear Double Resonance) [62] is more demanding on the instrumentation side. As the method requires efficient excitation of both the electron spin (typically microwave frequency region) and the nuclear spins (typically radio frequency region) a more sophisticated double resonance structure has to be employed. On the other hand, the obtained ENDOR spectrum is usually simpler to analyze because it only depends on allowed electron spin transitions. This method shows optimum performance at high external magnetic field strengths and can also be applied to radicals in liquid solution. In addition to CW ENDOR, several pulsed ENDOR sequences exist. In the case of large hyperfine couplings the Davies ENDOR sequence [63] is commonly used, whereas the pulsed Mims ENDOR experiment [64] is of advantage for detecting small hyperfine couplings arising from distant nuclei (Fig. 6). Similar to the 3-pulse ESEEM, Mims ENDOR has "blind spots", which can be circumvented by taking spectra with different pulse separation times. In all ENDOR experiments the change of the electron spin resonance signal as a function of the radiofrequency excitation frequency is observed. The spectra are rather easy to interpret: in the case of a hyperfine coupling smaller than the corresponding nuclear Larmor frequency the two ENDOR lines (for a nuclear spin $I = 1/2$) are centered at the nuclear Larmor frequency and split by the effective hyperfine coupling constant. This allows distinguishing and assigning different nuclei by their respective Larmor frequency. Again the separation between nuclear spins with similar gyromagnetic ratios becomes better at higher magnetic fields. Moreover, the high spin polarization at high magnetic fields reveals the sign of the hyperfine interaction [65]. Alternatively, advanced pulse and CW ENDOR methods also allow this at lower magnetic field strengths [66]. At lower magnetic fields the mixed nuclear spin states hamper the interpretation of ENDOR spectra and hinder the optimum performance. Additionally, an efficient excitation of the nuclear spin transitions at rf-frequencies below 5 MHz is cumbersome. In such cases the ESEEM techniques are more sensitive and successful. Therefore, both methods are mostly complementary and it is recommended to use both in order to obtain maximum hyperfine information.

3.3 *Pulsed Electron-Electron Double Resonance*

As explained in Sect. 3.1, CW EPR spectra of nitroxide radicals in solution reveal hyperfine splitting owing to the nitrogen nuclear spin interaction and can be used to monitor the molecular motion. If at least two spin labels are attached to the NA the splitting caused by the electron dipolar spin–spin interaction may also be detected for distances less than 2 nm, provided that the inhomogeneous line width is smaller than the dipolar coupling strength and is not averaged by fast rotational tumbling. For distances exceeding 2 nm the interaction strength is smaller than 10 MHz and is therefore not resolved any more in a CW EPR spectrum. In such cases pulsed electron-electron double resonance (PELDOR) can be used to increase the spectral resolution. The spectral resolution is only limited by the transverse relaxation time T_2 because the inhomogeneous linewidth is refocused. The PELDOR experiments are usually performed at low temperatures (~50K), where a frozen ensemble of molecular conformers contributes to the total signal. At this temperature the electron spin relaxation time is long enough to measure dipolar couplings in a corresponding distance range of up to 8 nm.

In a 4-pulse PELDOR experiment (see Fig. 6) the refocused echo signal of spin A of an A–B spin pair is measured via excitation with microwave frequency ν_A. The pulse sequence refocuses the inhomogeneous contributions to the linewidth. A pump pulse with frequency ν_B inverts the spin state of the second spin B. If A and B spins are dipolar coupled then a modulation of the refocused echo magnitude is observed depending on the time delay T between the Hahn echo position of spin A and the pump pulse (Fig. 7). Lengths and frequency difference of the microwave pump and probe pulses are chosen to avoid strong spectral overlap, but short enough to achieve sufficient excitation of the electron spins. This is important to achieve good detection sensitivity and signal modulation.

The PELDOR signal of a spin ensemble can be described as a sum of the products of individual pair interactions between spins k and j:

$$V(T) = \sum_{k=1}^{N} \prod_{j \neq k}^{N} m_k \left(1 - \lambda_j + \lambda_j \cos\left(\frac{DT}{R_{jk}^3} \left(1 - 3\cos^2\theta_{jk}\right) \right) \right) \tag{12}$$

Here m_k is the transversal magnetization of the kth spin at the time of the refocused echo and λ_j is the inversion efficiency of the jth spin by the pump pulse. R_{kj} and Θ_{kj} define the length and orientation with respect to the static magnetic field of the vector connecting both electron spins, respectively. D is the dipolar interaction constant, given as $(2\pi \times 52.18)$ MHz·nm^3 for two radicals.

In the case where the paramagnetic centers are randomly distributed within the sample the average of (12) leads to an exponential expression of the signal decay:

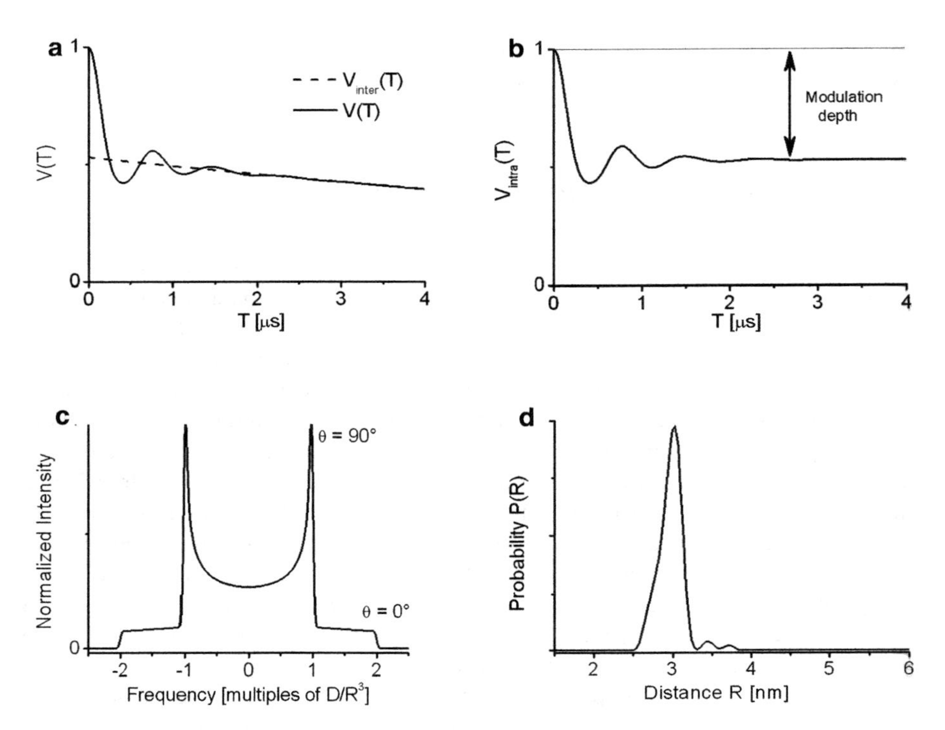

Fig. 7 PELDOR signal analysis. (**a**) Time domain PELDOR signal as a function of the delay time T of the pump pulse. The *dashed line* shows the exponentially decaying intermolecular dipolar contribution to the signal. (**b**) Time domain PELDOR signal after division of the original PELDOR time domain data by the fit-function representing the intermolecular decay. (**c**) Fourier transform of the PELDOR time trace (**b**) representing the dipolar Pake-pattern. (**d**) Distance distribution function obtained from the PELDOR time traces (**b**) by Tikhonov regularization. From the last representation the distances for spin pairs A-B can be the most easily extracted

$$V_{\text{inter}}(T) = V(0)\exp\left(-\frac{4\pi^2\lambda c}{9\sqrt{3}}DT\right) \tag{13}$$

where c is the spin concentration in the sample and λ is the fraction of spins excited by the inversion pulse. Typically, this formula is used to describe the intermolecular part of the PELDOR signal decay for spin labeled molecules (dashed curve in Fig. 7a). More interesting is the specific intramolecular interaction between the spin labels attached to one individual NA or a specific NA–protein complex. This intramolecular PELDOR signal for N spin labels per randomly oriented macromolecules in the sample can be calculated according to [67] by using

$$V_{\text{intra}}(T) = \frac{1}{N}\sum_{k=1}^{N}\prod_{j\neq k}^{N}\left(1-\lambda+\lambda\int_0^{\pi/2}\cos\left(\frac{DT}{R_{jk}^3}\left(1-3\cos^2\theta\right)\right)\sin\theta\,\mathrm{d}\theta\right) \tag{14}$$

The experimentally observed PELDOR signal is finally a product of the inter-molecular decay function V_{inter}, given by (13) and the intra-molecular function V_{intra}, described by (14).

In most EPR applications, the molecules under study are double-labeled, simplifying (14) for the intra-molecular decay function to

$$V(T) = 1 - \lambda + \lambda \int_0^\infty f(R) \int_0^{\pi/2} \cos\left(\frac{DT}{R^3}\left(1 - 3\cos^2\theta\right)\right) \sin\theta \, d\theta \, dR \qquad (15)$$

where $f(R)$ describes the distance distribution function between the two spin labels. This equation is most frequently used for the quantitative analysis of experimental PELDOR time traces. It allows a parameter free extraction of the distance distribution function $f(R)$ and therefore the determination of the average distance between the two unpaired spins of the spin-labeled macromolecule from PELDOR measurements. A typical example of such a data analysis is depicted in Fig. 7. In a first step the intermolecular decay is separated from the original signal. The experimental decay function at long times T is fitted with an exponential function and then the experimental PELDOR time trace is divided by this function (Fig. 7b). After that, the remaining PELDOR signal, which represents the intra-molecular interactions V_{intra}, is analyzed to determine the average distance R or the distance distribution function $f(R)$. The singularities of the dipolar Pake pattern, obtained after Fourier transformation, can be used to extract the average distance R (Fig. 7c), or the Fredholm equation of the first kind (15) can be solved using Tikhonov regularization methods to obtain the distance distribution function $f(R)$.

However, it should be mentioned that this procedure is only valid for spin-pairs ($N = 2$) with a random mutual orientation. If the two spin labels have a fixed relative orientation, the inversion efficiency depends on the specific orientation of the molecule with respect to the external magnetic field, leading to a more elaborate calculation of the PELDOR time trace:

$$V(T) = 1 - \int_0^\infty f(R) \int_0^{\pi/2} \lambda(\theta)\left(\cos\left(\frac{DT}{R^3}\left(1 - 3\cos^2\theta\right)\right) - 1\right) \sin\theta d\theta dR \qquad (16)$$

In general, the separation of the distance distribution function $f(R)$ from the part containing the mutual angular orientation between the two spin labels is more cumbersome and not straightforward [68–71]. However, it allows one to obtain additional valuable information regarding structure and flexibility of the macro-molecule under investigation, as will be shown in Sect. 4.2.2.

4 Applications to Nucleic Acids

4.1 CW EPR Spectroscopy

CW EPR spectroscopy is extensively used to probe local and global dynamics of nucleic acid molecules and to correlate these data with NA structures and functional aspects. This is particularly useful to investigate and monitor NA–protein or NA–ligand interactions. In the following we are going to highlight some recent applications of CW EPR to study RNA and DNA molecules.

4.1.1 CW EPR Studies of RNAs

Secondary structure elements, such as duplexes, single-stranded regions, hairpin loops, bulges, mismatches, internal loops, abasic sites, and junctions, have been thoroughly investigated by CW EPR spectroscopy [72]. For this purpose, single-labeled NA molecules carrying fairly rigid spin labels have been widely employed. In viscous media, such as buffer solutions containing 20–30 wt% sucrose at 0°C, the global motion of the NA molecules is strongly reduced and the CW EPR spectra are mostly sensitive to internal spin label mobility. As the spin label is rigidly connected to the nucleotide base, the nitroxide mobility reports on the nucleotide dynamics occurring on picosecond–nanosecond timescales, which are accessible via CW EPR. Thus nucleotides located in different secondary structure elements can easily be distinguished. As an illustration, in Fig. 8 we compare X-band (9.5 GHz) CW EPR data of a neomycin-responsive riboswitch containing spin-

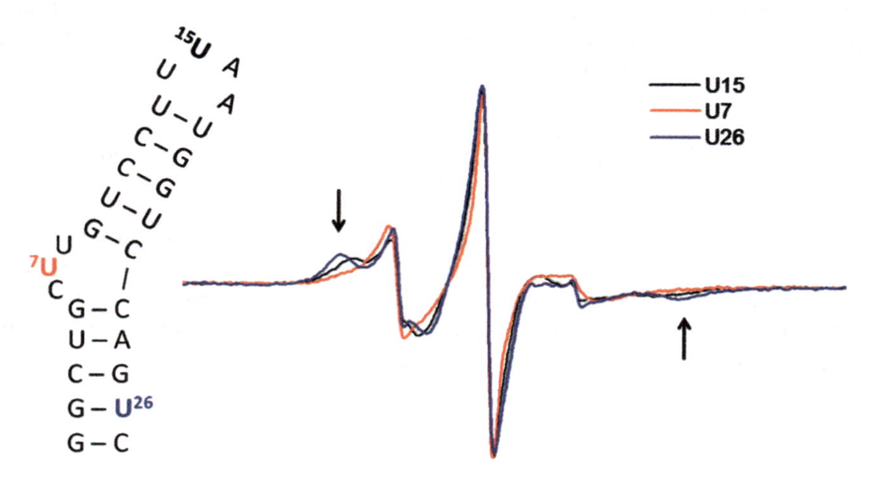

Fig. 8 The secondary structure of the neomycin-responsive riboswitch (*left*) and CW EPR data of the single-labeled RNA samples at 0°C in 20 wt% sucrose solution (*right*)

label **4** (Fig. 1) at either the duplex (U26), bulge (U7), or terminal loop region (U15) [73].

In each region the nitroxide displays a different spectral width, reflecting the different mobility of these parts of the RNA molecule. The most immobilized nucleotide is located in the closing stem (U26). The spectrum of this site is characterized by a splitting of both low field and high field peaks, due to partial averaging of the anisotropic hyperfine tensor. The nucleotide U15 resides in the terminal loop, which is a well known RNA motif called U-turn, defined by a sequence UNRN (N-any nucleotide, R-purine). The conformation of this motif is stabilized by hydrogen bonds between the imino proton of U and a phosphate oxygen of the last nucleotide in the motif and between 2′-OH of U and N7 in the purine base (R) as inferred from crystallographic data [74]. Thus, the spin label at U15 reveals only a slightly narrower hyperfine splitting due to the high order of the nucleotide U15 in the loop. A much higher mobility is expected for a bulge region, which is the least structured part of the riboswitch, as known from NMR [75]. Indeed, the spin label attached to U7 shows the highest level of hyperfine averaging and the narrowest EPR spectrum.

Conformational dynamics and tertiary structure of RNA molecules are essential for their functions as catalysts, regulatory elements, or structural scaffolds. Tertiary structural elements between helical and unpaired regions of the molecule, e.g., pseudoknots or "kissing" loops, are stabilized through van der Waals interactions, specific hydrogen bonds, metal ions, or π-stacking interactions. The motional dynamics of such motifs and the changes induced by their interaction with metal ions, small organic molecules, and peptides can be probed by CW EPR spectroscopy. Some illustrative examples of such applications will be described in the following.

The interaction between the trans-activation responsive (TAR) RNA of the human immunodeficiency virus (HIV) with the trans-activator of transcription (Tat) protein is essential for production of full-length RNA transcripts during viral replication and has been studied via CW EPR in detail by Sigurdsson and collaborators [47, 76–79]. The TAR RNA consists of two helical regions, connected by a tri-nucleotide bulge which acts as a joint between them. The motion between the two helices is strongly reduced upon addition of Ca^{2+} or cognate peptide, leading to a coaxial stacking. The changes in the spectral width of the CW EPR spectrum of spin label **21** (Fig. 2) were measured at four different labeling positions (Fig. 9) upon interaction with metal ions, inhibitors, and the Tat motif. From the dynamic signature obtained for each interaction partner, the authors concluded that the conformational change does not require divalent metal ions because Na^{+} induces the same dynamic response as Ca^{2+}. They suggested that TAR is able to adapt its conformation to accommodate binding of metal ions with different size and coordination properties in the bulge region [77]. The dynamic response predicts that Arginin 52 (R52) is essential for Tat binding and that argininamide has a binding motive similar to that of the wild type Tat protein [78, 79]. Further investigations from the same authors with point mutations at the C- and N-terminal also anticipated a possible role of R56 for the Tat protein

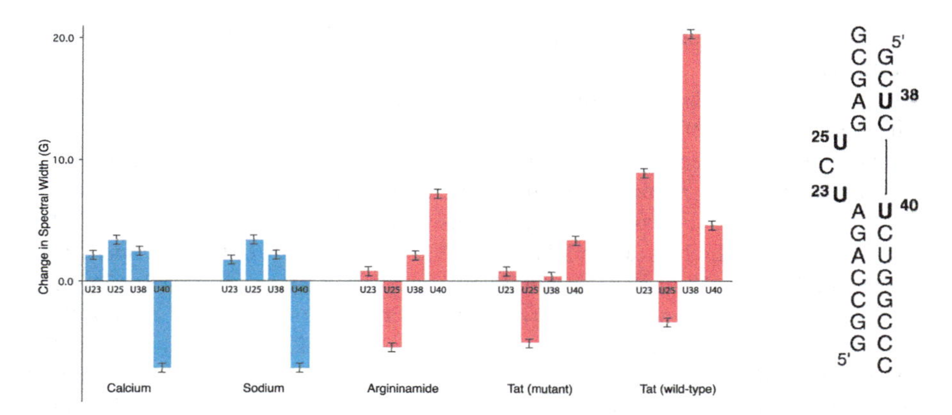

Fig. 9 Changes in CW EPR spectral width at four different labeling positions of TAR as a function of the interaction with calcium, sodium, argininamide, and Tat protein. The figure is from [79] with permission of the journal

binding. This point mutation affected the mobility of nucleotides U23 and U38, which are part of a triple base interaction with the Tat protein [76].

DeRose and coworkers have explored conformational changes of TAR RNA upon binding of divalent metal ions (Ca^{2+}) by measuring the dipolar coupling between two attached spin labels **20** using CW EPR (Fig. 2). The U25–U40 distances obtained from Fourier deconvolution methods are 11.9 ± 0.3 Å for TAR RNA in the absence of divalent metal cations and 14.2 ± 0.3 Å when 50 mM Ca^{2+} was added [45]. These results are in accordance with the proposed coaxial stacking of the two TAR helices upon addition of metal ions based on the X-ray crystal structure [80].

The group I intron of *Tetrahymena thermophila*, the first identified catalytic RNA molecule, is another example of a tertiary structured RNA which was investigated by CW EPR [39]. The exon substrate of this ribozyme forms a double-stranded region (duplex P1) with an internal guide sequence. Docking of P1 controls the substrate cleavage. Qin and coworkers have examined the nanosecond dynamics of the P1 duplex, with spin label R5a (bromo-substituted **18**; see Fig. 2) conjugated to the Sp-phosphorothioate diastereomer. CW EPR spectra of two spin-labeled RNA sequences, designated as S_C^{SL} and S_O^{SL} in Fig. 10, favoring either the open or closed conformation, revealed lower nitroxide mobility in the closed state of the ribozyme. The authors attributed this decreased P1 duplex dynamics to the requirement for proper positioning of the bound substrate for the catalytic reaction. Additionally, they have shown that in the open state of the ribozyme the length of the J1/2 junction can influence the P1 duplex mobility [39].

Interaction between the GNRA hairpin loop and the asymmetric internal loop (receptor) is one of the most abundant long-range tertiary interaction motifs in large RNAs, such as group I and II introns. Hubbell, Qin, and coworkers have investigated the hairpin–receptor complex formation in the presence of Mg^{2+} ions by CW EPR employing spin label **18** (Fig. 2) attached to an RNA hairpin containing

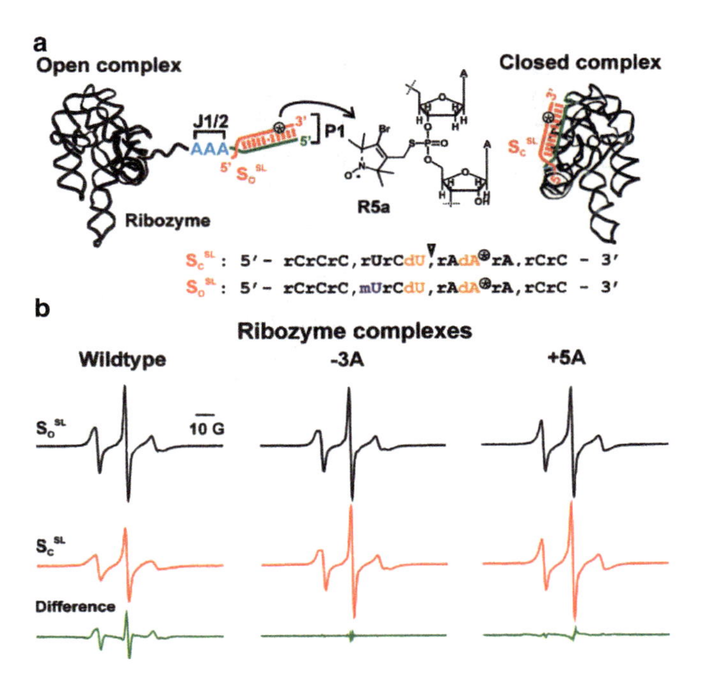

Fig. 10 CW EPR probing of P1 helix docking. (**a**) Schematic renditions of the open and closed ribozyme complex. The substrate sequences, the R5a spin label (*asterisk*), the cleavage site (triangle), and the J1/2 junction are marked. (**b**) EPR spectra of various substrate/ribozyme complexes. The figure is from [39] with permission of the journal

the GAAA tetraloop [41]. Monitoring the changes in the rotational correlation time of the labeled RNA tetraloop, the formation of tetraloop–receptor complex due to the increase in molecular size upon binding to the receptor has been detected. The presence of Mg^{2+} ions was required for the complex formation, and the binding constant was determined to be 0.40 ± 0.05 mM. A free energy of −4.6 kcal/mol has been calculated for this weak tetraloop–receptor interaction, assuming the absence of other tertiary constraints.

Using spin-labeled nucleobase **2** (Fig. 1) incorporated in a tetraloop receptor sequence (designated as TLR), Hubbell, Qin, and coworkers have studied conformational changes induced by the GAAA tetraloop (designated as TL) docking in the presence of Mg^{2+} [81]. As shown in Fig. 11, in the bound state of the receptor, the mobility of the U19 base significantly increases, indicating unstacking of the base upon tetraloop binding. This finding is consistent with the crystal structure of a group I intron ribozyme domain containing the tetraloop–receptor complex, in which the base equivalent to U19 is unstacked. On the other hand, spectral changes were not detected at U5 and U17 within the receptor sequence, suggesting no structural rearrangement in this local environment.

Most catalytically active RNAs require divalent metal ions, primarily Mg^{2+}, to accomplish full catalytic activity. CW EPR has been utilized to study the metal ion

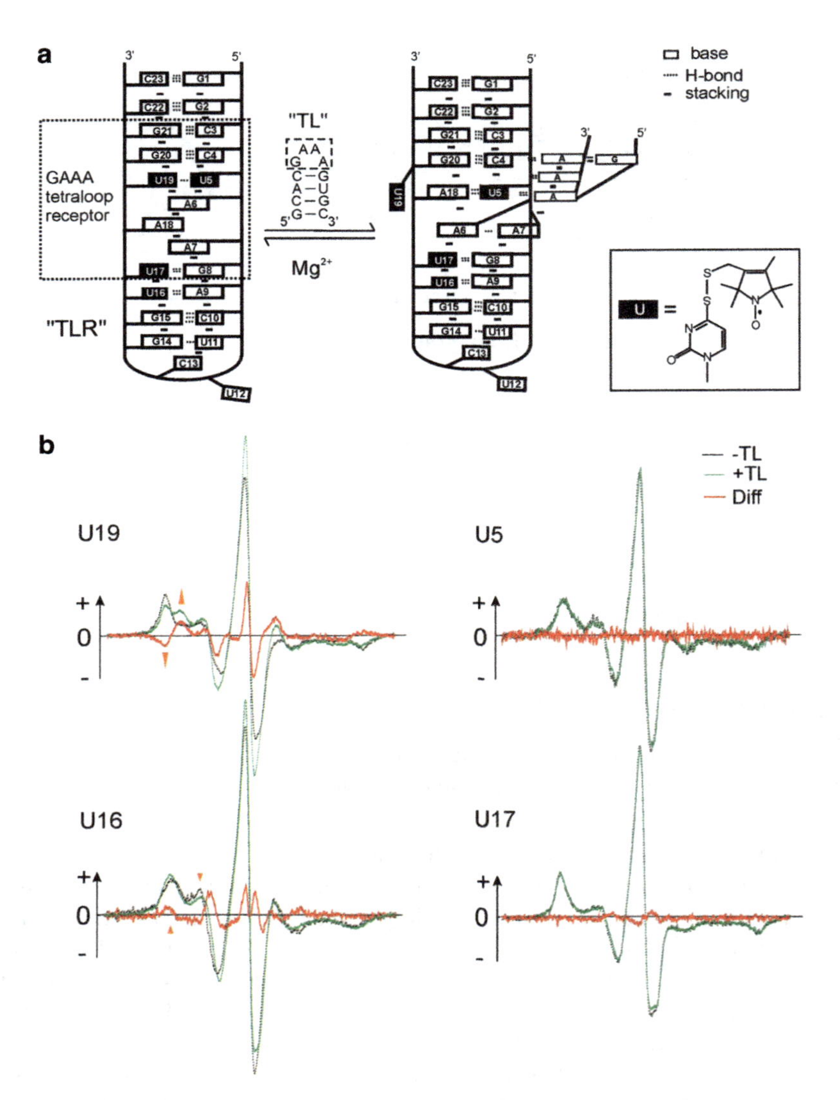

Fig. 11 Conformational changes of the tetraloop receptor upon interaction with GAAA tetraloop detected by CW EPR. (**a**) Schematic representations of receptor structures, with the *dotted box* indicating the 11 nucleotide GAAA tetraloop receptor motif. (**b**) Spectra observed in the absence (−TL, *black traces*) and presence (+TL, *green traces*) of tetraloop. The difference spectra (Diff, *red traces*) were obtained by subtracting the −TL spectra from the corresponding +TL spectra. This figure is from [81] with permission of the journal

binding affinities in the hammerhead ribozyme (HHRz) and the Diels–Alder ribozyme. For such studies, the naturally occurring diamagnetic Mg^{2+} has to be substituted by the paramagnetic Mn^{2+} ion ($S = 5/2$, $I = 5/2$). Despite the fact that the catalytic activity of ribozymes is reduced by metal exchange (even though both metal ions have similar ionic radii, charge, coordination geometry, and ligand exchange rates [82]), it has been shown by X-ray crystallography that Mn^{2+} occupies the same binding sites as Mg^{2+} in the minimal HHRz (mHHRz) [83, 84]. CW EPR spectroscopy can easily distinguish the free Mn^{2+} ions in a hexa–aqua complex in liquid solution from Mn^{2+} ions bound to the ribozyme, thus allowing one to count binding sites and to determine binding constants by titration of Mn^{2+} into the solution.

The mHHRz contains a single Mn^{2+} binding site with an affinity of 10 μM in 1 M NaCl solution, whereas at lower ionic strength (0.1 M NaCl) four binding sites are identified with an overall K_d of 4 μM [85]. The aminoglycoside antibiotic neomycin B inhibits the catalytic activity of this ribozyme [86]. Prisner, Schiemann, and coworkers could show that the bound Mn^{2+} ion is released upon binding of neomycin B ($K_d = 1.2$ μM) to the hammerhead ribozyme. The tertiary stabilized extended HHRz possesses a single high-affinity Mn^{2+} binding site with a $K_d \leq 10$ nM at an NaCl concentration of 0.1 M. This dissociation constant is more than two orders of magnitude smaller than the K_d determined for the single high-affinity Mn^{2+} site in the mHHRz [87]. The addition of two loops that interact between stems I and II leads to an enhanced cleavage activity of HHRz at low Mg^{2+} concentrations [88]. Kim et al. have probed the conformational dynamics of this structural element as a function of concentration of added Mg^{2+} by using the spin-labeled nucleotides **21** (Fig. 2) located in the substrate stem. The splitting of both low field and high field lines of the nitroxide spectra observed for the spin-labeled U1.6 (position 6 in stem I) are ascribed to the docking of stems I and II originating from the folding of the HHRz. Spectra recorded at different Mg^{2+} concentrations were fitted by two sites with distinctive mobility. Thus, it was concluded that the docking of stems I and II occurs at low Mg^{2+} concentrations ($\leq$1 mM) [88]. A detailed description of the structure of the metal binding site in HHRz has been achieved by using hyperfine spectroscopy as will be described in more detail in Sect. 4.2.1.

Similar CW EPR titration experiments have been performed by Schiemann and coworkers on the Diels–Alderase ribozyme [89]. It was demonstrated that five high-affinity Mn^{2+} binding sites with an upper K_d of 0.6 ± 0.2 μM exist in this ribozyme. From the spectral changes on stoichiometric mixtures of Mn^{2+} and Cd^{2+} the binding sites could be assigned to three different types: inner sphere, outer sphere, and a dimeric site.

RNA–protein interactions play a central role in cellular processes; therefore the investigation of structure, conformational dynamics of such complexes, and their relation to biological function is of major importance. Again, CW EPR spectroscopy can help to provide detailed understanding of such processes and we will state two selected examples in the following.

The interactions between a 22-amino acid long peptide (called N-peptide) and a nascent mRNA hairpin element (called boxB) are of a crucial function in transcription anti-termination. Qin and coworkers have studied this N-peptide/boxB complex by monitoring the nanosecond dynamics of nitroxide spin labels attached to different positions of the N-peptide [90]. Analysis of CW EPR spectra recorded in viscous aqueous solutions at 5°C indicates that the bound N-peptide exists in a dynamic equilibrium between two different conformational states, with the peptide C-terminus either stacking on the RNA loop or pointing away from the peptide/RNA interface. At the same time, the N-terminus of the N-peptide adopts a single well defined conformation, confirming that the arginine-rich motif accounts for boxB binding. Based on these findings, the authors suggested that the stacked state is responsible for the function of the N-peptide/boxB [90]. Another example is the interaction between a 20-mer HIV-1 RNA stem loop 3 and the HIV-1 nucleocapsid Zn-finger protein NCp7, investigated by Scholes and coworkers using a stopped-flow CW EPR setup, with a time resolution of about 4 ms [91]. In their experiments, spin-labeled RNA stem loop 3 was mixed with the protein NCp7 in a ratio of 4:1 and the decrease in amplitude of the central line of the nitroxide EPR spectrum recorded as a function of reaction time. The stopped-flow CW EPR signal showed that the main interaction of NCp7 with the RNA stem loop 3 occurs within the dead time of the apparatus (<4 ms); a second contribution with a time constant of ~30 ms and a much slower immobilization over seconds (possibly concomitant with large complex formation).

4.1.2 CW EPR Studies of DNAs

Similar to RNA, secondary structure elements of DNA (duplexes, hairpin loops, bulges, mismatches, junctions, and abasic sites) have been identified and characterized by CW EPR methods by several research groups, using both rigid and flexible spin labels [92–94].

Robinson and coworkers stated that the nitroxide mobility increases as spin label **14** (Fig. 1) is placed further away from the middle of the DNA duplex [95]. This was recently confirmed by Qin and coworkers employing the flexible spin label **18** (Fig. 2) attached to a specific phosphate group on the backbone [96]. In this case, the nitroxide internal motion depends on the three-dimensional structure of the DNA molecule, which enforces restrictions to the sterical accessible volume and therefore rotamer freedom of the nitroxide spin label. The specific base pair dependence of the internal conformational dynamics of duplex DNA molecules has been investigated thoroughly by Robinson and coworkers. Modeling their data on the basis of a weakly bending rod, they demonstrated that A–T base pairs are 20% more flexible than average base pairs in the 50-mer control sequence from *Drosophila melanogaster* TATA-box binding protein TFIID gene [32]. Simulations based on a model treating the spin-labeled duplex as a diffusing cylinder containing internal dynamics indicated that Z-DNA is more rigid in comparison to B-DNA [97]. Incorporation of an extremely rigid spin label **15** (Fig. 1) into double-stranded DNA enabled Sigurdsson and coworkers to detect the end-to-end helical stacking [35]. This was achieved by observation of the dipolar coupling between spin labels positioned at the 5′; end positions from CW EPR spectra. This additional dipolar line splitting diminished after addition of 10 equiv. of unlabeled duplex.

CW EPR spectra can also be used to characterize more flexible parts in secondary structure elements in DNA molecules: a spin-labeled G–T mismatch and a T bulge in long DNAs yielded EPR spectra clearly distinct from those of spin-labeled

Watson–Crick paired double-stranded DNAs, pointing towards increased mobility of these structural elements [17]. Single base mismatches in duplex DNA were successfully identified with the flexible spin label **11** (Fig. 1), whereas a rigid spin label **15** (Fig. 1) was not sensitive to changes in the motion induced by these impairing. Such spin labels conjugated to the exocyclic amino group of a cytosine base can not only be applied to detect mismatches but also to identify the base-pairing partner in DNA duplex [93]. The structural flexibility of single base bulges adjacent to a single base mismatch in duplex DNA was studied by Robinson and coworkers employing the same rigid spin label **15**. According to their findings the "bulge-mismatch" motif can be described by equilibrium between two possible conformations. The enthalpic and entropic contributions to the interconversion have been determined by analyzing the changes in population of the different states as a function of temperature [98].

The genetic information stored in DNA must be accessible to proteins that transcribe it into RNA or that assist in DNA replication. Gene expression is often regulated by proteins that activate or repress transcription by binding to short, specific DNA sequences. A detailed understanding of such DNA–protein interactions is therefore highly important. It is well accepted that protein–DNA recognition involves two processes:

1. Charge neutralization of the phosphate groups on the DNA backbone
2. Protein-induced DNA bending

Robinson and coworkers reported that the application of methylphosphonates as a model for an electrostatic neutralized phosphate backbone increased the backbone flexibility by 40% [99]. This entropically stabilized partially uncharged DNA duplex may be an intermediate along the protein-DNA binding pathway. Hubbell and coworkers have shown for a homodimeric motif in the DNA-binding transcription factor GCN4 (GCN4 bZip) that the protein backbone motion is attenuated when it is bound to DNA [100]. Steinhoff and Suess employed CW EPR distance measurements to study the molecular mechanism of gene regulation. They could reveal a twisting motion of the DNA reading heads of a Tet repressor upon interaction with a tetracycline antibiotic [101].

4.2 *Pulse-EPR applications*

Throughout the last few years, pulse EPR experiments have been increasingly applied to study NAs. As explained in Sect. 3, such methods can be subdivided into two major categories: hyperfine and dipolar spectroscopy. Hyperfine spectroscopy has so far been used mainly in the context of NAs to probe the binding of Mn^{2+} metal ions to RNA structures, whereas dipolar spectroscopy has found more widespread applications in the investigation of secondary and tertiary structure and conformational changes on NAs.

4.2.1 Characterization of Metal Ion Binding Sites by Hyperfine Spectroscopy

Hyperfine spectroscopy methods, such as 1D ESEEM, 2D HYSCORE, and ENDOR, have been employed to determine the coordination sphere of Mn^{2+} ions bound to the HHRz or to the Diels–Alder ribozyme. The HHRz was first investigated by X-band stimulated echo ESEEM by Britt and coworkers [102]. The ESEEM data revealed nitrogen hyperfine coupling of $A(^{14}N)$ ~ 2.3 MHz, indicating that the Mn^{2+} in HHRz is coordinated by a guanine. One postulated site in stem II of the ribozyme involves N7 of guanine G10.1 (the A9/G10.1 site). Definitive assignment of the high affinity site to this region was achieved by site-specific labeling of G10.1 with ^{15}N guanine [103]. ENDOR performed at Q-band frequencies (34 GHz) enabled DeRose and coworkers to observe the hyperfine coupling to ^{31}P ($A(^{31}P)$ ~ 4 MHz) providing evidence for a direct coordination (inner sphere) to a phosphodiester group in the hammerhead Mn^{2+} site [104] (Fig. 12). This finding was further supported by W-band ENDOR data on tertiary stabilized HHRz (tsHHRz) [105]. The W-band (95 GHz) proton ENDOR spectrum of the Mn-ribozyme sample shows features almost identical to those from $[Mn(H_2O)_6]^{2+}$, implying that the Mn^{2+} in the ribozyme contains one or more water molecules in its inner coordination sphere. Non-exchangeable proton(s) at ~ 3.6 Å were assigned to the C8-H of a coordinated purine base [104]. The coordinated

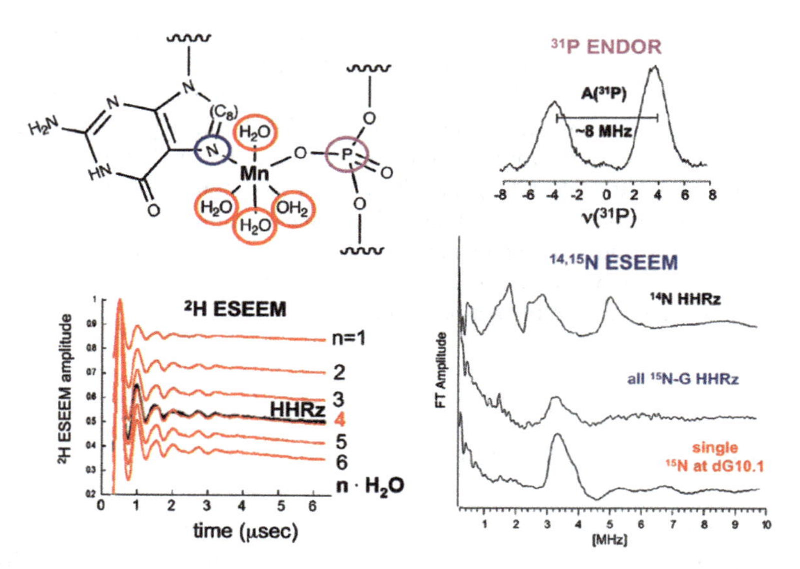

Fig. 12 Summary of hyperfine spectroscopy results on Mn^{2+} bound to HHRz. *Top right*: Q-band phosphor ENDOR showing a hyperfine splitting consistent with a inner-sphere coordination to the phosphor-diester oxygen. *Lower right*: Fourier transform of X-band ESEEM measurements showing the coordination to the nitrogen of dG10.1. *Lower left*: X-band ESEEM spectroscopy of HHRz in deuterated water. A quantitative comparison with simulations allows to determine the number of coordinated water molecules. Figure adapted from [106] with permission of the journal

metal ion retains four water ligands as measured by $^{1}H/^{2}H$ HYSCORE spectroscopy for tsHHRz and mHHRz [86, 87, 103].

Another application of ENDOR and HYSCORE techniques on NAs represents the study of the coordination environment of Cu^{2+} ions in poly (dG-dC)· (dG-dC) polymer reported by Jeschke and coworkers [107]. Two different copper species were detected and analyzed. Copper I was characterized by a weak ^{14}N coupling and a moderate ^{1}H coupling, which are assigned to a Cu^{2+} binding to N7 and H8 atoms of a single guanine. The authors proposed that species II forms coordinative bonds with N7 and H8 atoms of a guanine and with N3 and H5 of cytosine in a metal mediated Hoogsteen base pair.

4.2.2 Dipolar Spectroscopy to Study Structure and Conformational Changes of NAs

Within the last 10 years PELDOR [108–110] has demonstrated its applicability and reliability to provide structural information on NAs. In the following we want to highlight some applications of this method on double-helical and more complex doubly spin-labeled DNA and RNA molecules to determine distances and distance changes with high accuracy.

Secondary Structure of NAs

Several research groups demonstrated on double-labeled helical RNA and DNA molecules that very precise distances in a range of 2–7 nm could be obtained and that the structure of the NAs was not severely altered by the spin labeling procedure. The first PELDOR experiments on NA molecules were performed by Prisner, Schiemann, and coworkers on a self-complementary 12 bp long duplex RNA molecule double-labeled at the sugar moiety at the 2′; position by ureido-linked spin label **21** (Fig. 2) [111]. This enabled the measuring of an inter-spin distance of 3.5 ± 0.2 nm, in excellent agreement with the modeled structure. The development of the more rigid spin label **4** (Fig. 1) allowed extension of this work to distances between 1.9 and 5.2 nm, establishing a PELDOR-nanometer distance ruler for NAs [7, 19], as illustrated in Fig. 13. Results of MD simulations, using all-atom force field with explicit water, were in good agreement with the measured distances, indicating that the NAs retain their structure in frozen aqueous solution and that the attached spin labels induce only small local distortions [21]. The precision of the PELDOR experiment allowed the distinguishing of A-form RNA from B-form DNA duplex structures.

Hubbel, Qin, and coworkers exploited spin label **18** (Fig. 2) attached to the phosphate group of the backbone during chemical synthesis of DNA or RNA for PELDOR experiments. The extracted distances showed good correlation with the distance distributions predicted by the NASNOX computer program that they

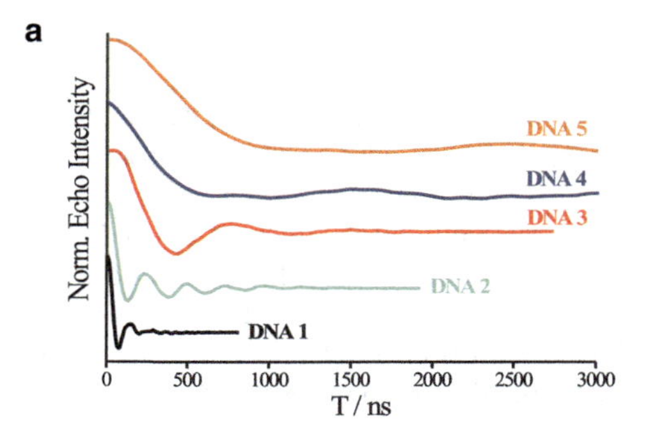

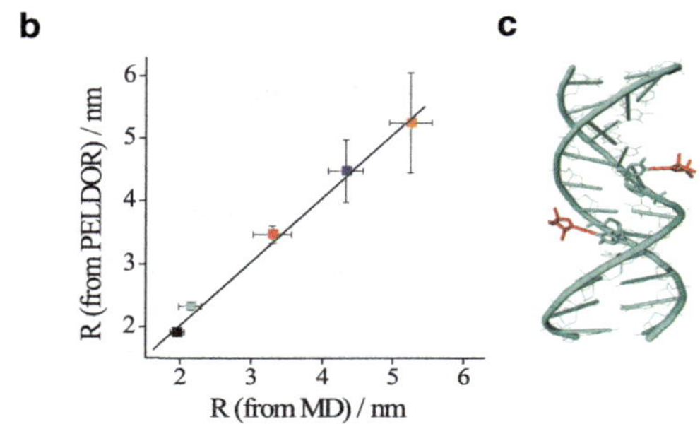

Fig. 13 (**a**) PELDOR time traces for double-stranded DNA molecules double spin-labeled at different positions. (**b**) Comparison of the distances measured by PELDOR and calculated by MD. (**c**) Modeled structure of the doubly spin-labeled DNA 2. Figure adapted from [19]

developed for calculating the phosphorothioate diastereomers and the allowable R5 conformations [40, 42, 43].

Ward et al. have employed the label **21** (Fig. 2) to investigate five double-labeled DNA duplexes individually and in mixtures [112]. The distances obtained in the range of 2.8–6.8 nm were in good agreement with B-form DNA. The deconvolution of multiple distances from simple heterogeneous samples was quantitative, but failed for more complicated mixtures. Seela and coworkers have applied the azide–alkyne "click" reaction to incorporate spin labels **9** and **10** (Fig. 1) into B-form DNA. Reporter groups have been introduced into the major groove and short distances measured by CW EPR (1.4 ± 0.2 nm) and PELDOR (1.8 ± 0.2 nm) have proven the general applicability of this approach to study structural aspects of DNA [26].

The novel nitroxide spin label **15** (Fig. 1) was covalently incorporated into DNA by Sigurdsson and coworkers [33]. It was integrated into the polycyclic fused ring system of a cytidine analog that forms a base pair with guanine. Therefore, the spin label cannot rotate around the interconnecting bonds leading to very precise distance determination without spin label rotamers. Moreover, the relative orientation of the two nitroxide spin labels with respect to the distance vector can be extracted from the PELDOR time traces taken with variable observer frequency because the out-of-plane axis of the nitroxide is fixed with respect to the DNA molecule (Fig. 14) [113].

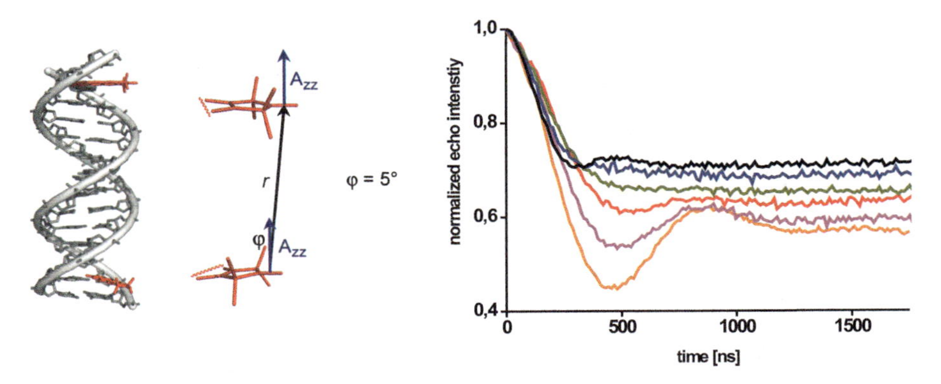

Fig. 14 Orientation selective PELDOR experiments on ds-DNA molecules with a rigid spin label **15** incorporated. The geometry between the out-of-plane vectors with respect to the interconnecting vector can be easily followed by measuring PELDOR as a function of the observer frequency, as shown on the *right side*. Figure adapted from [113]

This spin label acts as a rigid cantilever, attached to the DNA molecule, and therefore allows a very precise determination of the conformational flexibility of double-stranded DNA as stretching, twisting, and bending [114].

Selectively attached spin probe **8** (Fig. 1) was employed by Sicoli et al. to monitor quantitatively B-A conformational change of double-stranded DNA [115]. The B-A transition was induced by a high concentration of trifluoroethanol (higher than 70 vol.%) and very good agreement between the distances measured by PELDOR and the average distance obtained from MD simulations was achieved. These authors used the same spin label to study subtle structural changes in DNA induced by lesions as a model for DNA damaging. Distance measurements on a set of undamaged and damaged DNA duplexes that contain 8-oxoguanine, a nick, a gap, a bulge, an abasic site analog, and an anucleosidic site were supported by molecular dynamics studies [116].

Höbartner and coworkers have recently employed spin label **11** (Fig. 1) to detect simultaneously two competing structures of the incompletely self-complementary RNAs, the hairpin and duplex [30]. In the hairpin conformation the two spin labels are 6 bp apart, resulting in a distance of 1.8 nm (Fig. 15). Upon addition of complementary RNA strand the hairpin structure becomes disrupted and a continuous 20-bp duplex is formed. In the newly formed helical structure the two TEMPO groups are 11 bp apart, yielding a distance distribution centered at 3.1 nm. By increasing the amounts of complementary RNA the ratio between the two coexisting structures shifts completely towards the RNA duplex.

Tertiary Structure of NAs

Krstić et al. applied PELDOR spectroscopy for the first time to map the global structure of a tertiary folded RNA molecule, precisely the neomycin-responsive

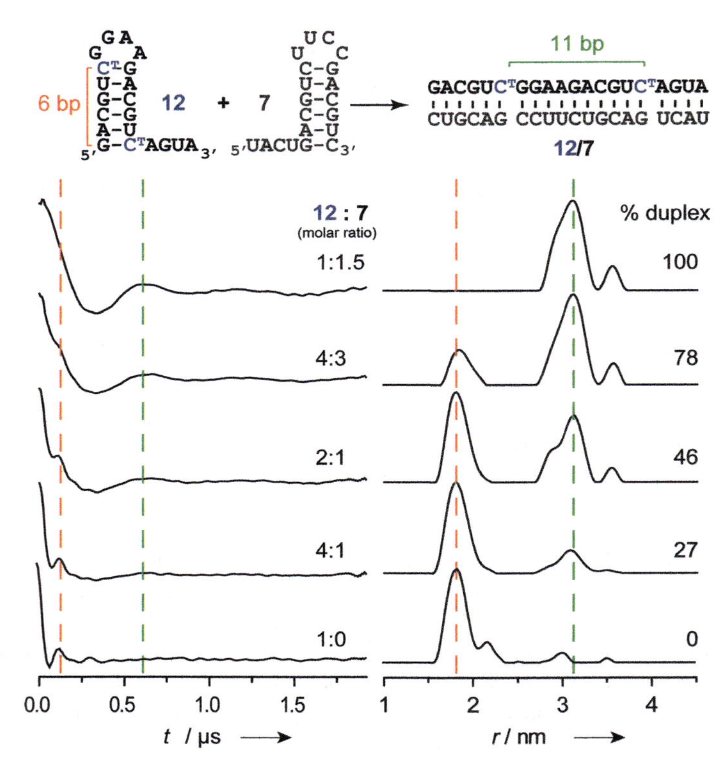

Fig. 15 Two RNA sequences, named 7 and 12, individually fold into a hairpin conformation but form an extended duplex when mixed. PELDOR experiments for different ratios of 7 and 12 are shown together with the distance distribution functions obtained from Tikhonov regularization of the PELDOR time traces. Integration of the peaks in the distance distribution allows a precise quantification of the different RNA structures in the mixture. Figure adapted from [30]

riboswitch [73]. It is an engineered riboswitch developed by combination of in vitro selection and in vivo screening [117]. Upon insertion into 5′; untranslated region of mRNA and binding the cognate ligand it is able to inhibit translational initiation in yeast. By enzymatic probing the secondary structure had been postulated comprising global stem-loop architecture with a terminal and an internal loop. The nucleotide labeling positions were chosen outside of the binding pocket and UV melting curves show that conjunction of the spin label **4** (Fig. 1) neither disturbs the secondary structure nor interferes with ligand binding. Efficient ligand binding was proven by thermal stabilization of 20.3 ± 3.3°C upon addition of neomycin, as well as by alteration of the CW EPR spectra. PELDOR time traces, as well as extracted distance distributions, for the aptamer with and without bound neomycin are shown in Fig. 16. The fact that there were no shifts in the measured distances upon addition of neomycin implied the existence of a prearranged tertiary structure of the neomycin-sensing riboswitch at low temperature, without a significant global conformational change induced by ligand binding. Measured distances were in very

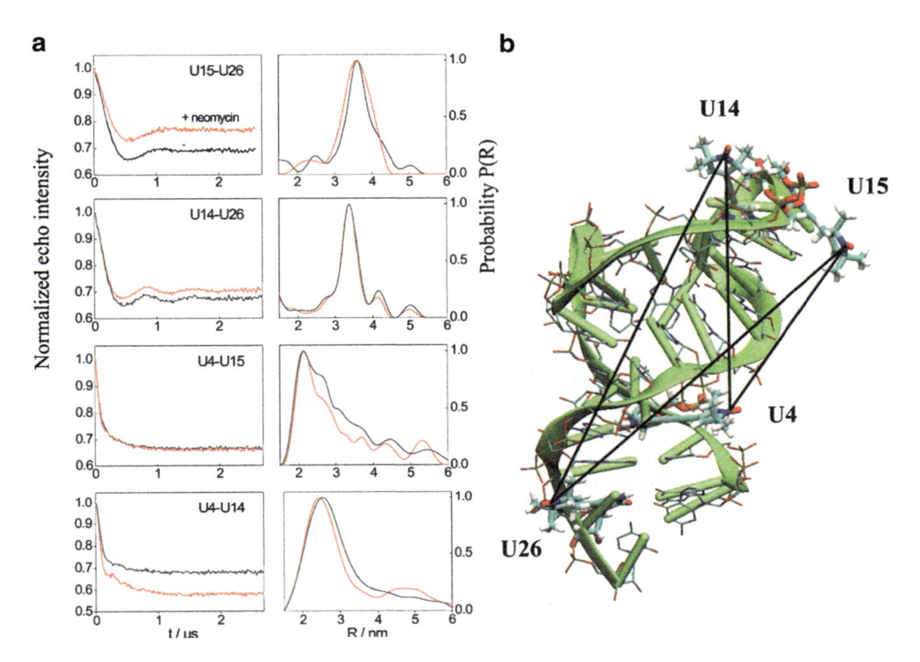

Fig. 16 (**a**) Background corrected PELDOR time traces and distance distribution functions of a double-labeled neomycin-responsive riboswitch in the presence (*red*) and absence (*black*) of the ligand. The measured distances are in good agreement with the NMR structure determined for the neomycin-bound state of the riboswitch. Figure adapted from [73]. (**b**) Model of labeled neomycin-sensitive riboswitch based on the NMR structure (pdb-code: 2kxm) with spin labels attached to C5-positions of U4, U14, U15, and U26

good agreement with the structure of the ligand-bound state of the riboswitch determined by NMR spectroscopy [73]. This indicates the intrinsic propensity of the global RNA architecture toward its energetically favored ligand-bound form [73].

The structure of human telomeric RNA repeat in the presence of K^+ ions has been investigated by Sicoli et al. [30]. Spin label **12** (Fig. 1) was introduced into the trinucleotide loop regions and efficient quadruplex formation was confirmed by UV melting and circular dichroism experiments. The distance of 3.7 ± 0.2 nm obtained from high quality PELDOR data is in accordance with the NMR structure of unmodified RNA, supporting the formation of a parallel-stranded G-quadruplex conformation. Drescher and coworkers have studied conformations of human telomeric DNA repeat in solutions containing different counter ions, employing spin label **7** (Fig. 1). They inferred that the quadruplex in the presence of K^+ ions experiences both the propeller and the basket conformation, whereas in Na^+ ions containing solution the antiparallel form is the only conformation present [118].

Spin probe **21** (Fig. 1) was introduced into stems I and II of the extended hammerhead ribozyme and folding process was investigated by measuring changes in spin-spin distances upon addition of Mg^{2+} ions [119]. DeRose and coworkers

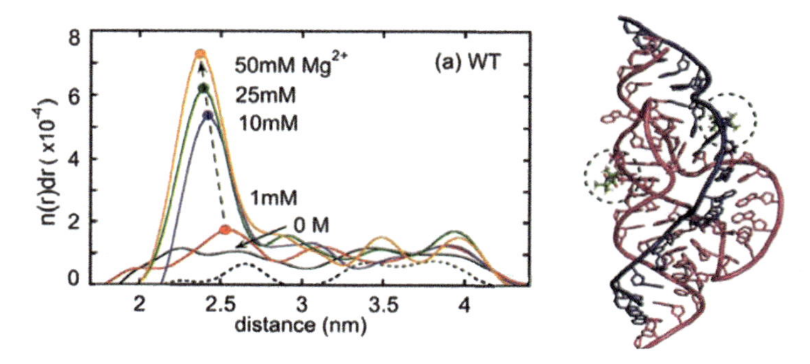

Fig. 17 Overlaid distance distribution functions for spin-labeled extended HHRz in 0.1 M NaCl, pH 7.0 at Mg^{2+} concentrations of 0 (*black*), 1 (*red*), 10 (*blue*), 25 (*green*), and 50 mM (*gold*). The *black dashed line* is for the isolated spin-labeled substrate in 1 mM Mg^{2+} and 0.1 M NaCl (*left*). Model of spin-labeled HHRz created from the crystal structure (PDB ID 2GOZ). Substrate is shown in *blue* and the enzyme is in *magenta*. The spin labels are shown in *green* with *dashed circles* (*right*). Figure adapted from [119]

reported that at low ionic strength the two stems of HHRz are randomly oriented resulting in very broad distance distribution. In contrast, at 10 mM and higher Mg^{2+} concentrations this RNA molecule is globally folded and the measured distance of ~ 2.4 nm is in accordance to a simple model constructed from a crystal structure of this tertiary folded ribozyme (Fig. 17).

Recently, Steinhoff and coworkers have used the same labeling procedure to study the conformational transition of a tetracycline aptamer upon ligand binding. They concluded that this synthetic riboswitch exhibits a thermodynamic equilibrium between two conformations in the ligand-free state and captures one conformation upon ligand binding [120].

5 Summary and Outlook

The global structure of nucleic acids depends on environmental conditions, such as concentration of ions and small molecules, molecular crowding, viscosity, and interactions with proteins. Lately, Krstić et al. performed for the first time nanometer distance measurements on NAs inside cells, laying the foundation for PELDOR application to study biological processes in cells, such as nucleic acid diffusion, interaction with proteins, and other factors or chemical reactions [121]. They investigated a double-labeled 12-base pair DNA duplex, the 14-mer cUUCGg tetraloop hairpin RNA, and the 27-mer neomycin-sensing riboswitch inside *Xenopus laevis* oocytes and compared the PELDOR distances obtained in cells with in vitro measurements. The reduced lifetime of nitroxide spin labels under cellular conditions has been a major obstacle in these measurements (Fig. 18). Investigation of nitroxide reduction kinetics in-cell have shown that the

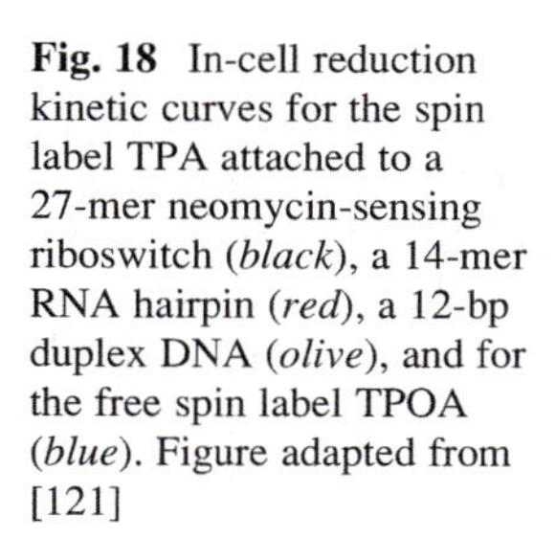

Fig. 18 In-cell reduction kinetic curves for the spin label TPA attached to a 27-mer neomycin-sensing riboswitch (*black*), a 14-mer RNA hairpin (*red*), a 12-bp duplex DNA (*olive*), and for the free spin label TPOA (*blue*). Figure adapted from [121]

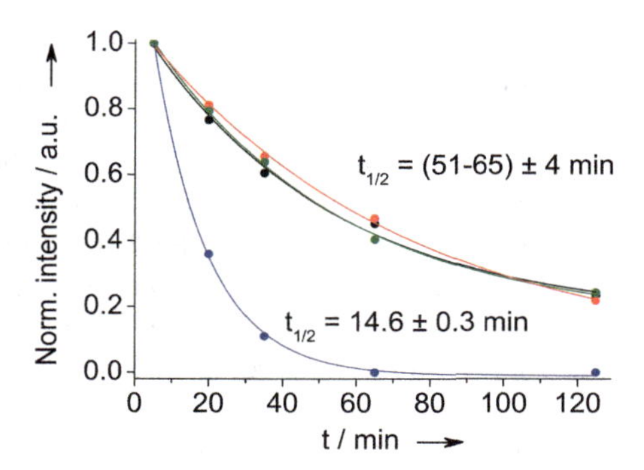

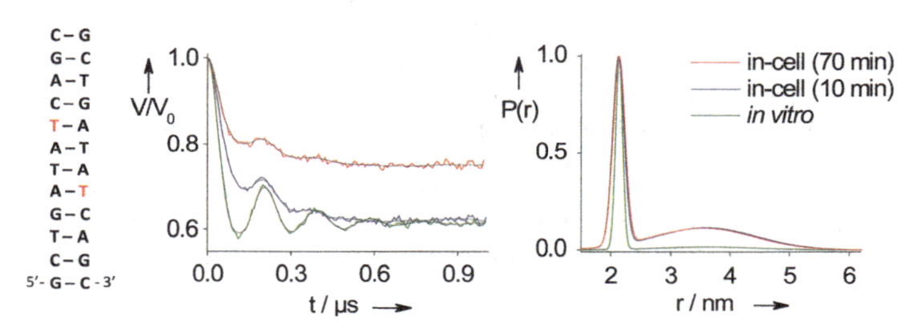

Fig. 19 Secondary structure (with spin-labeled nucleotides in *red*), baseline corrected PELDOR time traces and distance distribution for 12-bp double-labeled DNA in vitro (*green*) and in-cell (incubation time indicated in *legend*). Data were fitted with two Gaussian functions. Figure adapted from [121]

five-membered pyrrolidine and pyrroline rings are significantly more slowly reduced compared to six-membered piperidine ring based nitroxides. Due to prolonged lifetime of the spin probe **4** (Fig. 1) covalently attached to NA molecules, PELDOR signals could be measured with good signal-to-noise ratios of up to 70 min incubation time.

The partial loss of coupled spin labels due to nitroxide reduction only led to a decrease in the modulation depth upon increasing the incubation time (Fig. 19). No alterations in the measured distances between in vitro and in-cell experiments implies the existence of stable overall conformations of the 14-mer cUUCGg tetraloop hairpin RNA and the 27-mer neomycin-sensing riboswitch, whereas the 12-bp duplex DNA experiences stacking in-cell while retaining the secondary structure.

The variety of structural elements and the flexibility of NA molecules require both structural and dynamic methods for a functional characterization. As shown in this chapter, dipolar and hyperfine spectroscopy are important tools to determine short and long range distance constraints on such molecules. CW EPR can probe the dynamical behavior of NA molecules under physiological conditions and bridge the gap to the frozen-in conformational states, reflected by the distance distribution functions from low-temperature PELDOR experiments. The assortment of spin labels and spin-labeled positions in NAs offer the unique opportunity to place the spin markers in optimum positions to monitor specific dynamic modes or tertiary structure elements selectively and with high accuracy. On the other hand, the synthesis of spin-labeled NAs is still far from simple toolkit chemistry but requires high expertise in synthesis instead. Further work in the direction of rigid, redox-stable, or non-covalent spin labels, as well as the extension to larger NA molecules, will further improve and extend the applications of EPR to such systems in the future. Finally, the combination of NMR, MD, fluorescence spectroscopy, and EPR methods might allow one to achieve a full description of NA molecules on all length and time scales relevant for biological functions.

Acknowledgments Our own EPR work relied on the synthesis of spin-labeled RNA and DNA molecules performed by Nelly Piton and Olga Romainczyk from the group of Joachim W. Engels (Institute of Organic Chemistry and Chemical Biology, Goethe University Frankfurt) and the work by Pavol Cekan from the group of Snorri Th. Sigurdsson (University of Iceland) on the rigid spin labels for DNA. Vasyl Denysenkov is thanked for high-field G-band PELDOR experiments on DNA samples. Olav Schiemann (University of St. Andrews) is thanked for his major impact in the initial phase of this work as Habilitand in Frankfurt. Funding from the German Research Society (DFG) within the Collaborative Research Center 579 *RNA-Ligand Interaction* is gratefully acknowledged as well as support from the Center of Biomolecular Magnetic Resonance (BMRZ) and the Center of Excellence Frankfurt *Macromolecular Complexes* (DFG).

References

1. Fürtig B, Richter C, Wöhnert J, Schwalbe H (2003) NMR spectroscopy of RNA. Chembiochem 4(10):936–962. doi:10.1002/cbic.200300700
2. Varani G, Aboul-ela F, Allain FHT (1996) NMR investigation of RNA structure. Prog Nucl Magn Reson Spectr 29(1–2):51–127
3. Klostermeier D, Millar DP (2001) RNA conformation and folding studied with fluorescence resonance energy transfer. Methods 23(3):240–254
4. Roy R, Hohng S, Ha T (2008) A practical guide to single-molecule FRET. Nat Methods 5 (6):507–516. doi:10.1038/nmeth.1208
5. Wachowius F, Höbartner C (2010) Chemical RNA modifications for studies of RNA structure and dynamics. Chembiochem 11(4):469–480. doi:10.1002/cbic.200900697
6. Klare JP, Steinhoff HJ (2009) Spin labeling EPR. Photosynth Res 102(2–3):377–390
7. Schiemann O, Piton N, Plackmeyer J, Bode BE, Prisner TF, Engels JW (2007) Spin labeling of oligonucleotides with the nitroxide TPA and use of PELDOR, a pulse EPR method, to measure intramolecular distances. Nat Protoc 2(4):904–923. doi:10.1038/nprot.2007.97
8. Sowa GZ, Qin PZ (2008) Site-directed spin labeling studies on nucleic acid structure and dynamics. Prog Nucleic Acid Res Mol Biol 82:147–197. doi:10.1016/S0079-6603(08)00005-6
9. Berliner LJ (1998) Spin labeling: the next millenium. Biological magnetic resonance. Plenum Press, New York
10. Caron M, Dugas H (1976) Specific spin-labeling of transfer ribonucleic-acid molecules. Nucleic Acids Res 3(1):19–34
11. Sprinzl M, Kramer E, Stehlik D (1974) Structure of phenylalanine transfer-RNA from yeast – spin-label studies. Eur J Biochem 49(3):595–605
12. Sprinzl M, Scheit KH, Cramer F (1973) Preparation in-vitro of a 2-thiocytidine-containing yeast transfer-RNA Phe-A73-C74-S2c75-A76 and its interaction with para-hydroxymercuribenzoate. Eur J Biochem 34(2):306–310

13. Hara H, Horiuchi T, Saneyosh M, Nishimur S (1970) 4-Thiouridine-specific spin-labeling of E. coli transfer RNA. Biochem Biophys Res Commun 38(2):305
14. Mcintosh AR, Caron M, Dugas H (1973) Specific spin labeling of anticodon of Escherichia-coli transfer-RNA Glu. Biochem Biophys Res Commun 55(4):1356–1363
15. Qin PZ, Hideg K, Feigon J, Hubbell WL (2003) Monitoring RNA base structure and dynamics using site-directed spin labeling. Biochemistry-US 42(22):6772–6783. doi:10.1021/bi027222p
16. Ramos A, Varani G (1998) A new method to detect long-range protein-RNA contacts: NMR detection of electron-proton relaxation induced by nitroxide spin-labeled RNA. J Am Chem Soc 120(42):10992–10993
17. Spaltenstein A, Robinson BH, Hopkins PB (1989) Sequence-dependent and structure-dependent DNA-base dynamics – synthesis, structure, and dynamics of site and sequence specifically spin-labeled DNA. Biochemistry-US 28(24):9484–9495
18. Strube T, Schiemann O, MacMillan F, Prisner T, Engels JW (2001) A new facile method for spin-labeling of oligonucleotides. Nucleosides Nucleotides Nucleic Acids 20 (4–7):1271–1274
19. Schiemann O, Piton N, Mu YG, Stock G, Engels JW, Prisner TF (2004) A PELDOR-based nanometer distance ruler for oligonucleotides. J Am Chem Soc 126(18):5722–5729. doi:10.1021/ja0393877
20. Piton N, Schiemann O, Mu YG, Stock G, Prisner T, Engels JW (2005) Synthesis of spin-labeled RNAs for long range distance measurements by PELDOR. Nucleosides Nucleotides Nucleic Acids 24(5–7):771–775
21. Piton N, Mu YG, Stock G, Prisner TF, Schiemann O, Engels JW (2007) Base-specific spin-labeling of RNA for structure determination. Nucleic Acids Res 35(9):3128–3143. doi:10.1093/nar/gkm169
22. Spaltenstein A, Robinson BH, Hopkins PB (1988) A rigid and nonperturbing probe for duplex DNA motion. J Am Chem Soc 110(4):1299–1301
23. Gannett PM, Darian E, Powell J, Johnson EM 2nd, Mundoma C, Greenbaum NL, Ramsey CM, Dalal NS, Budil DE (2002) Probing triplex formation by EPR spectroscopy using a newly synthesized spin label for oligonucleotides. Nucleic Acids Res 30(23):5328–5337
24. Frolow O, Bode BE, Engels JW (2007) The synthesis of EPR differentiable spinlabels and their coupling to uridine. Nucleosides Nucleotides Nucleic Acids 26(6–7):655–659. doi:10.1080/15257770701490522
25. Hustedt EJ, Kirchner JJ, Spaltenstein A, Hopkins PB, Robinson BH (1995) Monitoring DNA dynamics using spin-labels with different independent mobilities. Biochemistry-US 34 (13):4369–4375
26. Ding P, Wunnicke D, Steinhoff HJ, Seela F (2010) Site-directed spin-labeling of DNA by the azide-alkyne ‘Click’ reaction: nanometer distance measurements on 7-deaza-2′-deoxyadenosine and 2′-deoxyuridine nitroxide conjugates spatially separated or linked to a ‘dA-dT’ base pair. Chem Eur J 16(48):14385–14396
27. Bobst AM, Kao SC, Toppin RC, Ireland JC, Thomas IE (1984) Dipsticking the major groove of DNA with enzymatically incorporated spin-labeled deoxyuridines by electron-spin resonance spectroscopy. J Mol Biol 173(1):63–74
28. Bobst AM, Pauly GT, Keyes RS, Bobst EV (1988) Enzymatic sequence-specific spin labeling of a DNA fragment containing the recognition sequence of ecori endonuclease. FEBS Lett 228(1):33–36
29. Okamoto A, Inasaki T, Saito I (2004) Nitroxide-labeled guanine as an ESR spin probe for structural study of DNA. Bioorg Med Chem Lett 14(13):3415–3418. doi:10.1016/j.bmcl.2004.04.076
30. Sicoli G, Wachowius F, Bennati M, Höbartner C (2010) Probing secondary structures of spin-labeled RNA by pulsed EPR spectroscopy. Angew Chem Int Ed Engl 49(36):6443–6447. doi:10.1002/anie.201000713
31. Miller TR, Alley SC, Reese AW, Solomon MS, Mccallister WV, Mailer C, Robinson BH, Hopkins PB (1995) A probe for sequence-dependent nucleic-acid dynamics. J Am Chem Soc 117(36):9377–9378

32. Okonogi TM, Alley SC, Reese AW, Hopkins PB, Robinson BH (2000) Sequence-dependent dynamics in duplex DNA. Biophys J 78(5):2560–2571
33. Barhate N, Cekan P, Massey AP, Sigurdsson ST (2007) A nucleoside that contains a rigid nitroxide spin label: a fluorophore in disguise. Angew Chem Int Ed Engl 46(15):2655–2658. doi:10.1002/anie.200603993
34. Lin KY, Jones RJ, Matteucci M (1995) Tricyclic 2′-deoxycytidine analogs – syntheses and incorporation into oligodeoxynucleotides which have enhanced binding to complementary RNA. J Am Chem Soc 117(13):3873–3874
35. Cekan P, Smith AL, Barhate N, Robinson BH, Sigurdsson ST (2008) Rigid spin-labeled nucleoside C: a nonperturbing EPR probe of nucleic acid conformation. Nucleic Acids Res 36 (18):5946–5954. doi:10.1093/nar/gkn562
36. Cekan P, Sigurdsson ST (2008) Single base interrogation by afluorescent nucleotide: each of the four DNA bases identified by fluorescence spectroscopy. Chem Commun (29):3393–3395. doi:10.1039/B801833b
37. Shelke SA, Sigurdsson ST (2010) Noncovalent and site-directed spin labeling of nucleic acids. Angew Chem Int Ed Engl 49(43):7984–7986. doi:10.1002/anie.201002637
38. Nagahara S, Murakami A, Makino K (1992) Spin-labeled oligonucleotides site specifically labeled at the internucleotide linkage – separation of stereoisomeric probes and EPR spectroscopical detection of hybrid formation in solution. Nucleos Nucleot 11(2–4):889–901
39. Grant GP, Boyd N, Herschlag D, Qin PZ (2009) Motions of the substrate recognition duplex in a group I intron assessed by site-directed spin labeling. J Am Chem Soc 131(9):3136–3137. doi:10.1021/ja808217s
40. Qin PZ, Haworth IS, Cai Q, Kusnetzow AK, Grant GPG, Price EA, Sowa GZ, Popova A, Herreros B, He H (2007) Measuring nanometer distances in nucleic acids using a sequence-independent nitroxide probe. Nat Protoc 2(10):2354–2365. doi:10.1038/nprot.2007.308
41. Qin PZ, Butcher SE, Feigon J, Hubbell WL (2001) Quantitative analysis of the isolated GAAA tetraloop/receptor interaction in solution: a site-directed spin labeling study. Biochemistry-US 40(23):6929–6936
42. Cai Q, Kusnetzow AK, Hideg K, Price EA, Haworth IS, Qin PZ (2007) Nanometer distance measurements in RNA using site-directed spin labeling. Biophys J 93(6):2110–2117. doi:10.1529/biophysj.107.109439
43. Cai Q, Kusnetzow AK, Hubbell WL, Haworth IS, Gacho GPC, Van Eps N, Hideg K, Chambers EJ, Qin PZ (2006) Site-directed spin labeling measurements of nanometer distances in nucleic acids using a sequence-independent nitroxide probe. Nucleic Acids Res 34(17):4722–4730. doi:10.1093/nar/gkl546
44. Flaender M, Sicoli G, Fontecave T, Mathis G, Saint-Pierre C, Boulard Y, Gambarelli S, Gasparutto D (2008) Site-specific insertion of nitroxide-spin labels into DNA probes by click chemistry for structural analyses by ELDOR spectroscopy. Nucleic Acids Symp Ser 52:147–148. doi:10.1093/nass/nrn075
45. Kim NK, Murali A, DeRose VJ (2004) A distance ruler for RNA using EPR and site-directed spin labeling. Chem Biol 11(7):939–948. doi:10.1016/j.chembiol.2004.04.013
46. Edwards TE, Sigurdsson ST (2007) Site-specific incorporation of nitroxide spin-labels into 2′-positions of nucleic acids. Nat Protoc 2(8):1954–1962. doi:10.1038/nprot.2007.273
47. Edwards TE, Okonogi TM, Robinson BH, Sigurdsson ST (2001) Site-specific incorporation of nitroxide spin-labels into internal sites of the TAR RNA; structure-dependent dynamics of RNA by EPR spectroscopy. J Am Chem Soc 123(7):1527–1528
48. Schweiger A, Jeschke G (2001) Principles of pulse electron paramagnetic resonance. Oxford University Press, USA
49. Atherton NM (1993) Principles of electron spin resonance. Ellis Horwood, New York
50. Abragam A (1961) The principles of nuclear magnetism. The international series of monographs on physics. Clarendon, Oxford
51. Freed JH, Fraenkel GK (1963) Theory of linewidths in electron spin resonance spectra. J Chem Phys 39(2):326

52. Kivelson D (1960) Theory of ESR linewidths of free radicals. J Chem Phys 33(4):1094–1106
53. Stoll S, Schweiger A (2006) EasySpin, a comprehensive software package for spectral simulation and analysis in EPR. J Magn Reson 178(1):42–55. doi:10.1016/j.jmr.2005.08.013
54. Schneider DJ, Freed JH (1989) Calculating slow motional magnetic resonance spectra: a user's guide. In: Berliner LJ, Reuben J (eds) Spin labeling: theory and applications. Biological magnetic resonance. Plenum Press, New York
55. Mims WB, Nassau K, Mcgee JD (1961) Spectral diffusion in electron resonance lines. Phys Rev 1(6):2059–2069
56. Rowan LG, Hahn EL, Mims WB (1965) Electron-spin-echo envelope modulation. Phys Rev 137(1A):61–71
57. Mims WB (1972) Electron spin echoes. In: Geschwind S (ed) Electron paramagnetic resonance. Plenum Press, New York
58. Hofer P, Grupp A, Nebenfuhr H, Mehring M (1986) Hyperfine sublevel correlation (HYSCORE) spectroscopy – a 2D electron-spin-resonance investigation of the squaric acid radical. Chem Phys Lett 132(3):279–282
59. Ichikawa T, Kevan L, Bowman MK, Dikanov SA, Tsvetkov YD (1979) Ratio analysis of electron spin echo modulation envelopes in disordered matrixes and application to the structure of solvated electrons in 2-methyltetrahydrofuran glass. J Chem Phys 71 (3):1167–1174
60. Flanagan HL, Singel DJ (1987) Analysis of ^{14}N ESEEM patterns of randomly oriented solids. J Chem Phys 87(10):5606–5616
61. Jeschke G, Rakhmatullin R, Schweiger A (1998) Sensitivity enhancement by matched microwave pulses in one- and two-dimensional electron spin echo envelope modulation spectroscopy. J Magn Reson 131(2):261–271
62. Feher G (1956) Observation of nuclear magnetic resonances via the electron spin resonance line. Phys Rev 103(3):834–834
63. Davies ER (1974) A new pulse endor technique. Phys Lett A 47(1):1–2
64. Mims WB (1965) Pulsed electron nuclear double resonance (E.N.DO.R.) experiments. Proc Roy Soc (London) 283(Ser. A;1395):452–457
65. Epel B, Manikandan P, Kroneck PMH, Goldfarb D (2001) High-field ENDOR and the sign of the hyperfine coupling. Appl Magn Reson 21(3–4):287–297
66. Cook RJ, Whiffen DH (1964) Relative signs of hyperfine coupling constants by a double ENDOR experiment. Proc Phys Soc Lond 84(6):845–845
67. Bode BE, Margraf D, Plackmeyer J, Durner G, Prisner TF, Schiemann O (2007) Counting the monomers in nanometer-sized oligomers by pulsed electron-electron double resonance. J Am Chem Soc 129(21):6736–6745. doi:10.1021/ja065787t
68. Marko A, Margraf D, Yu H, Mu Y, Stock G, Prisner T (2009) Molecular orientation studies by pulsed electron-electron double resonance experiments. J Chem Phys 130(6):064102. doi:10.1063/1.3073040
69. Margraf D, Cekan P, Prisner TF, Sigurdsson ST, Schiemann O (2009) Ferro- and antiferromagnetic exchange coupling constants in PELDOR spectra. Phys Chem Chem Phys 11 (31):6708–6714. doi:10.1039/b905524j
70. Marko A, Margraf D, Cekan P, Sigurdsson ST, Schiemann O, Prisner TF (2010) Analytical method to determine the orientation of rigid spin labels in DNA. Phys Rev E 81(2)
71. Margraf D, Bode BE, Marko A, Schiemann O, Prisner TF (2007) Conformational flexibility of nitroxide biradicals determined by X-band PELDOR experiments. Mol Phys 105 (15–16):2153–2160
72. Zhang XJ, Cekan P, Sigurdsson ST, Qin PZ (2009) Studying RNA using site-directed spin-labeling and continuous-wave electron paramagnetic resonance spectroscopy. Method Enzymol 469:303–328. doi:10.1016/S0076-6879(09)69015-7
73. Krstić I, Frolow O, Sezer D, Endeward B, Weigand JE, Suess B, Engels JW, Prisner TF (2010) PELDOR spectroscopy reveals preorganization of the neomycin-responsive riboswitch tertiary structure. J Am Chem Soc 132(5):1454–1455. doi:10.1021/ja9077914

74. Moore PB (1999) Structural motifs in RNA. Annu Rev Biochem 68:287–300. doi:10.1146/annurev.biochem.68.1.287
75. Duchardt-Ferner E, Weigand JE, Ohlenschlager O, Schtnidtke SR, Suess B, Wohnert J (2010) Highly modular structure and ligand binding by conformational capture in a minimalistic riboswitch. Angew Chem Int Ed Engl 49(35):6216–6219
76. Edwards TE, Robinson BH, Sigurdsson ST (2005) Identification of amino acids that promote specific and rigid TAR RNA-Tat protein complex formation. Chem Biol 12(3):329–337. doi:10.1016/j.chembiol.2005.01.012
77. Edwards TE, Sigurdsson ST (2003) EPR spectroscopic analysis of TAR RNA-metal ion interactions. Biochem Biophys Res Commun 303(2):721–725
78. Edwards TE, Sigurdsson ST (2002) Electron paramagnetic resonance dynamic signatures of TAR RNA-small molecule complexes provide insight into RNA structure and recognition. Biochemistry-US 41(50):14843–14847
79. Edwards TE, Okonogi TM, Sigurdsson ST (2002) Investigation of RNA-protein and RNA-metal ion interactions by electron paramagnetic resonance spectroscopy. The HIV TAR-Tat motif. Chem Biol 9(6):699–706
80. Ippolito JA, Steitz TA (1998) A 1.3-A resolution crystal structure of the HIV-1 trans-activation response region RNA stem reveals a metal ion-dependent bulge conformation. Proc Natl Acad Sci USA 95(17):9819–9824
81. Qin PZ, Feigon J, Hubbell WL (2005) Site-directed spin labeling studies reveal solution conformational changes in a GAAA tetraloop receptor upon Mg^{2+}-dependent docking of a GAAA tetraloop. J Mol Biol 351(1):1–8. doi:10.1016/j.jmb.2005.06.007
82. Feig AL (2000) The use of manganese as a probe for elucidating the role of magnesium ions in ribozymes. Met Ions Biol Syst 37:157–182
83. Pley HW, Flaherty KM, Mckay DB (1994) 3-Dimensional structure of a hammerhead ribozyme. Nature 372(6501):68–74
84. Scott WG, Murray JB, Arnold JRP, Stoddard BL, Klug A (1996) Capturing the structure of a catalytic RNA intermediate: the hammerhead ribozyme. Science 274(5295):2065–2069
85. Horton TE, Clardy DR, DeRose VJ (1998) Electron paramagnetic resonance spectroscopic measurement of Mn^{2+} binding affinities to the hammerhead ribozyme and correlation with cleavage activity. Biochemistry-US 37(51):18094–18101
86. Schiemann O, Fritscher J, Kisseleva N, Sigurdsson ST, Prisner TF (2003) Structural investigation of a high-affinity Mn^{II} binding site in the hammerhead ribozyme by EPR spectroscopy and DFT calculations. Effects of neomycin B on metal-ion binding. Chembiochem 4 (10):1057–1065. doi:10.1002/cbic.200300653
87. Kisseleva N, Khvorova A, Westhof E, Schiemann O (2005) Binding of manganese(II) to a tertiary stabilized hammerhead ribozyme as studied by electron paramagnetic resionance spectroscopy. RNA 11(1):1–6
88. Kim NK, Murali A, DeRose VJ (2005) Separate metal requirements for loop interactions and catalysis in the extended hammerhead ribozyme. J Am Chem Soc 127(41):14134–14135. doi:10.1021/ja0541027
89. Kisseleva N, Kraut S, Jaschke A, Schiemann O (2007) Characterizing multiple metal ion binding sites within a ribozyme by cadmium-induced EPR silencing. HFSP J 1(2):127–136
90. Zhang X, Lee SW, Zhao L, Xia T, Qin PZ (2010) Conformational distributions at the N-peptide/boxB RNA interface studied using site-directed spin labeling. RNA 16 (12):2474–2483. doi:10.1261/rna.2360610
91. Xi X, Sun Y, Karim CB, Grigoryants VM, Scholes CP (2008) HIV-1 nucleocapsid protein NCp7 and its RNA stem loop 3 partner: rotational dynamics of spin-labeled RNA stem loop 3. Biochemistry-US 47(38):10099–10110. doi:10.1021/bi800602e
92. Spaltenstein A, Robinson BH, Hopkins PB (1989) DNA structural data from a dynamics probe – the dynamic signatures of single-stranded, hairpin-looped, and duplex forms of DNA are distinguishable. J Am Chem Soc 111(6):2303–2305

93. Cekan P, Sigurdsson ST (2009) Identification of single-base mismatches in duplex DNA by EPR spectroscopy. J Am Chem Soc 131(50):18054–18056. doi:10.1021/ja905623k
94. Jakobsen U, Shelke SA, Vogel S, Sigurdsson ST (2010) Site-directed spin-labeling of nucleic acids by click chemistry: detection of abasic sites in duplex DNA by EPR spectroscopy. J Am Chem Soc 132(30):10424–10428. doi:10.1021/ja102797k
95. Okonogi TM, Reese AW, Alley SC, Hopkins PB, Robinson BH (1999) Flexibility of duplex DNA on the submicrosecond timescale. Biophys J 77(6):3256–3276. doi:10.1016/S0006-3495(99)77157-2
96. Popova AM, Kalai T, Hideg K, Qin PZ (2009) Site-specific DNA structural and dynamic features revealed by nucleotide-independent nitroxide probes. Biochemistry-US 48 (36):8540–8550. doi:10.1021/bi900860w
97. Keyes RS, Bobst AM (1995) Detection of internal and overall dynamics of a 2-atom-tethered spin-labeled DNA. Biochemistry-US 34(28):9265–9276
98. Smith AL, Cekan P, Brewood GP, Okonogi TM, Alemayehu S, Hustedt EJ, Benight AS, Sigurdsson ST, Robinson BH (2009) Conformational equilibria of bulged sites in duplex DNA studied by EPR spectroscopy. J Phys Chem B 113(9):2664–2675
99. Okonogi TM, Alley SC, Harwood EA, Hopkins PB, Robinson BH (2002) Phosphate backbone neutralization increases duplex DNA flexibility: a model for protein binding. Proc Natl Acad Sci USA 99(7):4156–4160
100. Columbus L, Hubbell WL (2004) Mapping backbone dynamics in solution with site-directed spin labeling: GCN4-58 bZip free and bound to DNA. Biochemistry-US 43(23):7273–7287
101. Steinhoff HJ, Suess B (2003) Molecular mechanisms of gene regulation studied by site-directed spin labeling. Methods 29(2):188–195
102. Morrissey SR, Horton TE, Grant CV, Hoogstraten CG, Britt RD, DeRose VJ (1999) Mn^{2+}-nitrogen interactions in RNA probed by electron spin-echo envelope modulation spectroscopy: application to the hammerhead ribozyme. J Am Chem Soc 121(39):9215–9218
103. Vogt M, Lahiri S, Hoogstraten CG, Britt RD, DeRose VJ (2006) Coordination environment of a site-bound metal ion in the hammerhead ribozyme determined by N-15 and H-2 ESEEM spectroscopy. J Am Chem Soc 128(51):16764–16770. doi:10.1021/Ja057035p
104. Morrissey SR, Horton TE, DeRose VJ (2000) Mn^{2+} sites in the hammerhead ribozyme investigated by EPR and continuous-wave Q-band ENDOR spectroscopies. J Am Chem Soc 122(14):3473–3481
105. Schiemann O, Carmieli R, Goldfarb D (2007) W-band P-31-ENDOR on the high-affinity Mn^{2+} binding site in the minimal and tertiary stabilized hammerhead ribozymes. Appl Magn Reson 31(3–4):543–552
106. DeRose VJ (2009) Characterization of nucleic acid-metal Ion binding by spectroscopic techniques. In: Hud NV (ed) Nucleic acid-metal ion interactions. The Royal Society of Chemistry, London
107. Santangelo MG, Antoni PM, Spingler B, Jeschke G (2010) Can copper(II) mediate Hoogsteen base-pairing in a left-handed DNA duplex? A pulse EPR study. Chemphyschem 11(3):599–606. doi:10.1002/cphc.200900672
108. Milov AD, Ponomarev AB, Tsvetkov YD (1984) Electron electron double-resonance in electron-spin echo - model biradical systems and the sensitized photolysis of decalin. Chem Phys Lett 110(1):67–72
109. Schiemann O, Prisner TF (2007) Long-range distance determinations in biomacromolecules by EPR spectroscopy. Q Rev Biophys 40(1):1–53. doi:10.1017/S003358350700460X
110. Jeschke G, Polyhach Y (2007) Distance measurements on spin-labelled biomacromolecules by pulsed electron paramagnetic resonance. Phys Chem Chem Phys 9(16):1895–1910. doi:10.1039/b614920k
111. Schiemann O, Weber A, Edwards TE, Prisner TF, Sigurdsson ST (2003) Nanometer distance measurements on RNA using PELDOR. J Am Chem Soc 125(12):3434–3435. doi:10.1021/ja0274610

112. Ward R, Keeble DJ, El-Mkami H, Norman DG (2007) Distance determination in heterogeneous DNA model systems by pulsed EPR. Chembiochem 8(16):1957–1964. doi:10.1002/cbic.200700245
113. Schiemann O, Cekan P, Margraf D, Prisner TF, Sigurdsson ST (2009) Relative orientation of rigid nitroxides by PELDOR: beyond distance measurements in nucleic acids. Angew Chem Int Ed Engl 48(18):3292–3295. doi:10.1002/anie.200805152
114. Marko A, Denysenkov VP, Margraf D, Cekan P, Schiemann O, Sigurdsson ST, Prisner TF (2011) Conformational flexibility of DNA. J Am Chem Soc. doi:10.1021/ja201244u
115. Sicoli G, Mathis G, Delalande O, Boulard Y, Gasparutto D, Gambarelli S (2008) Double electron-electron resonance (DEER): a convenient method to probe DNA conformational changes. Angew Chem Int Ed Engl 47(4):735–737. doi:10.1002/anie.200704133
116. Sicoli G, Mathis G, Aci-Sèche S, Saint-Pierre C, Boulard Y, Gasparutto D, Gambarelli S (2009) Lesion-induced DNA weak structural changes detected by pulsed EPR spectroscopy combined with site-directed spin labelling. Nucleic Acids Res 37(10):3165–3176. doi:10.1093/nar/gkp165
117. Weigand JE, Sanchez M, Gunnesch EB, Zeiher S, Schroeder R, Suess B (2008) Screening for engineered neomycin riboswitches that control translation initiation. RNA 14(1):89–97. doi:10.1261/rna.772408
118. Singh V, Azarkh M, Exner TE, Hartig JS, Drescher M (2009) Human telomeric quadruplex conformations studied by pulsed EPR. Angew Chem Int Ed Engl 48(51):9728–9730. doi:10.1002/anie.200902146
119. Kim NK, Bowman MK, DeRose VJ (2010) Precise mapping of RNA tertiary structure via nanometer distance measurements with double electron-electron resonance spectroscopy. J Am Chem Soc 132(26):8882–8884. doi:10.1021/Ja101317g
120. Wunnicke D, Strohbach D, Weigand JE, Appel B, Feresin E, Suess B, Müller S, Steinhoff HJ (2010) Ligand-induced conformational capture of a synthetic tetracycline riboswitch revealed by pulse EPR. RNA. doi:10.1261/rna.2222811
121. Krstić I, Hänsel R, Romainczyk O, Engels JW, Dötsch V, Prisner TF (2011) Long-range distance measurements on nucleic acids in cells by pulsed EPR spectroscopy. Angew Chem Int Ed Engl 50(22):5070–5074. doi:10.1002/anie.201100886

Top Curr Chem (2012) 321: 199–234
DOI: 10.1007/128_2011_303

Published online: 11 November 2011

New Directions in Electron Paramagnetic Resonance Spectroscopy on Molecular Nanomagnets

J. van Slageren

Abstract Recent developments and results in the area of electron paramagnetic resonance (EPR) in molecular nanomagnetism are reviewed. Emphasis is placed on unconventional measurement methods, such as frequency-domain magnetic resonance spectroscopy, interferometer-based Fourier-transform, terahertz spectroscopy, and terahertz time-domain spectroscopy. In addition, different methods to investigate EPR by monitoring the change in magnetization or magnetic torque in the presence of microwave radiation are discussed. Finally, an overview is given of application of pulse EPR in investigations of molecular nanomagnets.

Keywords Electron paramagnetic resonance · Electron spin resonance · Molecular nanomagnet · Quantum coherence · Single-molecule magnet · Zero-field splitting

Contents

J. van Slageren (✉)
Institut für Physikalische Chemie, Universität Stuttgart, Pfaffenwaldring 55, 70569 Stuttgart, Germany
e-mail: slageren@ipc.uni-stuttgart.de

1 Introduction: Overview of EPR in Molecular Nanomagnetism

Several decades worth of research in the field of molecular nanomagnetism has yielded an impressive list of highlights in this area, and electron paramagnetic resonance (EPR) methods have played a decisive role in the discovery and especially in the explanation of a great number of these findings. An early highlight was the discovery of slow magnetization dynamics in molecular clusters of exchange-coupled paramagnetic ions, as revealed by alternating-current magnetic susceptibility and Mössbauer measurements [1–3]. However, only with the help of high-frequency EPR measurements was it shown that the energy barrier between up and down orientation of the magnetic moment is due to magnetic anisotropy in the form of zero-field splitting (ZFS) of the orientations of the magnetic moment, which are characterized by the M_S quantum number [3].

Similarly, it was first demonstrated by means of magnetization measurements that the magnetic moment can also tunnel under the energy barrier, if two levels on both sides of the barrier are in resonance, leading to minima in the field dependence of the relaxation time of the magnetization [4], and striking steps in the magnetic hysteresis curve [5, 6]. However, again the intricacies of this tunneling process and the transverse interactions responsible could only be fully unraveled with the help of EPR studies [7, 8].

The molecular exchange-coupled clusters that show slow relaxation of the magnetization have since become known as single-molecule magnets (SMMs) [9], while the more general term molecular nanomagnet (MNM) [10] is often reserved for any exchange-coupled cluster with "interesting" magnetic properties. However, with the recent arrival of mononuclear complexes of lanthanide [11], actinide [12], and even transition metal ions [13], that show slow relaxation of the magnetization, this definition can be expanded to include any zero-dimensional molecular species with "interesting" magnetic properties.

Finally, a number of observations had been published of energy splittings between symmetric and antisymmetric superposition states [14–17] that were reported to be larger than the expected rate of quantum decoherence and hence taken as a signature of quantum coherence in these systems. However, the direct observation of quantum coherence in MNMs was possible only with the help of pulse EPR measurements [18]. This area has received great impetus from the realization that molecular magnets are not only excellent systems to study mesoscopic quantum coherence, but are also promising candidates for qubit implementation [19–21]. The area of pulse EPR studies of MNMs will be described in detail in Sect. 3.

Further notable highlights in molecular nanomagnetism include the discovery that a range of MNMs act as efficient cooling agents, based on the magnetocaloric effect, because of their large spin state degeneracy [22]. Because these magnetic cooling MNMs have small anisotropies to provide the necessary large degeneracies, EPR plays a less prominent role in this area, beyond materials characterization. A final area where EPR has not played a decisive role so far, but possibly could, is that of single-chain magnets (SCMs). SCMs are spin chains that combine ferro- [23] or ferrimagnetic [24] interactions with Ising-type anisotropy, resulting in slow relaxation of the magnetization that is governed by concerted action of both the exchange coupling and the zero-field splitting of the individual units.

Most of the EPR studies of MNMs have focused on the investigation of the magnetic anisotropy in the form of zero-field splitting, which is responsible for both the energy barrier towards relaxation of the magnetization and for the quantum tunneling process of the magnetization. Because ZFS is a field-independent interaction, it is advantageous to carry out EPR-measurements at a range of different frequencies. Extrapolation of the dependence of frequency on resonance field to zero field yields the splitting of the spin multiplet in zero field, i.e., the ZFS (see Fig. 1 for a recent example). Different aspects of the origin of ZFS in MNM have been described in various books [27–30]. The ZFS of the ground spin multiplet can

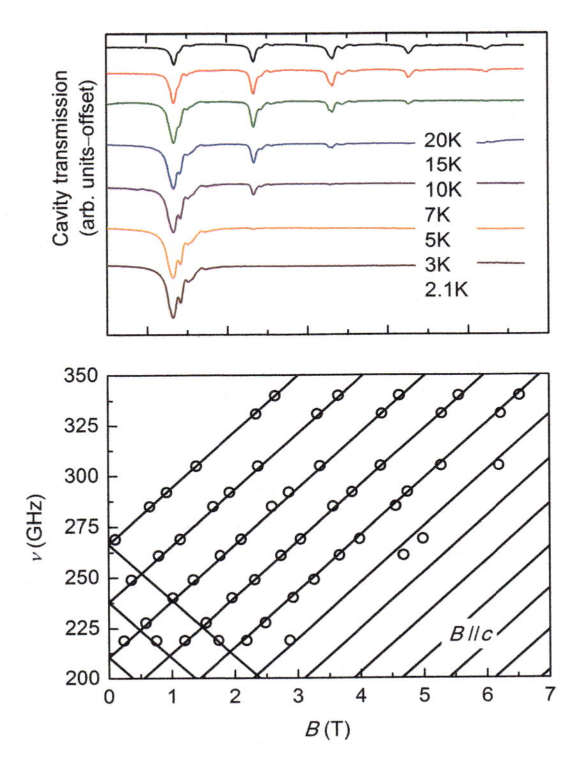

Fig. 1 *Top:* High-field EPR spectra recorded along the easy axis of a single crystal of $[Mn_6^{III}O_2(Et\text{-}sao)_6(O_2CPh(Me)_2)_2(EtOH)_6]$ at 331 GHz and different temperatures as indicated, and frequencies as indicated. *Bottom:* Diagram of microwave frequency as a function of resonance field. adapted from [25]. Used with permission. ©2009 Elsevier. See also Fig. 2

in most cases be described by a single-spin Hamiltonian of the type (Giant Spin Hamiltonian):

$$\mathcal{H} = \mu_B \mathbf{B} \cdot \mathbf{g} \cdot \hat{\mathbf{S}} + D\left[\hat{S}_z^2 - \tfrac{1}{3}S(S+1)\right] + E\left(\hat{S}_x^2 - \hat{S}_y^2\right) + \sum_{k \geq 4,q} B_k^q \hat{O}_k^q \quad (1)$$

where the first term is the Zeeman interaction between electron spin and magnetic field, and the second and third terms are the second order axial and transverse ZFS, respectively. Here, second order reflects the fact that the spin Hamiltonian contains the spin operators to the second power. The last term is the sum of higher (than second) order ZFS terms, which were shown to parametrize the effects of the ZFS-induced mixing between spin states on the level splittings within the ground multiplet (S-mixing) [31]. In the case of SMMs or, generally, MNMs with large, negative D-values, powder spectra or single crystal spectra recorded with the external field along the zero-field splitting quantization axis (the easy axis, if $D < 0$) are often not very informative regarding the transverse ZFS (the third term in (1) and the elements of the fourth term with $q \neq 0$), because the energies of the M_S states with large M_S quantum number are little affected by transverse ZFS. It is this transverse ZFS that is responsible for quantum tunneling of the magnetization. Much more informative in that regard are single-crystal EPR spectra that are recorded with the field perpendicular to the quantization axis (in the hard plane). For example, in a recent study on $[Mn_{12}O_{12}(t\text{BuCH}_2CO_2)_{16}(CH_3OH)_4]\cdot CH_3OH$, this approach allowed the determination of the transverse ZFS terms up to sixth order [32].

There have been a number of reviews of EPR studies of MNMs in the past 5 years [33–37] especially by Barra/Gatteschi and McInnes, and it is not the aim of the present chapter to repeat the content of those reviews in detail. From these reviews, it is clear that certain EPR studies of MNMs have become classics in their field and they are briefly outlined below. The reader is referred to the cited reviews and primary literature for further detail. In addition, the quasi-biannual reports by Collison and McInnes provide useful comprehensive overviews of EPR studies on exchange coupled clusters, including biological systems, for the relevant years [38–41].

Studies on a six-membered, ferromagnetically coupled copper(II) ring were an early triumph of high-frequency EPR in molecular nanomagnetism because of the textbook-quality, perfectly axial spectra [35, 42]. Rather more investigations have been carried out on *anti*ferromagnetic rings which, for an even number of equal metal ions, have $S = 0$ ground states. EPR measurements have been successfully employed to determine the ZFSs of the spin excited states in $[Fe_6(tea)_6]$ (tea^{3-} is triethanolaminate) [35, 43], $[Cr_8F_8Piv_{16}]$ (Cr_8, Piv = pivalate) [35, 44], and $[CsFe_8(tea)_8]Cl$ [45]. In contrast, the antiferromagnetic rings $(R_2NH_2)^+$ $[Cr^{III}7M^{II}F_8(Piv)_{16}]^-$, and (Cr_7M) have spin ground states of $S = 1/2$, 1, or 3/2 [46]. Again, EPR investigations were very fruitful and yielded a wealth of information on ground and excited states [36, 47, 48]. The ZFSs in Cr_8 were shown to result from a major contribution of the combined single-ion contributions and a minor, but important, contribution from the magnetic-dipolar interaction between the Cr^{III}-ions. Interestingly, in the Cr_7M systems, the anisotropic part of the exchange interaction (which has the same mathematical spin Hamiltonian form as the magnetodipolar interaction) was demonstrated to play a significant role in the ZFS of the spin states in these molecules. Only through the powerful combination of EPR spectroscopy and advanced calculations could this highly detailed understanding be obtained. Recent studies of linked rings have shown that EPR is an excellent tool to study the weak exchange interactions between the two rings [36, 49, 50]. These studies are of relevance in work towards implementing two-qubit gates utilizing MNMs (see Sect. 3).

Undoubtedly, the largest number of studies have been performed on the archetypical single-molecule magnet $[Mn_{12}O_{12}(OAc)_{16}(H_2O)_4]\cdot 2CH_3COOH$ ($Mn_{12}Ac$) and its derivatives [33–36]. Indeed, the progress in high-frequency EPR instrumentation over the years can be followed on the basis of studies of Mn_{12} [51]. Initially, these studies were aimed at the investigation of the sign and magnitude of the second order axial zero-field splitting, as well as a confirmation of the ground state spin of $S = 10$. A more detailed study several years later yielded more accurate ZFS splitting values, and this was probably also the first time fourth order ZFS parameters were used in the analysis of EPR spectra of MNMs [52]. Detailed hard-plane single-crystal investigations definitively determined the transverse anisotropy parameters of $Mn_{12}Ac$ [7, 53] and gave support to the now widely accepted model that ascribes local symmetry lowering to disorder of the cocrystallized acetic acid molecules [54]. Carefully recrystallized single-crystal samples of a derivative of Mn_{12} do possess cocrystallized solvent molecules, but these solvent molecules do not form hydrogen bonds with the cluster and thus do not influence its magnetic properties. The recent EPR study of this sample allowed the determination of parameter values of the ZFS up to sixth order, and also shed light on the origin of the transverse anisotropy through detailed analysis, which included considering the effect of tilting of the single-ion ZFS quantization axes on the higher-order transverse anisotropy [32].

A comprehensive series of investigations on tetranuclear metalloheterocubane nickel(II) clusters $[Ni(R'\text{-hmp})(ROH)Cl]_4$ (R',R = alkyl) gave detailed insight into the origin of ZFS and the efficient quantum tunneling of the magnetization observed in these systems [37, 55, 56]. The elegance of this work lies in the fact that the magnitude and the tensor orientation of the single ion ZFS were determined by studies on a diamagnetic isomorphous zinc(II) cluster, doped with a small amount of nickel(II) [57]. This then allowed one to fix a large number of parameters in the analysis of the data on the nickel cluster [8]. The analysis showed that in order to understand the physical origin of transverse anisotropy, one has to go beyond the Giant Spin Hamiltonian, and consider the isotropic exchange interaction and the single ion ZFS together.

Finally, the study of a range of star-shaped Fe_4 clusters has actually allowed devising a magnetostructural correlation for second order anisotropy terms $[Fe_4(L)_2\ (dpm)_6]$ (Hdpm = 2,2,6,6-tetramethylheptane-3,5-dione; H_3L = $R'OCH_2C(CH_2OH)_3$ with R' = alkyl and related moieties) [58, 59]. A clear correlation between the axial zero-field splitting value D and the helical pitch of the molecular structure was established.

To conclude this section, it is clear that over the past two decades EPR spectroscopic studies on MNMs have moved from sample characterization to detailed investigations of the origins of ZFS and the establishment of magnetostructural correlations for the ZFS. All these studies were carried out conventionally, in the sense that the external field was swept at constant radiation frequency, while monitoring the change in radiation intensity. There are a number of alternative measurement methods that can be broadly divided into those where the measurement is performed at fixed field and the frequency dependence of the transmitted or reflected radiation intensity is obtained directly, and, second, those where the magnetic resonance transitions are detected by a change in the thermodynamic properties of the sample, such as the magnetization or the magnetic torque. These methods are reviewed in Sect. 2. The third section of this review summarizes the pulse EPR studies that have been performed on MNMs. This is an exciting new area in molecular magnetism, where studies on MNMs can contribute to the understanding of quantum coherence and decoherence and their dependence on molecular and electronic structure and complexity.

2 Unconventional EPR Measurement Methods in Molecular Nanomagnetism

As the previous section documents, conventional EPR measurement methods have been extraordinarily successful in many aspects of molecular nanomagnetism. For the accurate determination of the zero-field splitting of high-anisotropy MNMs, best results are obtained from a multifrequency approach. This approach allows extrapolation of the resonance frequencies to zero-field, where the energy splittings directly correspond to the ZFS. The question then arises as to whether ZFSs can or should be measured by directly measuring the energy splittings in zero external fields. This section shows that this is indeed possible. Possible implementations of such a measurement scheme include frequency swept measurements at fixed or zero field (frequency domain magnetic resonance spectroscopy, Sect. 2.1), measurements using interferometer-based Fourier-transform techniques (far-infrared spectroscopy or Fourier-transform terahertz spectroscopy, Sect. 2.2), and measurements in the time-domain (terahertz time-domain spectroscopy, Sect. 2.3). Section 2.4 compares the methods described in Sects. 2.1–2.3 to each other and to complementary methods such as inelastic neutron scattering (INS) spectroscopy and conventional EPR spectroscopy (Sect. 1). Section 2.5 reviews measurement methods where the change in magnetization or magnetic torque of the sample is monitored, rather than the transmitted or reflected electromagnetic radiation.

2.1 *Frequency Domain Magnetic Resonance Spectroscopy*

In frequency domain magnetic resonance (FDMR) spectroscopy, the electromagnetic radiation is swept, while the external field is zero or kept at a fixed value. The transmitted or reflected radiation is then measured as a function of frequency, and the observed resonance frequencies allow direct determination of the ZFS. The advantage of this approach is that no external fields (which are often considerable in high-frequency ESR) are needed, which simplifies both the experiment and the analysis. For example, spin Hamiltonian terms that contain the magnetic field in higher powers may become significant, although their invocation appears not to have been necessary in studies published so far [60]. In addition, lineshapes in the field domain are not strictly equivalent to those in the frequency domain [61]. Earlier work was reviewed by Van Slageren [62] and selected examples by McInnes [34]. The spectral simulation strategy was outlined by Kirchner et al. [63].

2.1.1 Instrumental Aspects

The prerequisite for FDMR is the availability of sweepable radiation sources. Published FDMR data were all obtained on spectrometers, which employ backward-wave oscillators (BWOs) as radiation sources [64]. BWOs are vacuum tubes, in which accelerated electrons interact with a periodic metal structure. Because the dispersion curves for the electron beam and the periodic structure intersect at a point where the phase and group velocities of the periodic structure have opposite signs, energy that is extracted from the electron beam by the electromagnetic wave will flow in the opposite direction to the electron beam [65]. In recent years, the high-frequency BWOs ($\nu > 250$ GHz) have become scarce, and solutions based on lower frequency BWOs (e.g., 100–180 GHz), combined with broadband frequency doublers and triplers, are becoming more common. For such source-plus-multiplier-chain type solutions, sweepable solid state sources can also be used. The transmitted radiation is detected by means of Golay cells or bolometers. With the use of an interferometer, phase sensitive detection can be realized, allowing extraction of intensity and phase of the transmitted or reflected signal. The samples are inserted into a cryostat with optical (Mylar) windows, optionally equipped with a split-coil magnet. The sample rod is equipped with a sliding sample holder allowing consecutive measurements with and without sample in the beam which, after division, yields the transmission spectrum [62].

2.1.2 Zero-Field Splittings

The zero-field splittings of the spin ground states of a number of MNMs were determined by means of FDMR (Table 1). In most of these studies, the data were analyzed in the giant spin Hamiltonian framework, i.e., considering the ground spin

Table 1 Zero-field splittings of molecular nanomagnets determined by frequency domain magnetic resonance spectroscopy

Compound	S	D/cm^{-1}	E/cm^{-1}	$10^5 B_4{}^0$/cm^{-1}	$10^5 B_4{}^4$/cm^{-1}	Ref.
$[Mn_{12}O_{12}(OAc)_{16}(H_2O)_4]$	10	−0.461	0	−2.2	0	[66]
$[Mn_{12}O_{12}(O_2CCMe_3)_{16}(H_2O)_4]$	10	−0.464	0	−1.9	0	[67]
$[Mn_{12}O_{12}(O_2CPh)_{16}(H_2O)_4]$	10	−0.47	0	−2	0	[68]
$(PPh_4)[Mn_{12}O_{12}(O_2CEt)_{16}(H_2O)_4]$	19/2	−0.454	0	+1	0	[69]
$[(tacn)_6Fe_8O_2(OH)_{12}Br_7(H_2O)]Br(H_2O)_8$	10	−0.203	±0.00307	+0.004	0	[70]
$[Mn_9O_7(O_2CCH_3)_{11}(thme)(py)_3(H_2O)_2]$	17/2	−0.247	0	+0.46	0	[71]
$[Ni_4(MeOH)_4(sea)_4]$[a]	4	−0.93	0.023	−43	−210	[72]
$[Mn_6O_2(sao)_6(O_2CMe)_2(EtOH)_4]$	4	−2.12	0	+15	0	[26]
$[Mn_6O_2(Et\text{-}sao)_6(O_2CPh)_2(EtOH)_4]$	12	−0.386	0	−0.4	0	[26]
$[Mn_6O_2(Et\text{-}sao)_6(O_2CPhMe_2)_2(EtOH)_4]$	12	−0.362	0	−0.6	0	[26]
$[Ni(hmp)(MeOH)Cl]_4$	4	−0.592[b], −0.771	0	0	0	[73]
$[Ni(hmp)(dmb)Cl]_4$	4	−0.621, 0.631	0	0	0	[73]
$[Ni(hmp)(dmp)Cl]_4$	4	−0.610	0	0	0	[73]

The parameters D and $B_4{}^0$ are related to the parameters D_2 and D_4 of the spin Hamiltonian $\mathcal{H} = D_2\hat{S}_z{}^2 + D_4\hat{S}_z{}^4$ that is also often used to describe the ZFS, according to $B_4^0 = D_4/35$ and $D = D_2 - \frac{25}{35}D_4 + \frac{30}{35}S(S+1)D_4$ [74]

[a] Including inelastic-neutron-scattering data

[b] Resonance lines due to two different species were observed

multiplet only. The first (1998) such study was performed on pressed powder pellet samples of $Mn_{12}Ac$ [66]. The ZFS spin Hamiltonian parameter values of $D_2 = -0.389$ cm^{-1} and $D_4 = -7.65 \times 10^{-4}$ cm^{-1} (Table 1) were obtained. These parameters corresponded well to the parameters obtained by high-frequency EPR. Furthermore, FDMR measurements showed the axial ZFSs of other Mn_{12} derivatives, such as $[Mn_{12}O_{12}(O_2CCMe_3)_{16}(H_2O)_4]$ ($Mn_{12}Piv$, Piv = pivalate) [67] and $[Mn_{12}O_{12}(O_2CPh)_{16}(H_2O)_4]$ ($Mn_{12}Bz$, Bz = benzoate) [68], to be quite similar (Table 1). One-electron reduction of Mn_{12} causes reduction of one of the anisotropic Mn^{III} ions to give a more isotropic Mn^{II} ion, with a concurrent decrease in ZFS, evidenced by FDMR measurements which gave a slightly decreased D-value of $D = -0.454$ cm^{-1}, and, surprisingly, a change in sign in the case of $B_4{}^0$ (Table 1). The measurement also confirmed the ground state spin of $S = 19/2$ [69].

Another early FDMR investigation further validated the usefulness of the technique. In this study, four resonance lines were observed for $[(tacn)_6Fe_8O_2(OH)_{12}Br_7(H_2O)]Br(H_2O)_8$ (Fe_8) [70]. Here the often excellent agreement between resonance frequencies determined by FDMR and those determined by INS was also noted for the first time. An attempt was made to extract the second order transverse ZFS parameter from the data, and a value of $|E| = 0.0307$ cm^{-1} was obtained, which is lower than that obtained from the INS data. It was noted that the simulations are not very sensitive to the E-value, and not at all to the sign of E. FDMR measurements on MNMs with large spins and large negative D-values are generally not very sensitive to transverse ZFS, because transverse ZFS mostly affects levels with small M_S values, i.e., those close to the top of the energy barrier, and transitions between these are often not observed in FDMR (see Sect. 2.4). The good correspondence between INS and FDMR was also noted in the case of $[Mn_9O_7(O_2CCH_3)_{11}(thme)(py)_3(H_2O)_2]\cdot MeCN\cdot Et_2O$ (H_3thme = 1,1,1-tris(hydroxymethyl)-ethane) [71].

Magnetization measurements demonstrated that the metallocubane clusters $[Ni_4(ROH)_4(sea)_4]$ (H_2sea = salicylidene-2-ethanolamine; R = Me or Et) with approximate S_4 symmetry have $S = 4$ ground states and large negative D values, and yet the expected slow relaxation of the magnetization was not observed, even down to 40 mK [72]. A combined INS and FDMR study proved that this is due to large transverse ZFS both in second and fourth order, which causes mixing of the M_S-states and hence efficient quantum tunneling of the magnetization. The origin of the ZFS in S_4 symmetry clusters was the subject of a theoretical study, which proved that the first excited total-spin states belong to a degenerate representation of the S_4 point group, which corresponds to a quasi-orbital angular momentum [75]. First order spin–orbit coupling then leads to an effective antisymmetric exchange term in the spin Hamiltonian. An attempt was made to fit the INS spectra of $[Ni_4(MeOH)_4(sea)_4]$ using this approach, although the nonzero E-value shows that the effective cluster symmetry is lower than S_4. In EPR studies on other Ni_4 clusters, no antisymmetric exchange interactions were found [8].

Finally, a recent study reported FDMR and INS spectra of a series of hexanuclear manganese(III) clusters, with both $S = 4$ and $S = 12$ ground states (Fig. 2) [26]. Magnetically, these clusters consist of two exchange coupled triangles, and ground state spin value depends on the nature of the exchange interaction

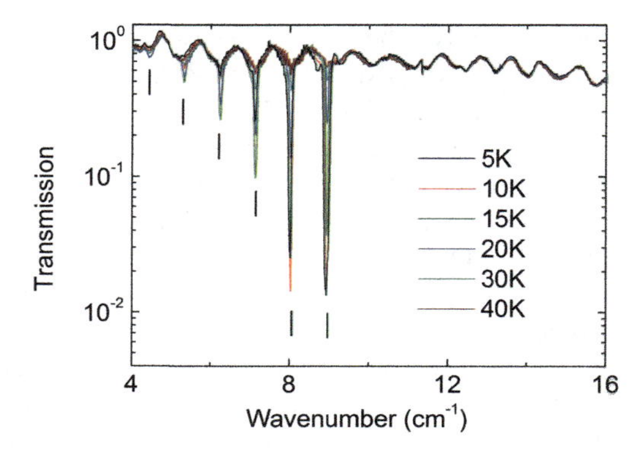

Fig. 2 FDMR-spectra recorded on a powder pellet of $[Mn_6O_2(Et\text{-}sao)_6(O_2CPhMe_2)_2(EtOH)_4]\cdot 4EtOH$ at different temperatures as indicated. Adapted from [26]. See also Fig. 1

(antiferromagnetic or ferromagnetic) within each triangle. The $S = 4$-ground-state of $[Mn_6O_2(sao)_6(O_2CMe)_2(EtOH)_4]\cdot 4EtOH$ (H_2sao = salicylaldoxime) has a very large D-value of $D = -2.12\ cm^{-1}$, which is much larger than the D-values for other MNMs. In contrast, the $S = 12$ ground states of the clusters $[Mn_6O_2(Et\text{-}sao)_6(O_2CPh)_2(EtOH)_4]\cdot 4EtOH$ and $[Mn_6O_2(Et\text{-}sao)_6(O_2CPhMe_2)_2(EtOH)_4]\cdot 4EtOH$ (Fig. 2) have much smaller D-values of ca. $-0.37\ cm^{-1}$, in spite of the, presumably, similar single-ion anisotropy. This is an elegant example of how the D-value of cluster ground state is related to the size of its spin. In the strong-exchange approximation, the D-value of a given spin state is a linear combination of the D-values of the single ions, where the coefficients become smaller for larger values of S [28].

2.1.3 Resonance Intensities

Because in the currently implemented FDMR spectrometers, measurements are performed by consecutive measurement of the detector intensity with and without sample, absolute transmission spectra are obtained. As a consequence, no scaling factor is, in principle, required to match experimental and simulated spectra. Hence, by comparison of the experimental and calculated intensities, further information on the properties of the material under study may be obtained. This opportunity has already been exploited in the first FDMR study on the $Mn_{12}Ac$ MNM, where careful examination of the temperature dependence of the line intensities suggested the presence of an excited state [66]. This state was subsequently identified as an $S = 9$ state by high-frequency EPR spectroscopy [32, 76]. In the case of $[Ni_4(ROH)_4(sea)_4]$, the calculated and experimental intensities matched very well, indicating the absence of low-lying excited states, in agreement with the lack of intermultiplet excitations below $22\ cm^{-1}$ in the INS spectra [72]. Finally, the resonance line intensity allowed determination of D, without investigation of the temperature dependence, in the mononuclear transition metal complex $[Ni(HIM2\text{-}py)_2NO_3]NO_3$ (HIM2-py = 2-(2-pyridyl)-4,4,5,5-tetramethyl-4,5-dihydro-1*H*-imidazolyl-1-hydroxy), giving $D = -10.0\ cm^{-1}$ [77].

2.1.4 Lineshapes

It was recognized early on that the lineshape formula that best describes the FDMR lines in spectra recorded on MNMs is usually the Gaussian lineshape rather than the Lorentzian [70]. This suggests that the linewidth is dominated by (Gaussian) distributions in the spin Hamiltonian parameters, especially in D. For $Mn_{12}Ac$ evidence for a discrete number of isomers rather than a continuous distribution was found [78] in agreement with high-frequency EPR results [7, 79]. In two cases, $(PPh_4)[Mn_{12}O_{12}(O_2CEt)_{16}(H_2O)_4]$ [69] and $[Ni_4(MeOH)_4(sea)_4]$ [72], the resonance lines were observed to be better described by Lorentzian functions. This suggests, that the linewidth may be determined by spin relaxation, in particular T_2 relaxation. This was supported by the fact that the linewidth increased towards higher temperatures. Assuming that the linewidth is determined by T_2 relaxation, an estimate of $T_2 = 50$–58 ps was obtained, which is much shorter than T_M times that were later obtained for other MNMs by pulsed EPR on dilute frozen solutions (Sect. 3.1). Temperature-dependent Gaussian linewidths were also found in high-frequency EPR measurements on Fe_8, which were attributed to the change in thermal fluctuations of the intermolecular magneto-dipolar interaction [80, 81]. The Lorentzian linewidth for $[Ni_4(ROH)_4(sea)_4]$ is largely temperature-independent, suggesting that the spin relaxation is temperature-independent in the investigated temperature range, or that the lineshape has a different origin. A priori, the lineshape of an FDMR line is not necessarily Lorentzian or Gaussian. However, when implementing a Mach–Zehnder interferometer, one can also measure the frequency dependence of the phase shift due to the sample (i.e., the dispersion, rather than the absorption). This allows direct determination of both real and imaginary parts of the magnetic permeability or dynamic magnetic susceptibility, without assumptions regarding the lineshape [62]. In a magnetized medium, the material parameters (dielectric permittivity and magnetic permeability) become tensorial in nature [82] and this can have an influence on the propagation of radiation in such media, leading to magneto-optical effects. FDMR measurements involve magnetic dipole transitions and, thus, only the magnetic permeability tensor needs to be taken into account. When the radiation propagation is perpendicular to the magnetization direction (Voigt geometry), the magneto-optical effects are rather small and were reported to lead to slight shifts in the apparent resonance frequency, as well as causing an asymmetry of the resonance lines, for measurements on magnetized single-crystal mosaics of $Mn_{12}Ac$ [78]. In HFEPR measurements in finite fields, asymmetric lines were also observed, due to distributions in the angle between molecular ZFS tensor and external field [83]. If the radiation propagation is parallel to the direction of the magnetization (Faraday geometry), the effects are much more pronounced. In this case, a strong rotation (>150°/mm sample thickness) of the plane of polarization of linearly polarized radiation was observed off-resonance (the familiar Faraday effect), again on magnetized $Mn_{12}Ac$ single-crystal mosaics (Fig. 3) [84]. At resonance, one circular component of the linearly polarized radiation is virtually fully absorbed, and the transmitted

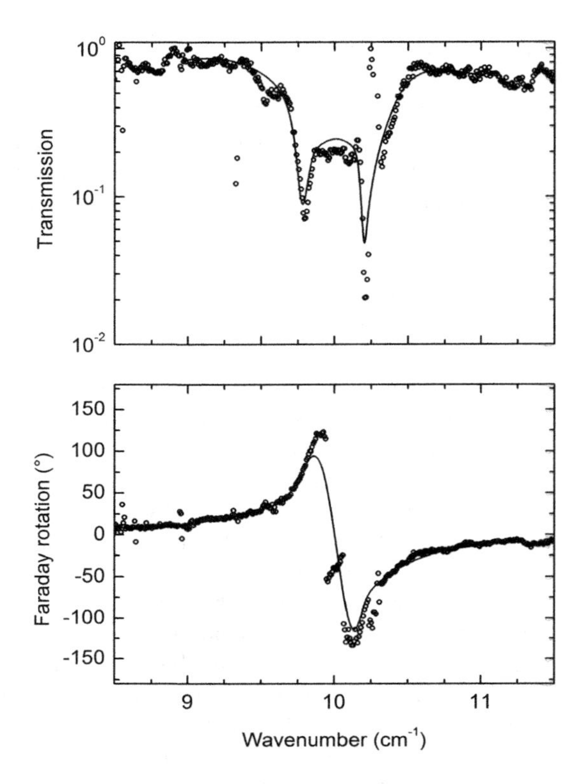

Fig. 3 *Top:* Transmission spectrum recorded at 1.77 K on a single crystals mosaic of $Mn_{12}Ac$ with the crystallographic *c*-axis (easy axis) parallel to the radiation propagation direction, using parallel polarizer and analyzer. *Bottom:* Frequency dependence of the Faraday rotation, derived from the transmission spectrum. Adapted from [84]

radiation is therefore circularly polarized of the opposite polarization. This effect is also exploited in induction-mode EPR [85].

2.1.5 Relaxation of the Magnetization

Because field and frequency can be changed at will, more elaborate measurements can be carried out. The relaxation of the magnetization in $Mn_{12}Ac$ was investigated in some detail, using the measurement procedure of cooling in a certain field, followed by inversion of the external field to a certain value in the opposite direction and subsequent recording of spectra at various time delays. FDMR measurements on powder samples after inverting the field from +0.45 to -0.45 T showed the gradual disappearance of one resonance line at 9.57 cm^{-1}, assigned to a transition from the metastable $M_S = +10$ state, and the appearance of a resonance line due to the transition from the -10 ground state. At the higher field of 0.9 T, the relaxation no longer proceeds homogeneously, but a pronounced hole develops in the resonance line, concurrently with the development of a peak at higher frequencies. These observations were attributed to quantum tunneling of the magnetization [86]. The relaxation of the magnetization was also performed on single-crystal mosaic samples of $Mn_{12}Ac$ [87]. The populations of the $M_S = \pm 10$ states were extracted from the intensities of the resonance lines, and from the time-dependence of these intensities the relaxation time was determined. These measurements were repeated at many different longitudinal fields. The field dependence of the relaxation time shows pronounced minima which were assigned to quantum tunneling of the magnetization occurring at those longitudinal fields where M_S levels on the left and right of the potential energy barrier are accidentally degenerate. The shift of the apparent resonance frequency during the relaxation process and dependence of this shift on the longitudinal field were discussed in terms of internal dipolar fields and quantum tunneling effects.

2.2 *Interferometer-Based Fourier-Transform Terahertz Spectroscopy*

Most FDMR investigations of MNMs (Sect. 2.1) were performed in the THz-Far Infrared (FIR) region of 90–600 GHz. This frequency range is also accessible by Fourier-transform infrared spectrometers, based on Martin–Puplett or Michelson interferometers. The advantage of this measurement method is that the entire spectrum can be acquired with a single radiation source, in contrast to current implementations of FDMR, which require about a dozen BWOs to cover the entire frequency range from 30 to 1,200 GHz. In addition, higher frequencies are easily accessed. Below ca. 300 GHz, the intensity of conventional far-infrared sources drops drastically, which decreases sensitivity. With the use of coherent synchrotron radiation, instead of a conventional mercury lamp, this disadvantage can be overcome [88]. To enhance sensitivity, lock-in detection was employed, locked to the pulse repetition rate of the synchrotron source (1.25 MHz). The FT THz method was successfully used to measure the ZFS in $Mn_{12}Ac$ [88]. The spectra were presented as the frequency dependence of detector signal at low-temperatures divided by that at high temperatures, to eliminate the frequency dependence of the source intensity, as well as absorption lines due to electric dipole transitions in air. Note that the physical principles of FT THz and FDMR spectroscopies are equal. FT-THz measurements on the SMM $(NEt_4)[Mn_2^{III}(5\text{-Brsalen})_2(MeOH)_2Ru^{III}(CN)_6]$ (5-Brsalen=*N,N′*-ethylenebis(5-bromosalicylidene) aminato anion) showed features at 9.0, 12.8, 18.6, and 19.3 cm^{-1}, of which only the first two were attributed to magnetic resonance transitions, because they split on application of a magnetic field [89]. The combination of these magnetic resonance measurements and susceptibility data allowed establishing of a consistent energy level diagram. The analogous chromium complex, $(NEt_4)[Mn_2^{III}(5\text{-Brsalen})_2(MeOH)_2Cr^{III}(CN)_6]$ was also investigated by FT-THz spectroscopy, as well as by INS, and gave similar results [90].

In addition to the studies cited above, Fourier Transform IR-spectrometers have been used for the study of magnetic excitations since the 1960s. Thus, collective excitations were observed in, among others, FeF_2 [91, 92], Fe^{III}dithiocarbamate

[93] and $LaMnO_3$ [94]. Crystal field splittings were observed for lanthanide ions [95]. The ZFS was studied in mononuclear complexes of Fe^{III} [96] Fe^{II} [97], and square planar Co^{III} [98, 99] among others. Even hemoglobin and myoglobin were successfully investigated in this manner [93, 100].

2.3 Terahertz Time-Domain Spectroscopy

A second broadband method that can be employed to study EPR excitations in MNM is THz time-domain spectroscopy (THz-TDS). This method employs mode-locked Ti:sapphire femtosecond lasers, where the fs pump pulse hits a photoconductive antenna across which a bias voltage is applied. The acceleration of the photogenerated charge carriers causes emission of a short (typically 2 ps) THz pulse [101]. This THz pulse interacts with the sample. For detection purposes, a similar antenna structure is employed, without bias voltage. A delayed fs probe pulse generates charge carriers, which are accelerated by the electric field of the THz pulse. By varying the delay between pump and probe pulses, the electric field of the THz pulse is mapped out. Fourier-transformation then gives the THz spectrum in the frequency domain. The advantage of this method is that both amplitude and phase are detected, while conventional FT-IR spectrometers only record the radiation intensity.

THz-TDS was also used to study the benchmark of molecular nanomagnetism, $Mn_{12}Ac$ [102–104], but has not been employed to study MNM since. These early studies demonstrated that the ZFS of MNMs can be obtained by THz-TDS. Parks et al. then investigated the linewidth in some detail, where they considered contributions from hyperfine interactions and intermolecular magneto-dipolar interactions on the linewidth and concluded that, in addition to these, there must also be a distribution in the D-parameter. These investigations also made use of the fact that both amplitude and phase of the THz electric field are obtained, which can be converted to the real and imaginary parts of the index of refraction.

2.4 Comparison of EPR Measurement Techniques

The following is the personal view of the author, based on his experience with the discussed techniques. The advantage of frequency-domain techniques, described in Sects. 2.1–2.3 is that no external field is required and ZFSs are obtained directly. The ability to apply a small field is useful, especially at frequencies above ca. 20 cm^{-1}, in order to distinguish magnetic resonance transitions from molecular vibrations and other phonon-type excitations. The advantage of field-swept techniques is that field-swept spectra tend to have flatter baselines, which increases sensitivity. Cavity and other resonator methods, which are only easily implemented in field-swept experiments, are much more sensitive. Therefore, single-crystal measurements are more easily carried out, which give access to a great deal of detail that is not easily available from polycrystalline samples.

The linewidths of the resonance lines in FDMR and FT-THz measurements (Fig. 4) are limited by the sample, although the intrinsic resolution of FDMR is higher. FT-THz has the advantage that a single period of the interferometer covers the entire frequency range available (typically up to 40–100 cm^{-1}, depending on source and detector), with the possibility of accumulating a number of scans (using a mercury lamp source typically dozens of scans are acquired). FDMR, on the other hand, is better suited to measurements below 10 cm^{-1}, where the optics of most IR spectrometers starts to cut off some of the intensity. THz TDS suffers from much

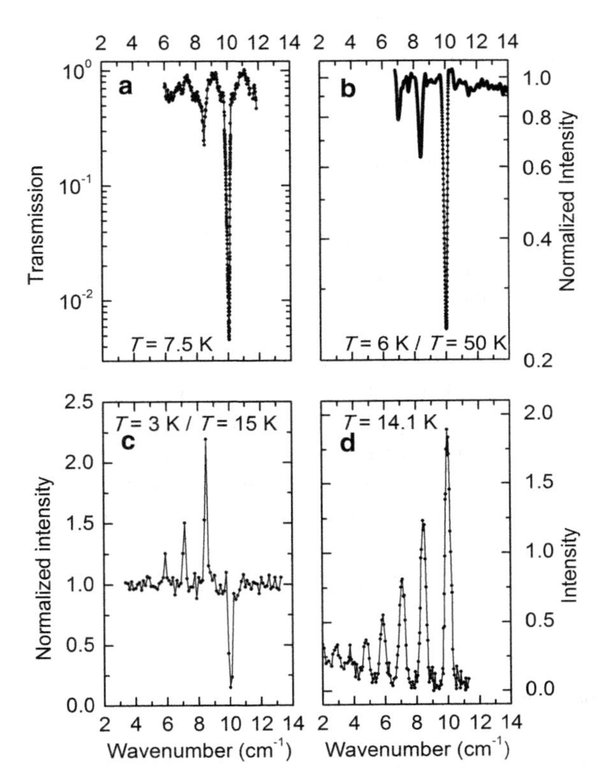

Fig. 4 Comparison of Zero-Field techniques to determine the Zero-Field Splitting in $Mn_{12}Ac$. (**a**) Frequency Domain Magnetic Resonance Spectroscopy [105]. (**b**) Frequency Domain Fourier-Transform Terahertz Spectroscopy [88] (Schnegg, Personal communication). (**c**) Terahertz Time-Domain Spectroscopy, adapted from [102]. Used with permission. ©2001 American Physical Society. (**d**) Inelastic Neutron Scattering, adapted from [106]. Used with permission. ©1999 American Physical Society. Readers may view, browse, and/or download material for temporary copying purposes only, provided these uses are for noncommercial personal purposes. Except as provided by law, this material may not be further reproduced, distributed, transmitted, modified, adapted, performed, displayed, published, or sold in whole or part, without prior written permission from the American Physical Society

lower resolution, which is limited by the length along which the delay stage can be moved. Its advantage lies in the ability to record both amplitude and phase and the inherent fast time resolution available, although both factors have yet to be exploited in molecular magnetism.

INS also has a lower resolution and also suffers from poor sensitivity. Its clear advantages (that should not be underestimated) are that transitions between spin multiplets are allowed and that both the frequency- and wavevector (Q)-dependence of the scattering intensity are recorded. The Q-dependence allows, in principle, the distinguishing of magnetic from nonmagnetic excitations and also gives information on the spin wavefunction [60]. Finally, because the scattering intensity only depends on the population of the initial state, INS is very useful at high temperatures, where the EPR intensities are low, because of limited population differences between initial and final states.

2.5 *Magnetization Detected EPR*

Not all molecular systems are amenable to EPR in the field-, frequency-, or time-domain. For example, fast spin relaxation can lead to excessively broad lines. Furthermore, the sensitivity of nonresonator techniques, including all non-field-swept measurement methods, is generally insufficient to allow for detailed single crystal measurements. Finally, during the wide-range field sweep of a field-swept, conventional (high-frequency) EPR measurement, the actual physical properties, including the spin ground state may change, e.g., in the case of field-induced spin ground state changes in antiferromagnetic (ring) systems. For these reasons, effort has been invested into the development of other detection schemes, based on the monitoring of a thermodynamic quantity of the system, such as the magnetization or the magnetic torque. Many of the instruments described below were devised to investigate the spin dynamics of MNMs, for which purpose pulse EPR spectrometers are also currently used (Sect. 3).

2.5.1 Implementations

An early implementation of magnetization-detected EPR by Wernsdorfer et al. employed a micro-SQUID detection system, employing field sweeps in the presence of microwave irradiation, at temperatures down to 40 mK, with the sample in vacuum [107]. The advantage of this method is the very high sensitivity of micro-SQUIDs that can detect the magnetic moment of 10^3 spins [108], with single-spin detection sensitivity predicted [109]. This setup was used to investigate magnetic resonance transitions in the molecule $K_6[V_{15}As_6O_{42}(H_2O)]\cdot 8H_2O$, abbreviated as V_{15}. Clear dips in the magnetization curves at temperatures down to 40 mK, attributed to $M_S = -1/2$ to $M_S = +1/2$ transition. Measurements at frequencies of up to 20 GHz allowed extraction of the g-value as $g = 2.02$, which is slightly higher

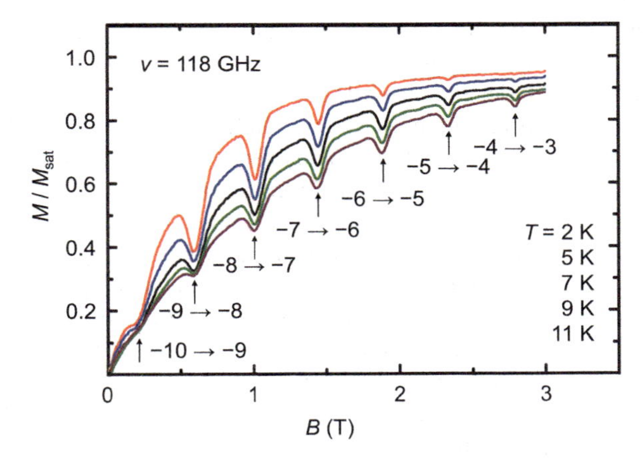

Fig. 5 Magnetization curves measured on Fe_8 in the presence of microwave radiation with $\nu = 118$ GHz and $T = 10$ K. The dips are due to transitions between the M_S states of the $S = 10$ ground state, as indicated. Adapted from [115]. Used with permission.

than previously reported values $g_{//c} = 1.98$ and $g_{\perp c} = 1.95$ [110] or $g_{//c} = 1.98$ and $g_{\perp c} = 1.968$ [111] (both X-band EPR), $g = 1.97$ (57.8 GHz) [112], $g_{//c} = 1.981$, and $g_{\perp c} = 1.953$ (FDMR) [113]. The same setup was used to detect EPR in $(H_2NMe_2)[Cr_7NiF_8(O_2CCMe_3)_{16}]$ (Cr_7Ni) [114] In later measurements, on Fe_8, a Hall bar was employed rather than a micro-SQUID (Fig. 5) [115].

A second instrument, developed by Friedman et al., employed a Hall bar sensor to measure the magnetization of the sample, which has the advantage over micro-SQUIDs that the operating magnetic field can be much higher [116, 117]. The authors used this and later setups to investigate magnetic resonance and spin dynamics of Fe_8 in a series of papers. High-power BWOs and a rectangular cavity were used to enhance the microwave field at the sample, and the Hall bar sensor was mounted outside the cavity [117]. Heating of the sample to 4 K, at a nominal cryostat temperature of 2 K by nonresonant heating of the resonator by the radiation was observed, with additional heating observed at resonance. The *D*-value of 0.2944 K, obtained from measurements at different frequencies between 127 and 136 GHz, is in good agreement with published values. Further measurements were performed on $[Mn_4O_3Cl(O_2CCH_3)_3(dbm)_3]$ (Hdbm = dibenzoylmethane) [118]. An improved setup employed a cylindrical rather than a rectangular cavity and a thin film pick-up loop rather than a Hall bar sensor for detection [119]. The pick-up coil makes the instrument well suited for the investigation of fast spin dynamics (see below) but less for spectroscopic purposes. The emf induced in the loop is measured with a SQUID voltmeter coupled to a room temperature amplifier.

A Hall bar sensor, but no resonator, was also used by del Barco, Kent et al. at temperatures down to 0.38 K with the sample in vacuum [120]. Measuring the longitudinal field dependence of the magnetization of $[Ni(hmp)(dmb)Cl]_4$ (Ni_4, Hhmp = 2-hydroxymethylpyridine, dmb = 3,3-dimethyl-1-butanol) with and without radiation (at single frequencies between 24 and 43 GHz) allowed the observation of magnetic resonance transitions, with extracted ZFS values in good agreement with published values. In transverse fields, the plots of resonance fields vs frequencies show a clear curvature near zero-field, which is a signature of the ground state being a quantum superposition of states with opposite magnetization, as is also required for quantum tunneling of the magnetization. This setup was later improved, when a strip line resonator was implemented on the chip with the Hall sensor, where different resonators allow for measurements between 10 and 30 GHz, and both the transmitted radiation and the magnetization can be measured [121, 122].

A commercial SQUID magnetometer was used for magnetization detection in combination with Klystron sources at 95 and 141 GHz, where the advantage lies in the ability to determine the absolute magnetization quantitatively [123]. Investigation of the power dependence allowed extraction of T_1 in copper sulfate. Direct time-resolved measurements appear to be difficult to implement in view of the slow measurement time scale of commercial SQUID magnetometers. Again, the method was used to investigate Fe_8, and clear magnetic resonances were observed [124].

The magnetic resonance transitions can also be observed by measuring the magnetic torque in magnetically anisotropic systems [125]. The setup employed BWO radiation sources in combination with an optical split coil superconducting magnet, allowing excitation frequencies between 30 GHz and 1 THz, and fields up to 7 T. A sample rotator is incorporated which allows measurements at arbitrary angles with the magnetic field. The use of an optical cryostat enables arbitrary polarization of the radiation with the external field. Preliminary measurements on a star-shaped Fe_4 single-molecule magnet in both field and frequency domains were presented. The combination of EPR and torque magnetometry was also used by Ohmichi et al., but has not yet been used for investigations of MNMs [126, 127].

2.5.2 Spin Dynamics

A number of publications report the investigation of spin dynamics of MNMs, where the change in magnetization is monitored following a pulse of microwave radiation. As recognized early on [107], the phonon-bottleneck effect, i.e., a long thermal equilibration time between phonons that are resonant with the spin system and the cryostat, plays a major role in these investigations. This effect depends on the crystal size, the coupling of the sample to the sample holder, and the sample surroundings (vacuum/exchange gas/liquid), and, as a consequence, a large range of time constants have been reported.

The most studied sample in this respect is Fe_8. In this sample, the magnetization relaxation, over or under the energy barrier, is a further dynamic process, in addition to spin–lattice relaxation and thermalization with the cryostat.

In measurements on small crystals at temperatures between 2 and 10 K, two processes were observed in the relaxation of the magnetization after 10 ms microwave pulses [128] with time constants in the range of 1–10 ms and 25–200 ms, respectively. Both processes were reported to be determined by the phonon-bottleneck equilibration time, with T_1 expected to be several orders of magnitude shorter. For larger crystals, the magnetization continues to change after the pulse, before relaxing back. This was ascribed to the slow equilibration of the sample with the lattice that was heated due to the microwave radiation or due to the phonons released by the spin after absorption of microwave photons by the spins [117, 118, 128]. This thermalization involves the magnetic moment crossing the energy barrier generated by the axial ZFS, or tunneling of the magnetic moment under the barrier. This interpretation is corroborated by the fact that measurements at a field where quantum tunneling is possible showed much smaller overshoots [117]. Even longer relaxation times, on the time scale of seconds, were reported for Ni_4 [120]. Later measurements gave relaxation times of 30 ms, attributed to phonon-bottleneck processes [122].

By means of pump–probe measurements, it proved to be possible to gain information on the spin–lattice relaxation time T_1 [129, 130]. These studies employed much shorter pulses of 20–50 ns. A first pulse excites the Fe_8 molecule from $M_S = -10$ to $M_S = -9$. After a delay, a second pulse resonant with the $M_S = -9$ to $M_S = -8$ transition is used. The effect of the latter pulse on the magnetization, measured by a Hall bar sensor, depends on the population of the $M_S = -9$ level at the time of the second pulse, which in turn depends on T_1. The delay between pulses was in the microsecond regime, which is much faster than the thermal equilibration effects described above. Although the measurements were hampered by the slow response of the Hall bar, the authors managed to extract T_1 times of the order of 2 μs at 1.6 K, decreasing to 400 ns at 15 K. By means of calculations using the generalized master equation for magnetization relaxation, estimates were extracted for the spin-phonon coupling. In addition, it was shown that phonon induced transitions with $\Delta M_S = \pm 2$ play an important role.

Bal et al. made extensive measurements on single-crystals of Fe_8 to disentangle phonon-bottleneck processes from true spin–lattice relaxation [131]. Resonant pulses 20 μs in length at 2.1 K induced a change in the magnetization, which continued on a long ca. 1 ms time scale after the end of the pulse. This was attributed to the lattice and the spins being out of equilibrium. On top of this, a faster decrease was observed with a time constant of about 5 μs, with no systematic dependence on temperature, field or radiation power, which was attributed to the phonon bottleneck. The authors note that this time constant is similar to that reported for T_1 by Bahr et al. (see above) [129]. Extensive simulations gave a much shorter time for T_1 in these measurements of $T_1 \sim 40$ ns [131].

No direct information on the phase memory or quantum coherence time T_M has yet been obtained from magnetization detected EPR investigations. In one study, a spin echo measurement was attempted on frozen solutions of $[Ni(hmp)(dmb)Cl]_4$ in toluene:dichloromethane (1:1) at 5.5 K, but no echoes were observed. An upper limit of $T_M = 50$ ns was deduced from this observation [132]. For homogeneously broadened lines, the half-width in CW-EPR is related to the real spin–spin relaxation time T_2, of which T_M is a lower limit, by $\Delta\nu = (2\pi T_2)^{-1}$, assuming $T_1 >> T_2$ [133]. Although EPR lines of MNMs are generally inhomogeneously broadened, a lower limit of T_M can still be obtained from the linewidth. This approach yielded values in the range of 10^{-10} to 10^{-9} s, including 0.2 ns [121], 0.5 ns [120], 2 ns [107], and 10 ns [114]. Pulse EPR measurements (Sect. 3) revealed that phase memory times can fortunately be much longer than these estimates.

3 Pulse EPR in Molecular Nanomagnetism

Much effort has been devoted to the study of the dynamics of the magnetization in single molecule magnets, where the axial magnetic anisotropy creates an energy barrier towards inversion of the magnetic moment [29]. The relaxation time is generally governed by an Arrhenius law $\tau = \tau_0\exp(-\Delta E/k_BT)$. Microscopically, relaxation of the magnetization is a multistep process, in which the molecule undergoes multiple phonon-induced transitions to increasingly higher lying M_S-states, until the highest-lying M_S state is reached. This is then followed by multiple downward transitions on the other side of the energy barrier. For direct spin-phonon interactions, the oscillating strain caused by a phonon causes modulation of the ZFS, which can induce transitions between M_S states with the selection rule $\Delta M_S = \pm 1, \pm 2$ [29, 74]. Because the phonon density decreases with decreasing energy, the direct spin-phonon interaction process becomes slowest close to the top of the energy barrier, where the spacing between M_S levels is smallest. The rate of the slowest spin-phonon transition is reflected by the Arrhenius prefactor τ_0. In EPR terms, these individual phonon-induced transitions are nothing but spin–lattice relaxation. In addition to the direct process mentioned above, in ESR-spectroscopy, two two-phonon processes, the Orbach and the Raman process, are generally considered, where in the former the intermediate state is real while in the latter it is virtual [74]. Although there is ample information on the spin–lattice relaxation in single-spin systems [74, 134, 135], much less is available on exchange coupled systems [28]. One important finding is that spin relaxation in polynuclear systems is generally faster, because of the presence of a number of additional relaxation pathways. Thus modulation of the exchange interaction can lead to Orbach-type spin–lattice relaxation, with an excited spin state as the intermediate state. For dimers only the nonisotropic exchange interactions are effective in this way, but in larger systems, modulation of the isotropic exchange interaction can lead to relaxation [28, 136].

The spin–spin relaxation T_2, or the experimentally determined phase memory time T_M, is a parameter of interest in molecular magnetism, because it is the time available for a quantum computation. It was recognized several years ago that MNMs may be used to implement quantum bits [137, 138]. There are currently essentially three proposals for using MNMs for quantum information processing [21, 139, 140]. The first is an elaborate scheme to use high-spin MNMs, such as

SMMs, where it is important that the M_S states are not equidistant. This is the case for a ZFS split ground state [19]. A combination of multifrequency pulses at high frequencies where the radiation magnetic field is perpendicular to the ZFS quantization axis, and a low-frequency pulse where the field is parallel to this axis, can create a simultaneous coherent population of many M_S states. In addition, the field strengths and phases of the high-frequency components must be judiciously controlled. If those conditions are met, Grover's algorithm, which, in certain cases, allows finding an entry in a database with one single query, can be implemented. For this scheme, entanglement of many-particle systems is not necessary [19]. A second proposal deals with more conventional two-level systems for use as qubits [139]. The advantages of MNMs for such purposes are generally listed as follows. The larger physical size of MNMs compared to single sites makes resolution requirements for local addressing less strict. Second, the excited spin states can, in carefully engineered systems, be used to implement effective switching of the interaction between two qubits [141]. Engineering the interaction between pairs of MNMs is currently receiving a great deal of attention [50, 142] but is far from trivial. The interaction should be much weaker than the interactions within the MNM qubit so that the two can be considered separate entities, but the coupling energy should be much larger than the thermal energy [139]. The third proposal concerns the electric control of the exchange interaction in triangular MNMs, where an electric field can couple the two $S = 1/2$ states, that can show a small energy splitting. These two states are characterized by their chirality [140, 143]. The advantage of electric field control is that electric fields can be applied much more locally. In all of these schemes, the quantum operations need to be performed within the lifetime of the generated coherent superposition states, i.e., within the quantum coherence time, for which the experimentally determined phase memory time T_M is a lower limit.

The spin dynamics are also critically important in biophysical EPR. This field revolves around the elucidation of geometric and electronic structures and dynamics of biological systems, by the determination of g-tensors, of distances between electron spins and of couplings between nuclear and electron spins. For many of the measurement techniques employed, sufficiently long T_M times are required for these techniques to be fruitfully applied.

It is clear from the above that the spin dynamics of MNMs is a subject of great current interest. The techniques to implement quantum operations and those of pulse EPR are essentially the same. In both cases, pulses of arbitrary length and phase, that are resonant with the energy splitting of the relevant states of the investigated system, are required. For a two-state system, the two states can be pictured as opposite poles on a sphere (the Bloch sphere), where all possible coherent superposition states are points on that sphere. Applying a resonant microwave pulse causes a rotation of the spin in a direction perpendicular to that of the microwave magnetic field [144]. The turning angle is given by the microwave field strength (B_1) and the pulse duration (t_p) according to $\theta = \omega_1 t_p$, with $\omega_1 = g\mu_B B_1/\hbar$. For a given turning angle, short, intense pulses excite spins in a larger frequency range than do long, less intense pulses, because of the Heisenberg uncertainty relation, where the FWHM is given by $\Delta\nu_{1/2} \approx 1.207\ t_p^{-1}$ for a rectangular pulse [144]. Applying series of pulses of judiciously chosen lengths and varying the interpulse delays allow determination of the spin relaxation time constants of MNMs, as detailed in the next two sections. The work by Takui et al. on pulse EPR spectroscopy on high-spin organic radicals is not reviewed here [145, 146].

3.1 Spin–Lattice Relaxation

Some of the earliest spin–lattice relaxation time measurements were carried out using superheterodyne detection by partial saturation of the magnetic resonance by a microwave pulse, and subsequent monitoring of the transient recovery on an oscilloscope [147], which is essentially a type of saturation recovery measurement [144]. In this manner, the temperature dependence of T_1 in iridium(IV) pairs in $(NH_4)_2PtCl_6$ and copper(II) pairs in $Zn(dtc)_2$ [dtc = bis(diethy1-dithiocarbamate)] was investigated in detail. These early measurements have been reviewed [28, 148] and are not discussed further here. Another version of the saturation-recovery experiment involves a long saturation pulse or sequence of pulses, followed after variable time delay T by the primary echo pulse sequence $\pi/2$–τ_{fix}–π–τ_{fix}–echo to detect the longitudinal magnetization. A third method to measure T_1 is by inversion recovery, using the pulse sequence π–T–$\pi/2$–τ_{fix}–π–τ_{fix}–echo. This last method is more sensitive to spectral and spin diffusion than the former [144]. Finally, stimulated echo decay can also be used to measure T_1, by variation of T in the pulse sequence $\pi/2$–τ_{fix}–$\pi/2$–T–$\pi/2$–τ_{fix}–echo. This method is even more sensitive to spin and spectral diffusion. In the following, we discuss spin–lattice relaxation times measured on MNMs and related systems (Table 2). We restrict ourselves to pulse EPR measurements, omitting measurements of the electron spin dynamics by NMR [134], or a.c. magnetic susceptibility [164].

The antiferromagnetic ring $(H_2NMe_2)[Cr_7NiF_8(O_2CCMe_3)_{16}]$ (Cr_7Ni) has a spin ground state of $S = 1/2$, and was the first MNM to be investigated by pulse EPR [18]. The spin–lattice relaxation time was investigated by means of inversion recovery measurements on frozen solutions in toluene at X-band. T_1 reaches a value of about 1 ms at the lowest temperature of 1.8 K (Table 2), but decreases rapidly with increasing temperatures, down to 10^2 ns at 9.5 K, suggesting interaction with phonons, i.e., Raman or Orbach processes. Measurements on (H_2NMe_2) $[Cr_7MnF_8(O_2CCMe_3)_{16}]$ (Cr_7Mn), which has an $S = 1$ ground state, gave similar T_1 values, where the authors note that the T_1 values show a variation of a factor of two at different points in the spectrum, indicating that the ZFS plays a role in the relaxation. W-Band (95 GHz) measurements on the same compound revealed that the spin-relaxation time is little dependent on microwave frequency, with $T_1 = 2.9$ μs (W-band, 4.5 K) [149] compared to $T_1 \approx 3.5$ μs (X-band, 4.5 K) [18].

The spin–lattice relaxation was studied in more detail in a ferric triangle, $[Fe_3(\mu_3\text{-}O)(O_2CPh)_5(salox)(EtOH)(H_2O)]$ (Fe_3salox, H_2salox = salicylaldehyde oxime) [150]. Inversion recovery measurements at X-band frequencies yielded

Table 2 Spin relaxation data of exchange-coupled metal ion species, determined by pulse EPR

Species	Abbrev.	S	T_1 / μs	T_M / μs	T / K	Ref
$(H_2NMe_2)[Cr_7NiF_8(O_2CCMe_3)_{16}]$	Cr_7Ni	1/2	10^3	0.55	1.8	[18]
$(H_2NMe_2)[Cr_7NiF_8(O_2CCMe_3)_{16}]$ (X-band)		1/2	~3.5	0.379	4.5	[18]
$(H_2NMe_2)[Cr_7NiF_8(O_2CCMe_3)_{16}]$ (W-band)		1/2	~2.0	0.357	5.0	[149]
$(H_2NMe_2)[Cr_7NiF_8(O_2CCMe_3\text{-}d_9)_{16}]$		1/2		3.8	1.8	[18]
$(H_2NMe_2)[Cr_7NiF_8(O_2CCMe_3\text{-}d_9)_{16}]$		1/2		2.21	4.5	[18]
$(H_2NMe_2)[Cr_7MnF_8(O_2CCMe_3)_{16}]$		1	10^3	~0.55	1.8	[18]
$[Fe_3(\mu_3\text{-}O)(O_2CPh)_5(salox)(EtOH)(H_2O)]$	Fe_3	1/2	693	2.6	5.5	[150]
$K_6[V_{15}As_6O_{42}(H_2O)]\cdot 8H_2O$	V_{15}	3/2		0.340	4	[151]
$[Fe_4(acac)_6(Br\text{-}mp)_2]$	Fe_4	5	1.06	0.63	4.3	[152]
$[(tacn)_6Fe_8O_2(OH)_{12}Br_7(H_2O)]Br(H_2O)_8$	Fe_8	10		0.093	1.93	[153]
$[(tacn)_6Fe_8O_2(OH)_{12}Br_7(H_2O)]Br(H_2O)_8$		10	9.5×10^2	0.714	1.27	[153]
$[Mn^{III}Mn^{IV}(\mu\text{-}O)_2bipy_4]ClO_4$		1/2	1,380		25	[154]
$[Mn^{III}Mn^{IV}(\mu\text{-}O)_2bipy_4]ClO_4$		1/2	11		75	[154]
$[Mn^{II}Mn^{III}(\mu\text{-}OH)(\mu\text{-}piv)_2(Me_3tacn)_2](ClO_4)_2$		1/2	35		4.3	[154]
$[Mn^{II}Mn^{III}(\mu\text{-}OH)(\mu\text{-}piv)_2(Me_3tacn)_2](ClO_4)_2$		1/2	0.42		9	[154]
$[Fe_2S_2(S_2\text{-}o\text{-}xylyl)_2]^{3-}$		1/2	~55×10^3		5	[155]
$\{Mn_4O_xCa\}$ Photosystem II S_0 state		1/2	9.4		4.3	[154]
$\{Mn_4O_xCa\}$ Photosystem II S_0 state		1/2	0.85		6.5	[154]
$\{Mn_4O_xCa\}$ Photosystem II S_2 state		1/2	1,220		4.2	[156]
$\{Mn_4O_xCa\}$ Photosystem II S_2 state		1/2		0.8	4.2	[157]
$[4Fe\text{–}4S]^{3+}$ (*Ectothiorhodospira halophila*)		1/2	20–23	0.68	5	[158]
$[4Fe\text{–}4S]^{3+}$ (*Chromatium vinosum*)		1/2	9.1–11.9	0.46	5	[159]
$[4Fe\text{–}4S]^{1+}$ (*Desulfovibrio gigas* hydrogenase)		1/2	5	0.2	4.2	[160]
$[4Fe\text{–}4S\,]^{x+}$ (*Bacillus stearothermophilus*)		1/2	5×10^3		5	[161]
Ir^{4+} pairs in $(NH_4)_2PtCl_6$		1	1.3×10^3		1.8	[162]
Cu^{2+} pairs in $Zn(dtc)_2$		1	18×10^3		1.8	[163]
Ti^{2+} pairs in SrF_2		2	1.1×10^2	1.3	10	[148]

T_1 times between 693 μs (5.5 K) and 0.55 μs (11.0 K). The temperature dependence of T_1 could be fitted equally well to the equations describing Orbach or Raman processes. The unrealistic obtained exponent value in case of the Raman process led the authors to prefer the Orbach process. The Orbach energy gap was derived to be 57 cm^{-1}. This energy gap corresponds to the energy of the first excited spin state, which allowed choosing between two sets of exchange coupling parameters that fitted the susceptibility curve equally well.

A similar strong temperature dependence of T_1 was found in inversion recovery measurements on frozen toluene solutions of $[Fe_4(acac)_6(Br\text{-}mp)_2]$ (Fe_4, acac = acetyl acetonate, or 1,3-pentanedioate, $Br\text{-}mpH_3$ = 2-(bromomethyl)-2-(hydroxy-methyl)-1,3-propanediol) [152] which has an $S = 5$ ground state [165], with $T_1 \approx 1.06$ μs at the lowest temperature ($T = 4.3$ K) employed. The strong temperature dependence suggests that the spin–lattice relaxation proceed via a two-phonon process. Although the limited temperature range available did not allow the drawing of definite conclusions, an Orbach fit gave a value of about 5 cm^{-1} for the Orbach energy gap, which is well within the ground spin multiplet.

Investigations of the spin–lattice relaxation time of single crystals of Fe_8 at $T = 1.27$ K, utilizing the stimulated echo sequence, revealed two relaxation processes with time constants of 1.0 μs and 0.95 ms, respectively [153]. The authors attribute the former time constant to spectral diffusion, and speculate that the latter may reflect true spin–lattice relaxation, although the time constant is two orders of magnitude longer than values found by magnetization-detected EPR (see above) [129, 131].

The results described above demonstrate that there is a large variation in T_1 times for different MNMs, as well as a strong temperature dependence of T_1 (Table 2). An extreme example of this variation in T_1 is given by the two $S = 1/2$ dimers $[Mn^{III}Mn^{IV}(\mu\text{-}O)_2bipy_4]ClO_4$ (bipy = 2,2′-bipyridine) and $[Mn^{II}Mn^{III}(\mu\text{-}OH)(\mu\text{-}piv)_2(Me_3tacn)_2](ClO_4)_2$ (Me_3tacn = N,N',N''-trimethyl-1,4,7-triazacyclononane) [154]. The spin–lattice relaxation of the former is very slow and characterized by a Raman process below 50 K, while that of the latter is best described by an Orbach process. Apart from the metal oxidation states involved, the main difference between these complexes is the strength of the exchange interaction which leads to energy gaps between ground and first excited state of $\Delta E = 450$ cm^{-1} and $\Delta E = 25.5$ cm^{-1}, respectively, for the two complexes. The Orbach energy gap ($\Delta \sim 26$ K) in the latter complex was indeed found to match the energy gap to the first excited state very well. Similar results (Table 2) are obtained for polynuclear clusters in biological systems, e.g., the $\{Mn_4O_xCa\}$ cluster in Photosystem II [154, 156, 157] and iron–sulfur clusters in different oxidation states [158–161].

3.2 *Phase-Memory Times*

Phase memory times in solids are usually determined by the Hahn or primary echo sequence, $\pi/2$–τ–π–τ–echo, by variation of τ [144]. For quantum computation purposes, the quantum coherence time must be sufficient to allow extensive quantum manipulations, including error corrections. The quantum coherence time is the same as the spin–spin relaxation time T_2, for which the experimentally determined phase memory time T_M is a lower limit.

For a long time it was unclear whether quantum coherence times in MNMs would be long enough to make MNMs viable qubit candidates. Indeed, this was the very question that the title of the first paper in this area asked [18]. The phase memory time was determined to be $T_M = 379$ ns by X-band Hahn echo measurements on frozen toluene solutions of Cr_7Ni, which is two orders of magnitude longer than previous estimates of the lower limit of T_M [18]. Interestingly, the related compound $(H_2NMe_2)[Cr_7MnF_8(O_2CCMe_3)_{16}]$, which has an $S = 1$ ground state, has a very similar T_M, which demonstrates that the ZFS or ground state spin plays a limited role in the decoherence process. The echo decay for Cr_7Ni displays a pronounced modulation when using short, intense microwave pulses. This effect is called electron spin echo envelope modulation (ESEEM), and is caused by formally forbidden nuclear spin flips by the π-pulse of the Hahn echo sequence [144]. The oscillation frequency, therefore, corresponds to the Larmor frequency of the nuclei that the electron spin couples to via superhyperfine coupling. The ESEEM in Cr_7Ni was found to be due to coupling to the ligand protons. No coupling to the bridging ^{19}F nuclei was reported. It is this coupling to proton nuclear spins that is the main decoherence pathway. By deuteration of the ligand it proved possible to enhance T_M by a factor of 6 ($T_M = 2.21$ μs at $T = 4.5$ K), in agreement with the smaller nuclear magnetic moment of D, compared to H. The phase memory time was found to increase with decreasing temperature (Table 2), but not as strongly as T_1 (see above). A very similar T_M of 379 ns was found in W-band measurements on Cr_7Ni [149]. W-band pulse ENDOR measurements on Cr_7Ni demonstrated that the coupling between electron spin and proton nuclear spin is dipolar in nature, and its strength is up to ~2 MHz [149].

In X-band Hahn echo measurements on frozen solutions of Fe_3 in acetone, clear spin echoes were observed, from which $T_M = 2.18$ μs at 7 K was extracted, which is clearly longer than for Cr_7Ni under similar conditions. The observation of ESEEM again demonstrated hyperfine coupling of the electron spin to nuclear spins. Interestingly, T_M becomes temperature independent below ca. 7 K, at which point T_M reflects the true spin–spin relaxation rate, whereas at higher temperatures it is influenced by spin–lattice relaxation. The magnetization can be rotated by a microwave pulse with length t_p around an arbitrary angle $\theta = \omega_1 t_p$, where ω_1 is the microwave field strength. After a delay time, which ensures the decay of all quantum coherences, the magnetization along the z-axis is measured. The measurement of the z-magnetization as a function of t_p is called a nutation measurement. The corresponding oscillations of the magnetization are the so-called Rabi oscillations. A nutation measurement performed on Fe_3 showed that part of the observed oscillations was due to nutation of the electron spin, but another part was attributed to ESEEM-like effects (Fig. 6). The oscillations due to transient nutation were observed to decay very quickly, presumably due to microwave field inhomogeneities.

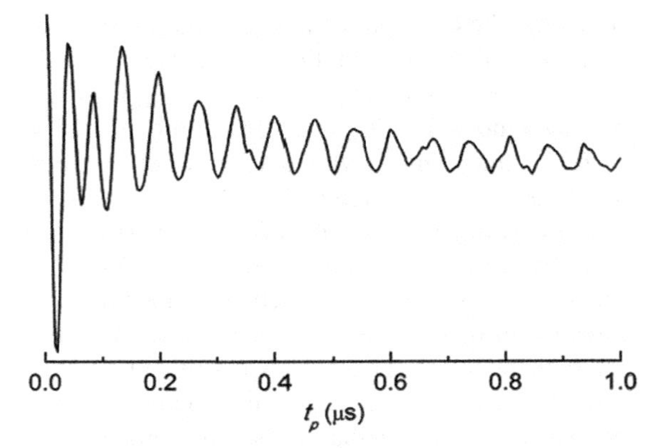

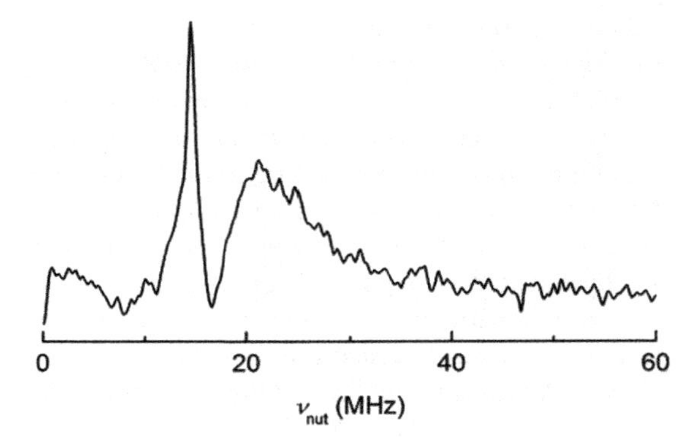

Fig. 6 *Left:* Echo-detected longitudinal magnetization of Fe_3 as a function of nutation pulse length. *Right:* Absolute-value Fourier Transform, showing contributions due to ESEEM-like effects (*sharp peak*) and Rabi oscillations (*broad peak*). Adapted from [150]. Used by permission of the PCCP Owner Societies

Use of a surfactant allows solubilization of the polyoxometalate cluster $K_6[V_{15}As_6O_{42}(H_2O)]\cdot 8H_2O$ (V_{15}) in the organic solvent chloroform. Spin echo measurements revealed a phase memory time of $T_M = 340$ ns, which was attributed to resonances in the $S = 3/2$ excited state of the cluster [166]. No quantum coherence was detected in the pair of $S = 1/2$ ground states [151]. By measurement of the z-magnetization after a nutation pulse, and a delay to ensure decay of all coherences, Rabi oscillations were observed. From the analysis of the different possible decoherence mechanisms, it was concluded that decoherence is almost entirely caused by hyperfine coupling to the ^{51}V nuclear spins.

The $S = 5$ ground state of the high spin cluster Fe_4 possesses a negative axial ZFS with $D = -0.342$ cm^{-1} [165]. As a consequence the zero-field energy gap between the $M_S = \pm 5$ and $M_S = \pm 4$ is 92 GHz, which is close to the frequency employed in a W-band EPR spectrometer (94 GHz). This fact was exploited in pulsed W-band EPR studies on frozen toluene solutions of Fe_4 in zero external field [152], where a phase memory time of $T_M = 307$ ns at $T = 4.3$ K was found in Hahn-echo measurements. The echo decay in an external field of 0.373 T shows ESEEM due to hyperfine coupling to protons. Interestingly, measurements on the protonated compound in deuterated toluene show ESEEM due to coupling to deuterium, proving that at this field the main interaction is with the nuclear spins of the solvent. In echo detected ESR spectra, ESEEM-like oscillations were observed at low field, which demonstrated that hyperfine couplings to the nuclear spins of the cluster are also present. Measurements in CS_2, which is largely nuclear-spin-free, show an increased phase memory time of Fe_4 of $T_M = 527$ ns at $T = 4.3$ K. The transient nutation measurement showed pronounced Rabi oscillations. These oscillations cannot be due to ESEEM-like effects, due to coupling to proton nuclei, because the measurements were performed in zero external field, where the proton nuclear Larmor frequency is virtually zero. On the other hand, quadrupole nuclei ($I > 1/2$) may exhibit nuclear ESEEM at zero field [167]. However, the only quadrupole nuclei in Fe_4 are the two bromine atoms of the Br-mp-ligands. Strong coupling of the electron spin to these bromine atoms is not expected because they are far away from the spin carrying iron ions.

Measurements of the quantum coherence are usually performed in dilute systems to prevent decoherence due to fluctuating intermolecular magnetic-dipolar electron–electron interactions. In SMMs these fluctuations can also be frozen out at low temperatures, below the blocking temperature of the magnetization. A single-crystal study on Fe_8 made use of this fact and, indeed, phase memory times of up to 712 ns were observed at 1.27 K [153]. Raising the temperature to 1.93 K results in a drastic reduction of T_M to 93 ns. Simulations showed that electron spin–electron spin interactions can account quantitatively for this behavior. A second decoherence process was identified from these simulations, with a decoherence time of about 1 μs, which was attributed to hyperfine-induced decoherence.

Spin–spin relaxation was also studied for polynuclear clusters in biomolecules, especially for Fe_4S_4 clusters. Interestingly a large range of values was found for chemically similar species (Table 2) [158–160].

3.3 *Future of Pulse EPR in Molecular Magnetism*

From the previous subsections it is clear that important first results from pulse EPR investigation of MNMs have been obtained. The phase memory times are generally two orders of magnitude longer than initially predicted, and were shown to increase significantly at very low temperatures. Much remains to be understood about the details of spin relaxation and decoherence in these materials, including the effects of spectral and spin diffusion. Furthermore, no attempts have yet been made to utilize nuclear spin in coherent spin manipulations. Indeed, no coherent manipulations beyond spin echo and nutation measurements (Rabi oscillations) have been reported. Here, the molecular magnetism community will be able to learn a great deal from interaction with the biophysical EPR and quantum information processing communities. Progress along these lines will depend in part on material development. To improve phase memory times, weakly coupled nuclear spins should be removed completely or moved as far as possible in space from the electron spin to limit dipolar hyperfine interactions. For detailed investigations of anisotropy (both ZFS and g-value anisotropy), single crystal measurements on dilute single crystals will be essential. This will require cocrystallization of the MNM of interest with a diamagnetic analog.

4 Emerging Trends and Outlook

Recently, a number of further directions in molecular magnetism have developed where EPR can or is starting to play a role of importance. These are very briefly outlined below, with some reference to recent and older literature without the aspiration to be comprehensive.

4.1 *Molecular Nanomagnets vs Magnetic Nanoparticles*

Most MNMs have well defined spin ground states. The splitting of the ground state by ZFS can be excellently studied by EPR techniques, as the many reviews on this subject attest. In several cases, this has proved to be impossible, often in systems that combine high molecular symmetry with competing exchange interactions, such as $Na_2[Mo_{72}{}^{VI}Fe_{30}{}^{III}O_{252}(CH_3COO)_{20}(H_2O)_{92}]\cdot$ca. 150 H_2O (Fe_{30}) [168] or $(C_5H_6N^+)_5[Fe_{13}F_{24}(OCH_3)_{12}O_4]\cdot CH_3OH\cdot 4H_2O$ (Fe_{13}) [169]. In such a case, in EPR spectra a single broad line is observed, which broadens and shifts downfield upon cooling [169, 170]. Such behavior is also observed in magnetic nanoparticles (MNP) [171] and exploitation of the classical models used in the MNP field may aid the description and analysis of the EPR spectra of the above-mentioned MNMs. The spin Hamiltonian framework seems to be unable to account for these observations, and for Fe_{30} a different nature of the low-lying excitations (spin waves) has been invoked, but only partial agreement with experimental data from INS was obtained [172]. Fittipaldi et al. have reviewed this area [173] and have recently published new results [174].

4.2 *Single Chain Magnets*

Research in the area of single-chain magnets (SCMs) continues to flourish. In these one-dimensional systems, the barrier for inversion of the magnetic moment is not only given by the ZFS of the building block, but also by the isotropic exchange interaction, which can potentially lead to much higher energy barriers [175]. These systems have been almost exclusively studied by magnetometric techniques. This is surprising, since EPR can yield information on both the ZFS and the exchange interaction in one-dimensional chains, depending on their relative magnitudes. Furthermore, one-dimensional spin chains show a very rich range of physical phenomena, including spin-Peierls transitions [176], field-induced transitions from quantum to classical physics [177], magnetic phase transitions, and soliton excitations [178], all of which can and have been investigated by EPR. EPR results on ferrimagnetic manganese(II)-radical chains could be related to the exchange interaction [179] and short range order [180]. In the SCM $[Mn_2(saltmen)_2Ni(pao)_2(py)_2](ClO_4)_2$ (saltmen=*N*,*N*-1,1,2,2-tetramethylethylene-bis(salicylideneiminate); pao= pyridine-2-aldoximate; py=pyridine), evidence for collective (spin-wave) excitations was obtained by HFEPR [181]. In addition, a Cu-Dy chain was studied, which shows the onset of slow relaxation of the magnetization at low temperatures [182, 183]. HFEPR measurements revealed that this system can be viewed as a coupled chain of Cu–Dy–Cu SMMs, and allowed quantification of the exchange interactions in the system.

4.3 *f-Element Molecular Nanomagnets*

Since the discovery of slow relaxation in lanthanide complexes [11], lanthanides [184, 185] and, more recently, actinides [12, 186] have become very popular in molecular magnetism. Most of the analysis of the magnetic properties of these complexes involved powder d.c. and a.c. susceptibility and magnetization measurements [187, 188]. Other techniques used for the investigation of lanthanide-containing MNMs have included single crystal susceptibility [189], Mössbauer spectroscopy [190], muon spin relaxation [191], and magnetic circular dichroism spectroscopy [192].

There have been very few EPR investigations on MNMs with f-elements, although EPR-based techniques have been used extensively to study lanthanides in extended lattices. Thus, EPR and ENDOR were used to study the excitations and spin dynamics within the lowest doublet in the crystal-field split ground multiplet [193, 194] and the crystal field splitting itself was investigated by far-infrared spectroscopy (FT-THz spectroscopy, see above) [95] and HFEPR [195]. Perpendicular- and parallel-mode X-band EPR spectroscopy was used to investigate the anisotropic exchange coupling in a series of lanthanide-transition metal ion dimers [196–198]. A similar approach was used in the Nd^{3+} dimer compound $\{[Nd_2(\alpha\text{-}C_4H_3OCOO)_6(H_2O)_2]\}_n$ [199] using perpendicular mode EPR at X- and Q-band. HFEPR studies were performed on the SMMs $[Dy_2Ni]$ and $[Dy_2Cu]$, but only signals due to the transition metal ion were found [182]. Extensive single-crystal EPR measurements were carried out for a series of complexes $[Ln(dmf)_4(H_2O)_3(\mu\text{-}CN)M(CN)_5]\cdot nH_2O$ (Ln = Ln^{3+}, Ce^{3+}, M = Fe^{3+}, Co^{3+}, dmf = *N*,*N*′-dimethylformamide) [200]. The low-energy electronic structure of the complexes are characterized in detail, and the authors were able to show that isotropic, anisotropic, and antisymmetric terms are required for a good description of the systems, where they profited from a detailed ligand field theoretical analysis.

Lanthanides may also come to play an important role in the investigation of quantum coherence in MNMs. Quantum coherence and coherent spin manipulations of lanthanide ions in lattices were recently reported, with phase memory times of $T_M = 50$ μs at 2.5 K for Er^{3+}:$CaWO_4$ (10^{-5} atom% Er) [201] and T_M values of up to 134 μs at 4 K for Yb^{3+}:$CaWO_4$ (0.0025% Yb) [202]. Indeed, some of the very first spin-echo experiments were carried out on lanthanides in lattices, e.g., Er^{3+}:$CaWO_4$ ($T_M \approx 14$ μs at ~1.8 K) [203] and Er^{3+}/Ce^{3+}:$CaWO_4$ ($T_M \approx 25$ μs at ~1.8 K for the Ce^{3+} ion) [204]. In molecular systems, T_M times of 430 ns were found for $\{[Nd_2(\alpha\text{-}C_4H_3OCOO)_6(H_2O)_2]\}_n$ [199].

References

1. Taft KL, Papaefthymiou GC, Lippard SJ (1993) Science 259:1302
2. Papaefthymiou GC (1992) Phys Rev B 46:10366
3. Caneschi A, Gatteschi D, Sessoli R, Barra AL, Brunel LC, Guillot M (1991) J Am Chem Soc 113:5873
4. Novak MA, Sessoli R (1995) In: Gunther L, Barbara B (eds) Quantum tunneling of magnetization – QTM'94. Kluwer Academic, Dordrecht
5. Thomas L, Lionti F, Ballou R, Gatteschi D, Sessoli R, Barbara B (1996) Nature 383:145
6. Friedman JR, Sarachik MP, Tejada J, Ziolo R (1996) Phys Rev Lett 76:3830
7. Hill S, Edwards R, Jones S, Dalal N, North J (2003) Phys Rev Lett 90:217204
8. Wilson A, Lawrence J, Yang EC, Nakano M, Hendrickson DN, Hill S (2006) Phys Rev B 74:140403(R)
9. Aubin SMJ, Wemple MW, Adams DM, Tsai HL, Christou G, Hendrickson DN (1996) J Am Chem Soc 118:7746
10. Cheesman MR, Oganesyan VS, Sessoli R, Gatteschi D, Thomson AJ (1997) Chem Commun 1677
11. Ishikawa N, Sugita M, Ishikawa T, Koshihara S, Kaizu Y (2003) J Am Chem Soc 125:8694
12. Rinehart JD, Long JR (2009) J Am Chem Soc 131:12558
13. Freedman DE, Harman WH, Harris TD, Long GJ, Chang CJ, Long JR (2010) J Am Chem Soc 132:1224
14. del Barco E, Vernier N, Hernandez JM, Tejada J, Chudnovsky EM, Molins E, Bellessa G (1999) Europhys Lett 47:722
15. Luis F, Mettes FL, Tejada J, Gatteschi D, de Jongh LJ (2000) Phys Rev Lett 85:4377
16. Hill S, Edwards RS, Aliaga-Alcalde N, Christou G (2003) Science 302:1015
17. Waldmann O, Dobe C, Güdel HU, Mutka H (2006) Phys Rev B 74:054429
18. Ardavan A, Rival O, Morton JJL, Blundell SJ, Tyryshkin AM, Timco GA, Winpenny REP (2007) Phys Rev Lett 98:057201
19. Leuenberger MN, Loss D (2001) Nature 410:789
20. Affronte M, Troiani F, Ghirri A, Carretta S, Santini P, Corradini V, Schuecker R, Muryn C, Timco G, Winpenny RE (2006) Dalton Trans 2810
21. Stepanenko D, Trif M, Loss D (2008) Inorg Chim Acta 361:3740
22. Evangelisti M, Candini A, Ghirri A, Affronte M, Brechin EK, McInnes EJL (2005) Appl Phys Lett 87:072504
23. Clérac R, Miyasaka H, Yamashita M, Coulon C (2002) J Am Chem Soc 124:12837
24. Caneschi A, Gatteschi D, Lalioti N, Sangregorio C, Sessoli R, Venturi G, Vindigni A, Rettori A, Pini MG, Novak MA (2001) Angew Chem Int Ed 40:1760
25. Datta S, Bolin E, Inglis R, Milios CJ, Brechin EK, Hill S (2009) Polyhedron 28:1788
26. Pieper O, Guidi T, Carretta S, van Slageren J, El Hallak F, Lake B, Santini P, Amoretti G, Mutka H, Koza M, Russina M, Schnegg A, Milios CJ, Brechin EK, Julià A, Tejada J (2010) Phys Rev B 81:174420

27. Kahn O (1993) Molecular magnetism. VCH, New York
28. Bencini A, Gatteschi D (1990) EPR of exchange coupled systems. Springer, Berlin
29. Gatteschi D, Sessoli R, Villain J (2006) Molecular nanomagnets. Oxford University Press, Oxford
30. Winpenny REP, McInnes EJL (2010) In: Bruce DW, O'Hare D, Walton RI (eds) Molecular materials. Wiley, Chichester
31. Carretta S, Liviotti E, Magnani N, Santini P, Amoretti G (2004) Phys Rev Lett 92:207205
32. Barra AL, Caneschi A, Cornia A, Gatteschi D, Gorini L, Heiniger LP, Sessoli R, Sorace L (2007) J Am Chem Soc 129:10754
33. Barra AL, Gatteschi D, Sessoli R, Sorace L (2005) Magn Reson Chem 43:S183
34. McInnes EJL (2006) Struct Bond 122:69
35. Gatteschi D, Barra A, Caneschi A, Cornia A, Sessoli R, Sorace L (2006) Coord Chem Rev 250:1514
36. McInnes EJL (2011) In: Winpenny REP (ed) Molecular cluster magnets. World Scientific, Singapore
37. Feng PL, Beedle CC, Koo C, Lawrence J, Hill S, Hendrickson DN (2008) Inorg Chim Acta 361:3465
38. Boeer AB, Collison D, McInnes EJL (2008) R Soc Chem Spec Period Rep 21:131
39. Collison D, McInnes EJL (2002) R Soc Chem Spec Period Rep 18:161
40. Collison D, McInnes EJL (2004) R Soc Chem Spec Period Rep 19:374
41. Collison D, McInnes EJL (2007) R Soc Chem Spec Period Rep 20:157
42. Rentschler E, Gatteschi D, Cornia A, Fabretti AC, Barra A-L, Shchegolikhina OI, Zhdanov AA (1996) Inorg Chem 35:4427
43. Pilawa B, Boffinger R, Keilhauer I, Leppin R, Odenwald I, Wendl W, Berthier C, Horvatic A (2005) Phys Rev B 71:184419
44. van Slageren J, Sessoli R, Gatteschi D, Smith AA, Helliwell M, Winpenny REP, Cornia A, Barra AL, Jansen AGM, Rentschler E, Timco GA (2002) Chem Eur J 8:277
45. Dreiser J, Waldmann O, Carver G, Dobe C, Güdel H-U, Weihe H, Barra A-L (2010) Inorg Chem 8729
46. Larsen FK, McInnes EJL, El Mkami H, Overgaard J, Piligkos S, Rajaraman G, Rentschler E, Smith AA, Smith GM, Boote V, Jennings M, Timco GA, Winpenny REP (2003) Angew Chem Int Ed 42:101
47. Piligkos S, Bill E, Collison D, McInnes EJL, Timco GA, Weihe H, Winpenny REP, Neese F (2007) J Am Chem Soc 129:760
48. Piligkos S, Weihe H, Bill E, Neese F, El Mkami H, Smith GM, Collison D, Rajaraman G, Timco GA, Winpenny REP, McInnes EJL (2009) Chem Eur J 15:3152
49. Timco GA, McInnes EJL, Pritchard RG, Tuna F, Winpenny REP (2008) Angew Chem Int Ed 47:9681
50. Timco GA, Carretta S, Troiani F, Tuna F, Pritchard RJ, Muryn CA, McInnes EJL, Ghirri A, Candini A, Santini P, Amoretti G, Affronte M, Winpenny REP (2009) Nat Nanotech 4:173
51. Barra AL (2008) Inorg Chim Acta 361:3564
52. Barra AL, Gatteschi D, Sessoli R (1997) Phys Rev B 56:8192
53. del Barco E, Kent AD, Hill S, North JM, Dalal NS, Rumberger EM, Hendrickson DN, Chakov N, Christou G (2005) J Low Temp Phys 140:119
54. Cornia A, Sessoli R, Sorace L, Gatteschi D, Barra AL, Daiguebonne C (2002) Phys Rev Lett 89:257201
55. Lawrence J, Yang EC, Hendrickson DN, Hill S (2009) Phys Chem Chem Phys 11:6743
56. Lawrence J, Yang EC, Edwards R, Olmstead MM, Ramsey C, Dalal NS, Gantzel PK, Hill S, Hendrickson DN (2008) Inorg Chem 47:1965
57. Yang EC, Kirman C, Lawrence J, Zakharov LN, Rheingold AL, Hill S, Hendrickson DN (2005) Inorg Chem 44:3827
58. Accorsi S, Barra AL, Caneschi A, Chastanet G, Cornia A, Fabretti AC, Gatteschi D, Mortalò C, Olivieri E, Parenti F, Rosa P, Sessoli R, Sorace L, Wernsdorfer W, Zobbi L (2006) J Am Chem Soc 128:4742

59. Gregoli L, Danieli C, Barra A-L, Neugebauer P, Pellegrino G, Poneti G, Sessoli R, Cornia A (2009) Chem Eur J 15:6456
60. Waldmann O, Güdel HU (2005) Phys Rev B 72:094422
61. Zhong YC, Pilbrow JR (1999) In: Poole CP, Farach HA (eds) Handbook of electron spin resonance, vol. 2. Springer, New York
62. van Slageren J, Vongtragool S, Gorshunov B, Mukhin AA, Karl N, Krzystek J, Telser J, Muller A, Sangregorio C, Gatteschi D, Dressel M (2003) Phys Chem Chem Phys 5:3837
63. Kirchner N, van Slageren J, Dressel M (2007) Inorg Chim Acta 360:3813
64. Kozlov GV, Volkov AA (1998) Top Appl Phys 74:51
65. Robson PN (1967) In: Martin DH (ed) Spectroscopic techniques for far infra-red submillimetre and millimetre waves. North-Holland, Amsterdam
66. Mukhin AA, Travkin VD, Zvezdin AK, Lebedev SP, Caneschi A, Gatteschi D (1998) Europhys Lett 44:778
67. El Hallak F, van Slageren J, Gómez-Segura J, Ruiz-Molina D, Dressel M (2007) Phys Rev B 75:104403
68. Carbonera C, Luis F, Campo J, Sánchez-Marcos J, Camón A, Chaboy J, Ruiz-Molina D, Imaz I, van Slageren J, Dengler S, Gonzalez M (2010) Phys Rev B 81:014427
69. Kirchner N, van Slageren J, Brechin EK, Dressel M (2005) Polyhedron 24:2400
70. Mukhin A, Gorshunov B, Dressel M, Sangregorio C, Gatteschi D (2001) Phys Rev B 63:214411
71. Piligkos S, Rajaraman G, Soler M, Kirchner N, van Slageren J, Bircher R, Parsons S, Güdel HU, Kortus J, Wernsdorfer W, Christou G, Brechin EK (2005) J Am Chem Soc 127:5572
72. Sieber A, Boskovic C, Bircher R, Waldmann O, Ochsenbein ST, Chaboussant G, Güdel HU, Kirchner N, van Slageren J, Wernsdorfer W, Neels A, Stoeckli-Evans H, Janssen S, Juranyi F, Mutka H (2005) Inorg Chem 44:4315
73. Moro F, Piga F, Krivokapic I, Burgess A, Lewis W, McMaster J, van Slageren J (2010) Inorg Chim Acta 363:4329
74. Abragam A, Bleany B (1986) Electron paramagnetic resonance of transition ions. Dover, New York
75. Kirchner N, van Slageren J, Tsukerblat B, Waldmann O, Dressel M (2008) Phys Rev B 78:094426
76. Petukhov K, Hill S, Chakov N, Abboud K, Christou G (2004) Phys Rev B 70 p 054426
77. Rogez G, Rebilly JN, Barra AL, Sorace L, Blondin G, Kirchner N, Duran M, van Slageren J, Parsons S, Ricard L, Marvilliers A, Mallah T (2005) Angew Chem Int Ed 44:1876
78. Vongtragool S, Mukhin A, Gorshunov B, Dressel M (2004) Phys Rev B 69:104410
79. Takahashi S, Edwards RS, North JM, Hill S, Dalal NS (2004) Phys Rev B 70:094429
80. Hill S, Maccagnano S, Park K, Achey R, North J, Dalal N (2002) Phys Rev B 65:224410
81. Park K, Novotny M, Dalal N, Hill S, Rikvold P (2002) Phys Rev B 66:144409
82. Lax B, Button KJ (1962) Microwave ferrites and ferrimagnets. McGraw-Hill, New York
83. Park K, Novotny MA, Dalal NS, Hill S, Rikvold PA (2002) J Appl Phys 91:7167
84. van Slageren J, Vongtragool S, Mukhin A, Gorshunov B, Dressel M (2005) Phys Rev B 72:020401
85. Smith GM, Lesurf JCG, Mitchell RH, Riedi PC (1998) Rev Sci Instrum 70:1787
86. Dressel M, Gorshunov B, Rajagopal K, Vongtragool S, Mukhin AA (2003) Phys Rev B 67:060405(R)
87. van Slageren J, Vongtragool S, Gorshunov B, Mukhin A, Dressel M (2009) Phys Rev B 79:224406
88. Schnegg A, Behrends J, Lips K, Bittl R, Holldack K (2009) Phys Chem Chem Phys 11:6820
89. Pedersen KS, Dreiser J, Nehrkorn J, Gysler M, Schau-Magnussen M, Schnegg A, Holldack K, Bittl R, Piligkos S, Weihe H, Tregenna-Piggott P, Waldmann O, Bendix J (2011) Chem Commun 6918
90. Dreiser J, Schnegg A, Holldack K, Pedersen KS, Schau-Magnussen M, Nehrkorn J, Tregenna-Piggott P, Mutka H, Weihe H, Bendix J, Waldmann O (2011) Chem Eur J 17:7492

91. Ohlmann RC, Tinkham M (1961) Phys Rev 123:425
92. Jensen MRF, Feiven SA, Parker TJ, Camley RE (1997) J Phys Condens Mater 9:7233
93. Brackett GC, Richards PL, Caughey WS (1971) J Chem Phys 54:4383
94. Talbayev D, Mihaly L, Zhou JS (2004) Phys Rev Lett 93:017202
95. Bloor D, Copland GM (1972) Rep Prog Phys 35:1173
96. Richards PL, Caughey WS, Eberspac H, Feher G, Malley M (1967) J Chem Phys 47:1187
97. Champion PM, Sievers AJ (1977) J Chem Phys 66:1819
98. Ray K, Begum A, Weyhermuller T, Piligkos S, van Slageren J, Neese F, Wieghardt K (2005) J Am Chem Soc 127:4403
99. van der Put PJ, Schilperoord AA (1974) Inorg Chem 13:2476
100. Champion PM, Sievers AJ (1980) J Chem Phys 72:1569
101. Schmuttenmaer CA (2004) Chem Rev 104:1759
102. Parks B, Loomis J, Rumberger E, Hendrickson DN, Christou G (2001) Phys Rev B 64:184426
103. Parks B, Loomis J, Rumberger E, Yang EC, Hendrickson DN, Christou G (2002) J Appl Phys 91:7170
104. Parks B, Vacca L, Rumberger E, Hendrickson DN, Christou G (2003) Physica B 329:1181
105. Vongtragool S (2004) Frequency-domain magnetic resonance spectroscopy on the mn12-acetate single-molecule magnet. PhD Thesis, Universität Stuttgart
106. Mirebeau I, Hennion M, Casalta H, Andres H, Güdel HU, Irodova AV, Caneschi A (1999) Phys Rev Lett 83:628
107. Wernsdorfer W, Müller A, Mailly D, Barbara B (2004) Europhys Lett 66:861
108. Jamet M, Wernsdorfer W, Thirion C, Mailly D, Dupuis V, Melinon P, Perez A (2001) Phys Rev Lett 86:4676
109. Cleuziou JP, Wernsdorfer W, Bouchiat V, Ondarcuhu T, Monthioux M (2006) Nat Nanotech 1:53
110. Barra AL, Gatteschi D, Pardi L, Müller A, Döring J (1992) J Am Chem Soc 114:8509
111. Ajiro Y, Itoh H, Inagaki Y, Asano T, Narumi Y, Kindo K, Sakon T, Motokawa M, Cornia A, Gatteschi D, Müller A, Barbara B (2002) In: Ajiro Y, Bouchier J-P (eds) French-Japanese symposium on quantum properties of low-dimensional antiferromagnets. Kyushu University Press, Fukuoka, Japan
112. Sakon T, Koyama K, Motokawa M, Ajiro Y, Müller A, Barbara B (2004) Physica B 346:206
113. Vongtragool S, Gorshunov B, Mukhin AA, van Slageren J, Dressel M, Müller A (2003) Phys Chem Chem Phys 5:2778
114. Wernsdorfer W, Mailly D, Timco GA, Winpenny REP (2005) Phys Rev B 72:060409
115. Petukhov K, Wernsdorfer W, Barra AL, Mosser V (2005) Phys Rev B 72:052401
116. Bal M, Friedman JR, Suzuki Y, Mertes KM, Rumberger EM, Hendrickson DN, Myasoedov Y, Shtrikman H, Avraham N, Zeldov E (2004) Phys Rev B 70:100408
117. Bal M, Friedman JR, Suzuki Y, Rumberger EM, Hendrickson DN, Avraham N, Myasoedov Y, Shtrikman H, Zeldov E (2005) Europhys Lett 71:110
118. Bal M, Friedman JR, Rumberger EM, Shah S, Hendrickson DN, Avraham N, Myasoedov Y, Shtrikman H, Zeldov E (2006) J Appl Phys 99:08D103
119. Bal M, Friedman JR, Tuominen MT, Rumberger EM, Hendrickson DN (2006) J Appl Phys 99:08D102
120. del Barco E, Kent AD, Yang EC, Hendrickson DN (2004) Phys Rev Lett 93:157202
121. de Loubens G, Chaves-O'Flynn GD, Kent AD, Ramsey C, del Barco E, Beedle C, Hendrickson DN (2007) J Appl Phys 101:09E104
122. Quddusi HM, Ramsey CM, Gonzalez-Pons JC, Henderson JJ, del Barco E, de Loubens G, Kent AD (2008) Rev Sci Instrum 79:074703
123. Cage B, Russek S (2004) Rev Sci Instrum 75:4401
124. Cage B, Russek SE, Zipse D, North JM, Dalal NS (2005) Appl Phys Lett 87:082501
125. El Hallak F, van Slageren J, Dressel M (2010) Rev Sci Instrum 81:095105
126. Ohmichi E, Mizuno N, Kimata M, Ohta H (2008) Rev Sci Instrum 79:103903

127. Ohmichi E, Mizuno N, Kimata M, Ohta H, Osada T (2009) Rev Sci Instrum 80:013904
128. Petukhov K, Bahr S, Wernsdorfer W, Barra AL, Mosser V (2007) Phys Rev B 75:064408
129. Bahr S, Petukhov K, Mosser V, Wernsdorfer W (2007) Phys Rev Lett 99:147205
130. Bahr S, Petukhov K, Mosser V, Wernsdorfer W (2008) Phys Rev B 77:064404
131. Bal M, Friedman JR, Chen W, Tuominen MT, Beedle CC, Rumberger EM, Hendrickson DN (2008) EPL 82:17005
132. de Loubens G, Kent AD, Krymov V, Gerfen GJ, Beedle CC, Hendrickson DN (2008) J Appl Phys 103:07B910
133. Rieger PH (2007) Electron spin resonance-analysis and interpretation. RSC, Cambridge
134. Bertini I, Martini G, Luchinat C (1994) In: Poole J, Farach HA (eds) Handbook of electron spin resonance. AIP, New York
135. Eaton SS, Eaton GR (2000) Biol Magn Reson 19:29
136. Tsukerblat BS, Botsan IG, Belinskii MI, Fainzilberg VE (1985) Mol Phys 54:813
137. Meier F, Levy J, Loss D (2003) Phys Rev Lett 90:047901
138. Meier F, Levy J, Loss D (2003) Phys Rev B 68:134417
139. Troiani F, Affronte M (2011) Chem Soc Rev 40:3119
140. Trif M, Troiani F, Stepanenko D, Loss D (2010) Phys Rev B 82:045429
141. Troiani F, Affronte M, Carretta S, Santini P, Amoretti G (2005) Phys Rev Lett 94:190501
142. Candini A, Lorusso G, Troiani F, Ghirri A, Carretta S, Santini P, Amoretti G, Muryn C, Tuna F, Timco G, McInnes EJL, Winpenny REP, Wernsdorfer W, Affronte M (2010) Phys Rev Lett 104:037203
143. Trif M, Troiani F, Stepanenko D, Loss D (2008) Phys Rev Lett 101:217201
144. Schweiger A, Jeschke G (2001) Principles of pulse electron paramagnetic resonance. Oxford University Press, Oxford
145. Sato K, Nakazawa S, Rahimi R, Ise T, Nishida S, Yoshino T, Mori N, Toyota K, Shiomi D, Yakiyama Y, Morita Y, Kitagawa M, Nakasuji K, Nakahara M, Hara H, Carl P, Höfer P, Takui T (2009) J Mater Chem 19:3739
146. Nakazawa S, Sato K, Shiomi D, Yano M, Kinoshita T, Franco MLTMB, Lazana MCRLR, Shohoji MCBL, Itoh K, Takui T (2011) Phys Chem Chem Phys 13:1424
147. Ruby RH, Benoit H, Jeffries CD (1962) Phys Rev 127:51
148. Hoffmann SK, Lijewski S, Goslar J, Ulanov VA (2010) J Magn Reson 202:14
149. Schlegel C, van Slageren J, Timco G, Winpenny REP, Dressel M (2011) Phys Rev B 83:134407
150. Mitrikas G, Sanakis Y, Raptopoulou CP, Kordas G, Papavassiliou G (2008) Phys Chem Chem Phys 10:743
151. Bertaina S, Gambarelli S, Mitra T, Tsukerblat B, Müller A, Barbara B (2010) Nature 466:1006
152. Schlegel C, van Slageren J, Manoli M, Brechin EK, Dressel M (2008) Phys Rev Lett 101:147203
153. Takahashi S, van Tol J, Beedle CC, Hendrickson DN, Brunel L-C, Sherwin MS (2009) Phys Rev Lett 102:087603
154. Kulik LV, Lubitz W, Messinger J (2005) Biochemistry 44:9368
155. Beardwood P, Gibson JF, Bertrand P, Gayda JP (1983) Biochim Biophys Acta 742:426
156. Lorigan GA, Britt RD (1994) Biochemistry 33:12072
157. Britt RD, Zimmermann JL, Sauer K, Klein MP (1989) J Am Chem Soc 111:3522
158. Dilg AWE, Mincione G, Achterhold K, Iakovleva O, Mentler M, Luchinat C, Bertini I, Parak FG (1999) J Biol Inorg Chem 4:727
159. Dilg AWE, Capozzi F, Mentler M, Iakovleva O, Luchinat C, Bertini I, Parak FG (2001) J Biol Inorg Chem 6:232
160. Guigliarelli B, More C, Fournel A, Asso M, Hatchikian EC, Williams R, Cammack R, Bertrand P (1995) Biochemistry 34:4781
161. Bertrand P, Gayda JP, Rao KK (1982) J Chem Phys 76:4715
162. Harris EA, Yngvesson KS (1968) J Phys C 1:990

163. Altshuler SA, Kirmse R, Solovev BV (1975) J Phys C 8:1907
164. Sanakis Y, Pissas M, Krzystek J, Telser J, Raptis RG (2010) Chem Phys Lett 493:185
165. Schlegel C, Burzurí E, Luis F, Moro F, Manoli M, Brechin EK, Murrie M, van Slageren J (2010) Chem Eur J 16:10178
166. Bertaina S, Gambarelli S, Mitra T, Tsukerblat B, Müller A, Barbara B (2008) Nature 453:203
167. Jeschke G, Schweiger A (1996) Chem Phys Lett 259:531
168. Müller A, Luban M, Schröder C, Modler R, Kögerler P, Axenovich M, Schnack J, Canfield P, Bud'ko S, Harrison N (2001) Chemphyschem 2:517
169. van Slageren J, Rosa P, Caneschi A, Sessoli R, Casellas H, Rakitin Y, Cianchi L, Giallo F, Spina G, Bino A, Barra A-L, Guidi T, Carretta S, Caciuffo R (2006) Phys Rev B 73:014422
170. Barra AL, Hassan AK, Gatteschi D, Sessoli R, Müller A (2001) GHMFL Annual Report, p 74
171. Berger R, Bissey J-C, Kliava J (2000) J Phys Condens Mater 12:9347
172. Waldmann O (2007) Phys Rev B 75:012415
173. Fittipaldi M, Sorace L, Barra AL, Sangregorio C, Sessoli R, Gatteschi D (2009) Phys Chem Chem Phys 11:6555
174. Fittipaldi M, Innocenti C, Ceci P, Sangregorio C, Castelli L, Sorace L, Gatteschi D (2011) Phys Rev B 83:104409
175. Coulon C, Miyasaka H, Clérac R (2006) Struct Bond 122:163
176. Demishev SV, Semeno AV, Sluchanko NE, Samarin NA, Vasil'ev AN, Leonyuk LI (1997) JETP 85:943
177. Kashiwagi T, Hagiwara M, Kimura S, Honda Z, Miyazaki H, Harada I, Kindo K (2009) Phys Rev B 79:024403
178. Motokawa M, Ohta H, Nojiri H, Kimura S (2003) J Phys Soc Jpn 72:1
179. Caneschi A, Gatteschi D, Rey P, Sessoli R (1988) Inorg Chem 27:1756
180. Caneschi A, Gatteschi D, Renard JP, Rey P, Sessoli R (1989) Inorg Chem 28:3314
181. Oshima Y, Nojiri H, Asakura K, Sakai T, Yamashita M, Miyasaka H (2006) Phys Rev B 73:214435
182. Okazawa A, Nogami T, Nojiri H, Ishida T (2008) Inorg Chem 47:9763
183. Okazawa A, Nogami T, Nojiri H, Ishida T (2008) Chem Mater 20:3110
184. Sessoli R, Powell AK (2009) Coord Chem Rev 253:2328
185. Rinehart JD, Fang M, Evans WJ, Long JR (2011) Nat Chem 3:538
186. Mills DP, Moro F, McMaster J, van Slageren J, Lewis W, Blake AJ, Liddle ST (2011) Nat Chem 3:454
187. Benelli C, Gatteschi D (2002) Chem Rev 102:2369
188. Sorace L, Benelli C, Gatteschi D (2011) Chem Soc Rev 40:3092
189. Bernot K, Luzon J, Bogani L, Etienne M, Sangregorio C, Shanmugam M, Caneschi A, Sessoli R, Gatteschi D (2009) J Am Chem Soc 131:5573
190. Akhtar MN, Mereacre V, Novitchi G, Tuchagues J-P, Anson CE, Powell AK (2009) Chem Eur J 15:7278
191. Salman Z, Giblin S, Lan Y, Powell A, Scheuermann R, Tingle R, Sessoli R (2010) Phys Rev B 82:174427
192. Gonidec M, Davies ES, McMaster J, Amabilino DB, Veciana J (2010) J Am Chem Soc 132:1756
193. Baker JM (1993) R Soc Chem Spec Period Rep 13B:131
194. Altshuler SA, Kozyrev BM (1974) Electron paramagnetic resonance in compounds of transition elements. Wiley, New York
195. Goiran M, Klingeler R, Kazei Z, Snegirev V (2007) J Magn Magn Mater 318:1
196. Tangoulis V, Figuerola A (2007) Chem Phys 340:293
197. Tangoulis V, Estrader M, Figuerola A, Ribas J, Diaz C (2007) Chem Phys 336:74
198. Figuerola A, Tangoulis V, Sanakis Y (2007) Chem Phys 334:204
199. Mingalieva LV, Voronkova VK, Galeev RT, Sukhanov AA, Melnik S, Prodius D, Turta KI (2009) Appl Magn Reson 37:737
200. Sorace L, Sangregorio C, Figuerola A, Benelli C, Gatteschi D (2009) Chem Eur J 15:1377

201. Bertaina S, Gambarelli S, Tkachuk A, Kurkin IN, Malkin B, Stepanov A, Barbara B (2007) Nat Nanotech 2:39
202. Rakhmatullin R, Kurkin I, Mamin G, Orlinskii S, Gafurov M, Baibekov E, Malkin B, Gambarelli S, Bertaina S, Barbara B (2009) Phys Rev B 79:172408
203. Mims WB, Nassau K, McGee JD (1961) Phys Rev 123:2059
204. Mims WB (1968) Phys Rev 168:370

Index